Forage Crops in the Omics Era

Faheem Shehzad Baloch
Muhammad Azhar Nadeem
Rüştü Hatipoğlu • Nafiz Çeliktaş
Muhammad Aasim • Abdulqader Jighly
Ajit Singh
Editors

Forage Crops in the Omics Era

Springer

Editors
Faheem Shehzad Baloch
Department of Biotechnology, Faculty of Sciences
Mersin University
Mersin, Türkiye

Muhammad Azhar Nadeem
Department of Biotechnology, Faculty of Sciences
Mersin University
Mersin, Türkiye

Rüştü Hatipoğlu
Department of Field Crops
Kırşehir Ahi Evran University
Kırşehir, Türkiye

Nafiz Çeliktaş
Department of Field Crops
Hatay Mustafa Kemal University
Hatay, Türkiye

Muhammad Aasim
Faculty of Agricultural Sciences and Technologies
Sivas University of Science and Technology
Sivas, Türkiye

Abdulqader Jighly
SuSTATability Statistical Solutions
Melbourne, Victoria, Australia

Ajit Singh
School of Biological and Environmental Sciences
University of Nottingham Malaysia
Semenyih, Selangor, Malaysia

ISBN 978-981-95-9248-7 ISBN 978-981-95-9249-4 (eBook)
https://doi.org/10.1007/978-981-95-9249-4

This Springer imprint is published by the registered company Springer Nature Singapore Pte Ltd.
The registered company address is: 152 Beach Road, #21-01/04 Gateway East, Singapore 189721, Singapore

Preface

The world faces a daunting challenge in increasing global food production while conserving natural resources to feed an estimated ten billion people by 2050. This goal can be achieved through better utilization of land, smart irrigation technologies, reduced chemical fertilizers utilization, higher labor costs, exploring novel genetic resources, involvement of OMICS approaches, and necessitating more efficient resource utilization. At the same time, environmental protection must be prioritized amid climate change and increasing public demand for sustainable practices. Forage crops will be crucial to deal with the climate change scenarios we are currently facing, as they can allow the efficient use of marginal lands while providing sufficient quantities of safe, high-quality, animal-based products. While humans do not consume forage crops directly, they serve as the foundation for protein and fat sources, including meat, eggs, milk, and dairy products, through livestock such as cattle, sheep, goats, and poultry. Numerous plant species have the potential to become valuable forage crops under specific regional climatic conditions. Additionally, perennial forage crops could serve as renewable energy sources, helping reduce reliance on fossil fuels. They will also play a larger role in sustainable farming systems, including crop rotations, cover cropping, and erosion control along waterways. Integrating perennial legumes into agricultural practices will not only protect soil health and support biodiversity but also fix nitrogen for future crops.

To meet growing demands for animal-based protein, expanding livestock populations require increasing quantities of fodder as their primary feed source. More forage crops are needed to explore and check their adoption to specific climatic conditions so that they should be used a feed for the increasing animal population. However, forage crop production encounters significant hurdles from biotic and abiotic factors, particularly drought. The increasing frequency and intensity of drought events threaten the yield and quality of forage crops.

This book, entitled Forage Crops in the Omics Era, edited by Faheem Shehzad BALOCH, Muhammad Azhar NADEEM, Rüştü HATİPOĞLU, Nafiz ÇELİKTAŞ, Muhammad AASIM, Abdulgader JIGHLY, Ajit SINGH, aimed to address the above-discussed challenges by presenting a comprehensive resource for leveraging molecular tools to improve the tolerance of forage crops to biotic and abiotic stresses. Marker-assisted breeding (MAB) and conventional and advanced genomic techniques have revolutionized crop enhancement initiatives. Marker-assisted breeding expedites the breeding process by facilitating accurate identification and selection of preferred attributes, enabling the development of resilient cultivars adapted to a changing climate.

This book targeted a diverse reader, comprising academics, plant breeders, graduate students, and policymakers. Researchers will discover the most recent insights into genomics-based breeding strategies, while plant breeders will gain from practical case studies the use of marker-assisted selection (MAS) in diverse forage crops. Graduate students will acquire a robust foundation in molecular breeding techniques in forage crops, while policymakers will comprehend the wider implications of implementing these strategies for achieving food and nutritional security.

This book is offered to the readership in a set of 17 chapters. Chapters 1–7 describe the global importance, agronomy, adaptability, and genetic improvement by conventional as well as biotechnological methods of forage crops. Chapters 8–16 give up-to-date information on breeding and genomics of some major forage legume and grass species as well as of some minor forage legume and grass species. Chapter 17 presents the future perspectives for forage breeding.

This book would not have been possible without the invaluable contributions of leading experts in the breeding of forage crops. Each chapter reflects the authors' extensive experience and dedication to advancing agricultural science. We express our profound appreciation to the contributors for imparting their experience and to the reviewers for their insightful criticism, which improved the quality of this work.

I am assured that this book will serve as an indispensable resource for individuals committed to enhancing the resilience and productivity of forage crops, thus significantly contributing to the future of sustainable agriculture.

Kırşehir, Turkey Rüştü Hatipoğlu

Contents

Editors and Contributors

About the Editors

Faheem Shehzad Baloch is Professor of Plant Genetics and Breeding at Mersin University. He received his PhD on QTL mapping in wheat from Çukurova University, Turkey, in 2012. With over 12 years of experience in plant genetics, genomics, and biotechnology, he focuses on NGS, molecular markers, GWAS and genomic prediction, especially in legume-cereal. He led national and international projects, and has supervised MSc, PhD, and post-doc students. He has participated in international conferences, over 150 publications, co-edited 13 books, and more than 5300 citations with an h-index of 38.

Muhammad Azhar Nadeem is Associate Professor of Plant Breeding and Genetics at the Department of Biotechnology, Mersin University, Türkiye. His research focuses on genetic diversity and GWAS in legumes, particularly on bean-lentil to identify genomic regions linked to agronomic and quality traits. He has over 90 SCI/SCIE publications and more than 10 chapters. He is participating in various national and international projects. He also serves on the editorial boards of international journals and actively involved in various COST actions. He has edited 6 books and has over 4000 citations with an h-index of 30.

Rüştü Hatipoğlu is Professor of Forage Crops and Plant Biotechnology at Kırsehir Ahi Evran University, Türkiye. He has extensive experience in forage crop production, breeding, and genomics. His research focuses on agronomy and breeding of forage crops, especially forage grasses, using conventional and biotechnological methods. He has published over 50 SCI/SCIE articles and has more than 3000 citations with an h-index of 25.

Nafiz Çeliktaş is a Professor in the Department of Field Crops at Hatay Mustafa Kemal University, Türkiye, specializing in forage crops and pasture management. He earned his PhD from Çukurova University with a focus on sainfoin tissue culture. Starting his academic career in 2002, he became Associate Professor in 2007 and Professor in 2012. He also spent 3 years working in Kyrgyzstan. His research focuses on plant tissue culture, haploidization, cryopreservation of endangered species, pasture ecosystem dynamics, and bioenergy farming, including second-generation bioethanol production.

Muhammad Aasim is a Professor at the Department of Plant Protection, Sivas University of Science and Technology, Turkey, since 2020. He earned his PhD with honors from Ankara University in 2010. He has worked as Assistant and Associate Professor at several Turkish universities. His research focuses on plant biotechnology, tissue culture, genetic transformation, phytoremediation, biopesticides, and nanoparticles. He has published over 100 SCIE articles, edited 3 books, and written 25 chapters and has an h-index of 29 with over 2500 citations.

Abdulqader Jighly is co-founder of AgriSapiens Statistical Solutions and a guest professor at Qingdao Agricultural University, China. With over 14 years of experience, he developed innovative statistical models combining diverse data types to improve agricultural selection decisions. Previously, he led the Computational Biology team at Agriculture Victoria, Australia, managing over 25 researchers and initiating global collaborations. He teaches at the University of Melbourne and serves as Associate Editor for *The Plant Genome* journal. He mentors many PhD and master's students in applying statistical tools.

Ajit Singh is Associate Professor of Agronomy at the University of Nottingham Malaysia. He earned his PhD from the University of Ibadan, Nigeria, focusing on legume-cereal rotations and soil fertility. His research covers urban agriculture with LED-integrated hydroponics, precision farming, nitrogen fixation in underutilized legumes, and converting food waste into resources. He has published 159 works, including 74 journal articles and 6 chapters. He supervised postgraduate students in PhD, MRes, and MSc programs.

Contributors

Banzeer Ahsan Abbasi Department of Botany, Rawalpindi Women University, Rawalpindi, Pakistan

Hadiqa Afzal Department of Botany, Division of Science and Technology, University of Education, Lahore, Punjab, Pakistan

Sadia Afzal Department of Botany, Lahore College for Women University, Lahore, Pakistan

Naveed Alam Institute of Forest Sciences, University of Swat, Charbagh Swat, Pakistan

Ahmad Ali Center for Plant Sciences and Biodiversity, University of Swat, Charbagh Swat, Pakistan

Amjad Ali Faculty of Agricultural Sciences and Technologies, Sivas University of Science and Technology, Sivas, Turkey

Arshad Ali Center for Plant Sciences and Biodiversity, University of Swat, Charbagh Swat, Pakistan

Zeeshan Ali Department of Plant Breeding and Genetics, Faculty of Agriculture and Environment, The Islamia University of Bahawalpur, Bahawalpur, Pakistan

Muhammad Tanveer Altaf Department of Field Crops, Faculty of Agriculture and Natural Sciences, Recep Tayyip Erdoğan University, Pazar, Rize, Turkey

Muhammad Arslan Ashraf Department of Botany, Government College University, Faisalabad, Pakistan

Lalarukh Azhar Department of Botany, Lahore College for Women University, Lahore, Pakistan

Swapnil Shamrao Baraskar Professor Jayashankar Telangana Agricultural University (PJTAU), Hyderabad, Telangana, India

Mehmet Bedir Faculty of Agricultural Sciences and Technologies, Sivas University of Science and Technology, Sivas, Turkey

Rashmi Bonoriya Dr. Rajendra Prasad Central Agricultural University, Pusa, Samastipur, Bihar, India

Maya Chandra Dr. Rajendra Prasad Central Agricultural University, Pusa, Samastipur, Bihar, India

Anuradha Chetukuri Professor Jayashankar Telangana Agricultural University (PJTAU), Hyderabad, Telangana, India

Sakshi Lingamurti Chinta Department of Microbiology, HPT Art's and RYK Science, Nashik, Maharashtra, India

D. Dinesh Varma Dr. Rajendra Prasad Central Agricultural University, Pusa, Samastipur, Bihar, India

Umar Farooq Department of Plant Breeding and Genetics, Faculty of Agricultural Sciences and Technology, University of Layyah, Layyah, Pakistan

Nadia Ghaffar Department of Botany, Lahore College for Women University, Lahore, Pakistan

Dipali Shantaram Gosavi Department of Microbiology, Sanjivani Art's, Commerce and Science College, Kopargaon, Maharashtra, India

Shumaila Ijaz Department of Burn and Plastic Surgery, Shenzhen Institute of Translational Medicine, The First Affiliated Hospital of Shenzhen University, School of Biomedical Engineering, Shenzhen University Medical School, Shenzhen University, Shenzhen, China

Madiha Imtiaz Department of Botany, Bacha Khan University, Charsadda, Khyber, Pakhtunkhwa, Pakistan

Aina Inam Department of Botany, Lahore College for Women University, Lahore, Pakistan

Javed Iqbal Department of Botany, Bacha Khan University, Charsadda, Khyber Pakhtunkhwa, Pakistan

Rashid Iqbal Department of Agronomy, Faculty of Agriculture and Environment, The Islamia University of Bahawalpur, Bahawalpur, Pakistan

Department of Life Sciences, Western Caspian University, Baku, Azerbaijan

Rehman Iqbal Department of Botany, Division of Science and Technology, University of Education Lahore, Punjab, Pakistan

Sumera Javad Department of Botany, Lahore College for Women University, Lahore, Pakistan

Monika Kalia Department of Entomology, Dr YS Parmar University of Horticulture & Forestry, COH & F Neri, Hamirpur, Himachal Pradesh, India

Sonika Kalia Department of Biotechnology, School of Applied and Life Sciences, Uttaranchal University, Dehradun, Uttarakhand, India

Mahmut Kaplan Department of Field Crops, Faculty of Agriculture, University of Erciyes, Kayseri, Turkey

Inam Mehdi Khan Department of Botany, Division of Science and Technology, University of Education Lahore, Punjab, Pakistan

Shoaib Khan Department of Chemistry, Abbottabad University of Science and Technology (AUST), Abbottabad, Pakistan

Kağan Kökten Faculty of Agricultural Sciences and Technologies, Sivas University of Science and Technology, Sivas, Turkey

Aslı Küçükrecep Department of Biotechnology, Faculty of Science, Mersin University, Mersin, Turkey

Venkata Chanda Sameer Kumar Professor Jayashankar Telangana Agricultural University (PJTAU), Hyderabad, Telangana, India

Sayanti Mandal Department of Chemistry & Biochemistry, Sharda School of Engineering & Science, Sharda University, Greater Noida, Uttar Pradesh, India

Parnaz Mortazavi Faculty of Agricultural Sciences and Technologies, Sivas University of Science and Technology, Sivas, Turkey

Mahlka Mukhtiar Department of Plant Breeding and Genetics, Faculty of Agricultural Sciences and Technology, MNS University of Agriculture Multan, Multan, Pakistan

Sidratul Muntaha Center for Plant Sciences and Biodiversity, University of Swat, Charbagh Swat, Pakistan

Ghulam Murtaza School of Agriculture, Yunnan University, Kunming, Yunnan, China

Zemran Mustafa Faculty of Agricultural Sciences and Technology, Sivas University of Science and Technology, Sivas, Turkey

Muhammad Naeem Department of Plant Breeding and Genetics, Faculty of Agriculture and Environment, The Islamia University of Bahawalpur, Bahawalpur, Pakistan

Shanvi Nautiyal Department of Biotechnology, School of Applied and Life Sciences, Uttaranchal University, Dehradun, Uttarakhand, India

Muhammad Nosherwan Department of Plant Breeding and Genetics, Faculty of Agriculture and Environment, The Islamia University of Bahawalpur, Bahawalpur, Pakistan

Selim Özdemir Field Crops Program, Department of Crop and Animal Production, Food, Agriculture and Livestock Vocational School, Bingol University, Bingol, Turkey

Pragati Institute of Plant and Microbial Biology, Academia Sinica, Taipei, Taiwan

Molecular and Biological Agricultural Sciences, The Taiwan International Graduate Program, Academia Sinica, Taipei, Taiwan

Graduate Institute of Biotechnology, National Chung-Hsing University, Taichung, Taiwan

Sarmad Ali Qureshi Faculty of Agricultural Sciences and Technologies, Sivas University of Science and Technology, Sivas, Turkey

Patlolla Ramakanth Reddy Professor Jayashankar Telangana Agricultural University (PJTAU), Hyderabad, Telangana, India

Shashi Ranjan Amrita Centre for Nanosciences and Molecular Medicine, Amrita Vishwa Vidyapeetham, Kochi, Kerala, India

Rizwan Rasheed Department of Botany, Government College University, Faisalabad, Pakistan

Indra Rautela Department of Biotechnology, School of Applied and Life Sciences, Uttaranchal University, Dehradun, Uttarakhand, India

Ali Raza Department of Plant Breeding and Genetics, Faculty of Agriculture and Environment, The Islamia University of Bahawalpur, Bahawalpur, Pakistan

Abdul Rehman Department of Plant Breeding and Genetics, Faculty of Agriculture and Environment, The Islamia University of Bahawalpur, Bahawalpur, Pakistan

Ruchi Department of Biotechnology, School of Applied and Life Sciences, Uttaranchal University, Dehradun, Uttarakhand, India

Muhammad Saqib Department of Agronomy, Faculty of Agriculture and Environment, The Islamia University of Bahawalpur, Bahawalpur, Pakistan

Meliha Feryal Sarıkaya Faculty of Agricultural Sciences and Technologies, Sivas University of Science and Technology, Sivas, Turkey

Sedat Severoglu Department of Field Crops, Faculty of Agriculture, University of Atatürk, Erzurum, Turkey

Vinay Sharma Center of Excellence in Genomics and Systems Biology, International Crops Research Institute for the Semi-Arid Tropics (ICRISAT), Patancheru, Hyderabad, India

Hassan Sher Center for Plant Sciences and Biodiversity, University of Swat, Charbagh Swat, Pakistan

Ajit Singh School of Biological and Environmental Sciences, University of Nottingham Malaysia, Semenyih, Selangor, Malaysia

Sabeen Rehman Soomro Faculty of Agricultural Sciences and Technologies, Sivas University of Science and Technology, Sivas, Turkey

Salma Naimatullha Soomro Faculty of Agricultural Sciences and Technologies, Sivas University of Science and Technology, Sivas, Turkey

Noopur Srivastava Department of Chemistry & Biochemistry, Sharda School of Engineering & Science, Sharda University, Greater Noida, Uttar Pradesh, India

Rada Sucur University of Novi Sad, Faculty of Agriculture, Novi Sad, Serbia

Muhammed Tatar Faculty of Agricultural Sciences and Technologies, Sivas University of Science and Technology, Sivas, Turkey

Dilek Tekdal Department of Biotechnology, Faculty of Science, Mersin University, Mersin, Turkey

Anamika Thakur ICAR-Vivekananda Parvatiya Krishi Anusandhan Sansthan, Hawalbhag, Almora, Uttarakhand, India

Niranjan Ravindra Thakur Vasantrao Naik Marathwada Krishi Vidyapeeth (VNMKV), Parbhani, Maharashtra, India

Vinod Anand Ubarhande Professor Jayashankar Telangana Agricultural University (PJTAU), Hyderabad, Telangana, India

Wahid Ullah Center for Plant Sciences and Biodiversity, University of Swat, Charbagh Swat, Pakistan

Zahid Ullah Center for Plant Sciences and Biodiversity, University of Swat, Charbagh Swat, Pakistan

Zakir Ullah Department of Plant Sciences, Faculty of Biological Sciences, Quaid-i-Azam University, Islamabad, Pakistan

Vaishnavi Dilip Umbarkar Department of Microbiology, Sanjivani Art's, Commerce and Science College, Kopargaon, Maharashtra, India

M. Vennela Dr. Rajendra Prasad Central Agricultural University, Pusa, Samastipur, Bihar, India

Priya Yadav Department of Chemistry, Amity Institute of Applied Sciences, Amity University, Noida, Uttar Pradesh, India

İlker Yüce Faculty of Agricultural Sciences and Technologies, Sivas University of Science and Technology, Sivas, Turkey

Sadia Zafar Department of Botany, Division of Science and Technology, University of Education Lahore, Punjab, Pakistan

Muhammad Abu Bakar Zia Land Information and Management System Center of Excellence (LIMS-COE) Green Pakistan Initiative, Rawalpindi, Pakistan

General Prospective of Forage Crops and World Status

Umar Farooq, Mahlka Mukhtiar, and Muhammad Abu Bakar Zia

Abstract

Forage crops play a pivotal role in global agriculture, serving as a primary source of nutrition for livestock. These crops, comprising grasses, legumes, and other herbaceous species, contribute significantly to sustainable livestock farming by ensuring a continuous supply of feed. Forage can be utilized in various forms, including direct grazing, hay, silage, and haylage, to meet seasonal feed demands. The importance of forage crops extends beyond animal nutrition, as they also enhance soil fertility, prevent erosion, and contribute to carbon sequestration, making them an essential component of agroecosystems. Globally, forage crop production varies due to climatic conditions, agronomic practices, and livestock management systems. While developed nations have well-established forage production systems supported by advanced breeding programs and mechanization, developing countries face challenges such as low yield potential, lack of improved cultivars, and inadequate storage facilities. With the increasing demand for animal products, optimizing forage crop production through improved genetics, sustainable agronomic practices, and efficient feed management strategies is critical. This chapter provides an overview of the significance of forage crops,

U. Farooq
Department of Plant Breeding and Genetics, Faculty of Agricultural Sciences and Technology, University of Layyah, Layyah, Pakistan

M. Mukhtiar
Department of Plant Breeding and Genetics, Faculty of Agricultural Sciences and Technology, MNS University of Agriculture Multan, Multan, Pakistan

M. A. B. Zia (✉)
Land Information and Management System Center of Excellence (LIMS-COE) Green Pakistan Initiative, Rawalpindi, Pakistan
e-mail: abzia@ul.edu.pk

F. S. Baloch et al. (eds.), *Forage Crops in the Omics Era*,
https://doi.org/10.1007/978-981-95-9249-4_1

their global status, and future prospects in ensuring food security and sustainable livestock production.

Keywords

Forage crops · Livestock nutrition · Sustainable agriculture · Feed management · Global forage production

1 Introduction

Deliberately grown to provide nutrition for livestock, these crops include a variety of grasses, legumes, and other herbaceous plants (Hazra 2014). In maintaining a crucial role in maintaining the balance between feed supply and demand, forage plays a crucial role throughout the season (Ogbu and Ilo 2021). For livestock farming, forage crops are essential, providing a primary source of nutrients (Chand et al. 2022). Animals either graze on the crop or feed on them directly when crops are harvested and kept as either hay, silage, or haylage (Taran 2019).

To the agricultural economy, by supporting the livestock sector, cultivated forage crops contribute significantly (Franzluebbers and Martin 2022). They improve the profitability of diary, meat, and wool industries by enhancing the productivity and health of livestock (Nosirov and Sobirova 2022). Through their root systems, forage crops improve soil structure and fertility, which helps to increase soil erosion, enhance water retention, and increase organic matter content (Haruna et al. 2020). This leads to more agricultural practices and healthier soils. By providing habitat for various species of insects, birds, and other wildlife, these crops support biodiversity (Kremen 2020). By capturing and storing atmospheric carbon dioxide, these crops play a role in carbon sequestration, helping mitigate climate change (Nazir et al. 2024).

Since ancient times, forage crops have been integral to agricultural systems (Lemaire et al. 2023). Necessitated the cultivation of plants, especially for animal feed, through the domestication of livestock (Brewer et al. 2024). By domesticated animals, early forage crops were likely wild grasses and legumes that were adapted to grazing (Perić et al. 2024). To improve their nutritional value, yield, and adaptability to different climates and soil types, the selection of breeding crops has evolved (Fita et al. 2015). To meet the needs of various livestock species, advances in agricultural practices and scientific research have led to the development of high-quality forage species (Hazra 2014). The cultivation of forage crops is a sophisticated practice in modern agriculture that involves selecting appropriate species, optimizing planting and harvesting times, and employing sustainable management techniques (Sarıkaya et al. 2023; Shah and Wu 2019). While contributing to the overall sustainability of agriculture systems, these practices ensure that forage crops continue to provide essential nutrients for livestock (Rao et al. 2015).

2 Classification of Forage Crops

In agriculture system, forage crops are essential as they provide feed for livestock and wildlife. On the basis of plant types and lifecycle characteristics, these crops can be categorized (Kremsa 2021). The classification includes grasses, legumes, and other forage types such as brassicas and forbs (Mitchell and Nelson 2017). In animal nutrition and pasture management, each category possesses its own distinctive characteristic, rendering them either applicable to a specific purpose (Allen et al. 2011). Classification understanding helps in optimizing forage use and improving the sustainability of agricultural practices. Based on their plant types, forage crops can be classified into three main categories.

Grasses: Forage crops used in agriculture system to feed livestock are called grasses. Characterized by narrow leaves and hollow stems, these plants are monocotyledons, and they are important for their fiber content and palatability (Joshi et al. 2023). Surviving one year or less, grasses can be annual or perennial, which live for multiple years, are essential in various climatic conditions, making them versatile (Mackiewicz-Walec et al. 2024).

2.1 Legumes

These are a diverse group of plants belonging to fabaceae family, known for their ability to fix atmospheric nitrogen through a symbiotic relationship with bacteria in their nodules (Raza et al. 2020; Ali et al. 2022). This process increases soil fertility, as well as making legumes crucial in sustainable agriculture. Examples of legumes include beans, chickpeas, peanuts, lentils, alfalfa, and clover. Legumes are necessary for their nutritional benefits, being rich in vitamins, fiber, and minerals (Maphosa and Jideani 2017; Nadeem and Baloch 2023). Forage helps in food security and agricultural production, as it is consumed by humans, and livestock is fed on it (Michalk et al. 2019).

2.2 Brassica

Genus of plants in the Brassicaceae family, commonly known as the cabbage or mustard family, is brassicas (Shankar et al. 2019). This group includes different crops such as broccoli, cauliflower, cabbage, turnips, and mustard (Moorthy et al. 2022). Due to the cross-shaped arrangement of their flowers, these plants are also referred to as cruciferous vegetables (Raza et al. 2020). Brassicas, which can be used as vegetables, oilseeds, and fodder, are grown primarily for their edible parts. They are known for their forest resistance, making them valuable crops in various climates and thriving in cool-season environments (Kijowska-Oberc et al. 2020). In grazing systems, brassicas play a significant role in providing high-quality forage with crude protein and digestible nutrients (TDN) levels if harvested before heading (Powell 2020).

2.3 Forbs

Herbaceous flowering plants that are not graminoid (grasses, rushes, sedges) are called forbs (Johnston and Balster 2020). They are nonwoody broadleaf plants always found in meadows and rangelands (Spaeth Jr and Spaeth Jr 2020). Many wildflowers, such as goldenrod, gentian, and firewood. Orbs have a more diverse structure for their significant ecological roles, unlike grasses, including providing food for pollinators and wildlife (Landsman et al. 2020).

3 Agronomic Characteristics of Forage Crops

Plants cultivated primarily for consumption by livestock are called forage crops (Chand et al. 2022). Ensuring high yield and quality agronomic characteristics of forage crops is crucial (Liliane and Charles 2020). The key characteristics of forage crops are as follows:

3.1 Growth Habits

Forage crops show diverse growth habits (El Sabagh et al. 2022). Like the legumes often have a spreading habit, some, such as grasses, grow in dense tufts (Trivedi 2010). How the crops are managed and harvested is influenced by growth habit (Connor et al. 2011).

3.2 Soil and Climatic Characteristics

Soil and climatic needs are varying by different forage crops. For example, legumes often require more fertile soil with good moisture retention, while the grasses thrive in well-drained soils and moderate climates (Delgado et al. 2021). For optimizing high yield and quality, adaptability to local conditions is crucial (Brito et al. 2021).

3.3 Nutritional Value and Palatability

Including proteins, minerals, and fiber, forage crops are valued for their nutritional content (Capstaff and Miller 2018). For ensuring livestock health and productivity, high nutritional value and palatability are essential. The nutrient balance affects reproduction, growth, and overall performance (Kubkomawa 2019).

3.4 Resistance to Pests and Diseases

Pest and disease resistance is a critical agronomic characteristic (Mundt 2014). To minimize losses and reduce the need for chemical interventions, forage crops must be resilient (Roberts and Mattoo 2018). To ensure sustainable production, breeding programs often focus on enhancing this trait (Phocas et al. 2016).

3.5 Global Production and Distribution

3.5.1 Major Forage Crops by Region

North America: Prominent forage cops are alfalfa, clover, ryegrass, and corn silage (Ranck 2018).

South America: Tropical grasses are Brachiaria, alfalfa, and clover (Solomon 2022).

Europe: Crops used for silage are ryegrass, clover, and different cereal crops (Reheul et al. 2017).

Africa: Sorghum, alfalfa, and Napier grass (Mwendia et al. 2017).

Australia: Alfalfa, clover, and ryegrass (McElroy et al. 2016).

Leading countries in forage crop production:

The United States: Especially for the dairy and beef industries, the largest producers of alfalfa and corn silage (Drouillard 2018).

Brazil: Supporting its extensive livestock center, notable for its large-scale production of tropical grasses and alfalfa (Neres et al. 2021).

China: Producer of sorghum, millet, and forage crops (Zhang et al. 2023).

India: Producer of sorghum and millet used as forage crops (Aruna et al. 2018).

Argentina: Renowned for the production of alfalfa and rye grass (Ojeda et al. 2016).

Case studies of successful forage crop practices:

The United States: Advanced irrigation techniques to increase yield and quality, through the use of genetically improved alfalfa varieties (Putnam 2021).

Brazil: To improve soil health and livestock productivity, implementation of integrated crop-livestock systems (Marchão et al. 2024).

Australia: To enhance forage production and maintain ecosystem health, the adoption of sustainable practices such as rotational grazing and mixed species pastures (Jaramillo et al. 2021).

4 Forage Crop Production and Management Practices

Significant experimentation has been directed at the standard development of forage crops. There are plentiful effective procedures for planting forage crops, but all utilize various essential doctrines that ensure a supportive biome for seed sprouting and seedling propagation. These comprise suitable seed positioning, nutrient-rich soil interface, optimal planting season, effective seed application proportion, diminishing challenges from alternative breeds (e.g., weeds), and defense from bugs, insects, rodents, pests, and maladies (Hazra 2014; Manoj and Shekara 2020).

Integration traditions and genetic upgrades have offered to flourish startups. Hub pivot irrigation supplies an efficient, standardized water delivery strategy for watering forage crops. Customarily, minimal pressure agricultural water management for twirls demands approximately half of the pressure rating regulation system and has performance optimization around 80–85%, judged against values of 60–70% for system integration. While deliberating, the twain lowered pump capacity compulsion demanded and elevated share of supplied water virtually supplied to the topsoil with pivots, energy usage for pivots ought to be 50 percent less for the process (Brik et al. 2022; Brown 2008). Nevertheless, precaution is required in system engineering and administration to evade water overflow and consequently deep grooving that may happen with pivots. Watering aims to supply an appropriate water volume on time to optimize fodder production (Hussein et al. 2024). Irrigation water utilized by this system will be at a pace that refuses to cultivate overland flow and will be arranged to lessen water flux beneath the root region.

Nitrogen and phosphorus symbolize the fewest macronutrients in sultry region forage grasses are extremely reactive to N-fertilizers, their implementation can result in mounting protein content, stem number, and foliage, and therefore solid content yield, due to other output variables being unlimited. N-fertilizers in forage grasses have displayed poor utilization in the tropical biome. N-fertilizers are typically implemented in rigorous supervision networks and lead to high NH2-N declines, triggering less N utilization capability. It positively impacts overall growth in fodder and forages (Singh and Sumeriya 2012; Al-Gazhal 2021). Utilizing N-fertilizers in forage grass needs to be carefully reflected. Inflation of N-fertilizers can advance the surplus of forage biomass beyond the capability of intake of fauna foraging; therefore, the grazing capacity must be tailored consistently with forage accessibility. We should recall that the answer to N-fertilizers is carefully associated with the sufficient content of P, K, and other nutrients obtainable in soil (Khan et al. 2014; Almodares et al. 2009). Only N use in forage cannot result in acceptable enhancement in plant material yield, if there are limitations of additional nutrients in the soil, less NUE is triggered.

To boost N-fertilizers' effectiveness in forage grasses, we need to recall that the notion of best management practices (BMPs) for fertilizer use demands adherence, which unites the Right time, Right place, Right rate, and Right source (4R) nutrient management (Fixen 2020). These notions are globally exercised, but the modification of 4Rs proposal rests on location-specific situations; therefore, the utilization of particular nutrient amounts for one site is inapplicable for the rest of the regions. Built on the preceding demonstration, this chapter highlights the model of 4R to clear BMPs to enhance NUE in forage grasses in order to unite all exercises ahead of the best commercial, ecological, and cultural situations.

4.1 Role of Forage Crops in Sustainable Agriculture

There are 1700 breeds of legumes worldwide, out of which around 1500 species can be utilized as fuel for livestock; in contrast, approximately 60 breeds of forage legumes have been cultivated and extensively utilized as cultured forages (Kumar et al. 2020). Legume forages have a broad spectrum of species from brief annuals to sustainable, persistent trees and minuscule non-woody species to tree-like species. Forage legumes are utilized as the basis of the husbandry animals' forage in the form of forage harvesting and pasturing, such as livestock. In the diet of milking animals, the addition of legumes in animal nutrition provides the feed with a nutritious and palatable diet (Vanzant et al. 2020). Similar to legumes, forage legumes can improve soil nitrogen, which differentiates them from other plants in terms of soil health betterment. Moreover, forage legumes are further capable of supplying a wide range of significant fertility compensation, explicitly optimizing soil organic matter, enhancing soil permeability, recycling dietary supplements, enhancing soil physical health, lessening soil acidity (pH), enhancing microbial life in the land, and preventing disease development and pest issues of fodder crops (Obrycki and Karlen 2020).

4.1.1 Forage Crops and Soil Sustainability

Soil sustainability relates to the prolonged conservation of soil productivity and output, which improves cultivation over time. Legumes (forages/grain) discharge nitrogen from disintegrating blades, roots, and lumps, leading to increased soil quality and efficiency; thus, they have an enormous possibility to refine soil environmentalism or sustainability (Stagnari et al. 2017). The vital function of legumes is enrichment in soil organic carbon, soil rehabilitation, and concentration of nitrogen, which influence the output of accomplishing crop sequencing and soil sustainability (Kumar et al. 2020).

4.1.2 Forage Crops and Crop Productivity

Prolonged agriculture primarily depends on the exploitation of chemical fertilizers and pest control agents, besides an efficient watering method. Nevertheless, it is clear that over a prolonged basis, the use of fertilizers will undermine production, and soon after initiation of rigorous agriculture, it overt output structure with a decline in yield. The introduction of legume crops into the cultivation systems is therefore considered one of the significant features that help to sustain the productivity of the soils and increase their potential to produce.

4.1.3 Forage Crops and Crop Rotations

Forage legumes provide different methods to fulfill the dietary needs of cyclical crops. Therefore, these crops are useful in decreasing the unintended consequences in Pakistan, which led to prolonged use of single crop rotation with cereals (wheat-rice) in a specific zone. Leguminous forage crops such as berseem can mend climatic nitrogen and transform it into useful parts of plants. This convertible nitrogen is not only utilized by the berseem but also increased the generation of the sequential crops in rotation (Takawale et al. 2022). Ghosh et al. (2020) recorded that the output of rice and various successive crops amplified considerably following leguminous crops compared to alternative crops. The boosted output of mustard grown with chickpea has been noted by the researchers (Singh et al. 2019). The rice-wheat cultivation framework is an extremely nutrient-depleting method, and the lengthy base roots create a scarcity of both micro- and macronutrients, which results in soil exhaustion. According to the rules of crop rotation, a curative crop is essential to add to an exhaustive framework to balance the fertility rate.

4.1.4 Forage Crops Mitigating Climate Change

Agriculture is considerably delicate to climate alterations, i.e., the alterations in thermic and moisture cycles, augmented regularity of intense climate fluctuations, and water accessibility (Sarıkaya and Tatar 2023). Climate fluctuation has remarkably impacted human communities and natural ecosystem's global warming by altering the atmospheric abundance of greenhouse gases (GHG) such as methane, carbon dioxide, and nitrous oxide (Wang et al. 2018). It is a renewed reality that the development of every kilogram of ammonium nitrate utilizing the Haber-Bosch procedure requires a substantial amount of energy (58 MJ) and releases an ample amount of greenhouse gases in the form of N2O (8.6 kg CO2 equivalent). Additionally, the application of each 100 kg of nitrogenous fertilizers is accepted to discharge 1.0 kg N2O, a greenhouse gas possessing 300 times more global warming capability than CO2 (Ahlgren et al. 2012). The addition of forage legumes in savannas and cultured forages minimizes N2O effluxes. (i) Mutually fixed N inside the nodules of legumes is restrictively accessible in the soil in a responsive state. (ii) In grass-forage legume-rooted combinations, the grass elements occupy nitrogen acquired from legumes and mineralization of soil biological material. N2O exhaust is highest in N-fertilized grass swards compared to varied grass-forage clover swards. The N2O dispersions triggered by the development of legume crops might be forecasted exclusively as an activity of the surface level and underground N inputs from forage remains during grassland rejuvenation. Effluents of N2O from legumes occur as a consequence of the decay of remains from leguminous flora, but the extent of this sort of emissions remains vague (Jeuffroy et al. 2013).

4.1.5 Forage Crops in Animal Husbandry

Pakistan is the fourth biggest milk supplier in the world. The persistent shortage of excellent feeds is a substantial restriction to financial and reasonable animal production. To utilize the maximum capacity of domesticated animals, it is necessary to breed them according to the necessity for reproduction and output. Food expenses of domesticated animals could be lessened significantly by interchanging the pricey concentrates with quality green fodder. Legume fodders are valuable sources of protein as they fix the environmental nitrogen, assisted by nodules in their roots (Lazali and Drevon 2023). Berseem and Lucerne are the basic legume fodder crops cultivated in the rabi season; concurrently, cowpea and guar are legume fodders grown in the kharif period. Moreover, these cultured fodders, a few legume trees, and undergrowth also supply nourishing fodder, especially for small ruminants.

5 Challenges of Improved Forage Crop Production

Despite a rich heritage and the existence of a few exceptional forage generation exercises, the participation of refurbished forage in all feed means is still less than one percent (Tulu et al. 2023). Owing to its low acceptance rate in the agrarian society due to distinct parameters, in the agriculture network. From the handful of research that has been conducted to recognize elements that impact the implementation of the innovations, the uplands plain faces identical limitations to welcome forage advancements. Soil scarcity, shortage of funds to buy inputs for better forage seeds, insufficiency, exorbitant planting supplies, ignorance and inexperience, crop-based cultivation framework, and inadequate outreach were recognized as the main hindrances to upgrading forage development (Mijena 2022). Land unavailability is an obstacle owing to the reluctance of agrarians to distribute land for advanced forage production relative to growing food crops cultivation, and is likely to turn intensifying as the inhabitants proceed to expand. Inefficiency of forage breed practiced after paying numerous reserves by cultivator was an additional obstacle, and employees also cited that the minimum graded trial of optimized forage fabrication elevated by the agrarians (Mengistu et al. 2021).

The scarcity of employment roles in whether agrarians accept forges is not due to the workforce getting engaged in instances when employees are needed for forges, as they are already working on other projects or fields. Therefore, labor shortage could go on to be an aspect affecting the approval of upgraded forges. Approach to agricultural outreach aid, contribution in forage workshops, and increased earnings, male population workforce, and fertilization cast an optimistic impact on embracing forage innovations. Household members, land area, geography, and distance from cultivators' abode to cultivators (Baba et al. 2011). Workshop sites are crucial aspects that substantially influence the likelihood of the adoption of enhanced forage technologies. This suggests that holding bigger land size stimulates cultivators to utilize regionally attainable grasses and explore more than the cultivators who possess smaller land. Agrarians possessing large land sizes own personal pastures and approach to get harvest remnants and stockpile them for the forthcoming drought interval to surmount food scarcity. Ensuring this itself adversely impacts enhanced forage utilization, so it is feasible to advance refined forage adoption via consciousness fabrication of agrarians with enormous land size. Empowering centers in the proximity to cultivators and fabricating agrarians' knowledge and explaining centers are considered to boost forage production, exercises, application-enhanced efficiency, producing domestic animals industry, and it is also achievable to speed up the inclusion of advanced forage in blended agrarian societies (Gebremedhin et al. 2003).

6 Future Prospects

The accessibility of genomics and bioinformatics has innovated all genetics and biotechnology, as well as repositories, which extend to comprise more varieties, breeds, and hybrid species. This knowledge can help forage producers to advance crop yield. By contrasting hybrid species series reports and employing GWAS for characteristics, for example, abundant growth tissues, amount of protein, NUPE, and particular trace elements, the dietary value and output of forage crops can be boosted. A few SNP genes that have been identified in research plants can be the destinations of genome engineering methods (Belhaj et al. 2015). Seedbed lines have also been used in a lot of forage crop research to understand gene activity. Additionally, succession data is helpful for a pattern of PCR primers that are utilized for tissue analysis. These experiments can be utilized to quickly recognize the overall wellness and nutritional adequacy of crops and particular microbes. One obstacle is probable to transmit the emerging genetic facts into forage crops. GM forage crops can be more permissible to the people because if fed to livestock and pets, their admission to the man's food supply chain is indirect. The utilization of CRISPR/Cas9 machinery could supply a satisfactory path for coercion like this and like multiple harvest practicality. Research is undertaken in forage crops; the alterations of the *Medicago sativa* squamosal promoter (DNA) Binding Protein like 9 (SPL9) have been tried and authenticated, despite defective genome modification. Performance is restricting progress currently (Subedi et al. 2023). Numerous applicant genes have been detected that could be rapidly converted into forage crops, but the technology for modification is constraining the growth of this upgraded flora. In the future, gene modification could become more embraced, especially for animal fodder crops.

References

Ahlgren S, Baky A, Bernesson S, Nordberg Å, Norén O, Hansson PA (2012) Consequential life cycle assessment of nitrogen fertilisers based on biomass–a Swedish perspective. Insciences J 2(4):80–101

Al-Gazhal SA (2021) The response of some forage grass crops to nitrogen fertilizer. Mesopotam J Agric 49(4):48–58

Ali A, Altaf MT, Nadeem MA, Karaköy T, Shah AN, Azeem H et al (2022) Recent advancement in OMICS approaches to enhance abiotic stress tolerance in legumes. Front Plant Sci 13:952759

Allen VG et al (2011) An international terminology for grazing lands and grazing animals. Grass Forage Sci 66(1):2

Almodares A, Jafarinia M, Hadi MR (2009) The effects of nitrogen fertilizer on chemical compositions in corn and sweet sorghum. Am-Eurasian J Agric Environ Sci 6(4):441–446

Aruna C et al (2018) Improving production and utilization of sorghum in Asia Indian Institute of Millets Research, India. In: Achieving sustainable cultivation of sorghum, vol 2. Burleigh Dodds Science Publishing, Philadelphia, pp 209–232

Baba SH, Wani MH, Shaheen FA, Zargar BA, Kubrevi SS (2011) Scarcity of agricultural labour in cold-arid Ladakh: extent, implications, backward bending and coping mechanism. Agric Econ Res Rev 24:391–400

Belhaj K, Chaparro-Garcia A, Kamoun S, Patron NJ, Nekrasov V (2015) Editing plant genomes with CRISPR/Cas9. Curr Opin Biotechnol 32:76–84

Brewer DJ et al (2024) Domestic plants and animals: the Egyptian origins. Oxbow Books, Oxford

Brik M, Guerrah A, Atia A (2022) Water-energy nexus: a systematic analysis and evaluation of a center-pivot irrigation system. Energy Sources Part A Recovery Util Environ Eff 44(3):8299–8313

Brito L, Bedere N, Douhard F, Oliveira HR, Arnal M, Peñagaricano F, Schinckel AP, Baes CF, Miglior F (2021) Genetic selection of high-yielding dairy cattle toward sustainable farming systems in a rapidly changing world. Animal 15:100292

Brown P (2008) Flood vs. pivot irrigation for forage crops: what are the advantages and disadvantages. In: Proceedings, California alfalfa & forage symposium and western seed conference, San Diego, CA

Capstaff NM, Miller AJ (2018) Improving the yield and nutritional quality of forage crops. Front Plant Sci 9:535

Chand S et al (2022) Agronomical and breeding approaches to improve the nutritional status of forage crops for better livestock productivity. Grass Forage Sci 77(1):11–32

Connor DJ et al (2011) Crop ecology: productivity and management in agricultural systems. Cambridge University Press, Leiden

Delgado JA et al (2021) Potential use of cover crops for soil and water conservation, nutrient management, and climate change adaptation across the tropics. Adv Agron 165:175–247

Drouillard JS (2018) Current situation and future trends for beef production in the United States of America—a review. Asian Australas J Anim Sci 31(7):1007

El Sabagh A et al (2022) Phytohormones as growth regulators during abiotic stress tolerance in plants. Front Agron 4:765068

Fita A et al (2015) Breeding and domesticating crops adapted to drought and salinity: a new paradigm for increasing food production. Front Plant Sci 6:978

Fixen PE (2020) A brief account of the genesis of 4R nutrient stewardship. Agron J 112(5):4511–4518

Franzluebbers AJ, Martin G (2022) Farming with forages can reconnect crop and livestock operations to enhance circularity and foster ecosystem services. Grass Forage Sci 77(4):270–281

Gebremedhin B, Ahmed MM, Ehui SK (2003) Determinants of adoption of improved forage technologies in crop-livestock mixed systems: evidence from the highlands of Ethiopia. Trop Grassl 37:262–273

Ghosh PK, Hazra KK, Venkatesh MS, Praharaj CS, Kumar N, Nath CP, Singh U, Singh SS (2020) Grain legume inclusion in cereal–cereal rotation increased base crop productivity in the long run. Exp Agric 56(1):142–158

Haruna SI et al (2020) Improving soil physical properties through the use of cover crops: a review. Agrosyst Geosci Environ 3(1):e20105

Hazra C (2014) Feed and forage resources for sustainable livestock development. Range Manag Agrofor 35(1):1–14. https://www.finance.gov.pk/survey_2022.html

Hussein MA, Haileslassie A, Derseh MB, Assefa TT, Riga FT, Adie A, Tebeje AK, Jones CS, Tilahun SA (2024) Enhancing irrigated forage crop production through water and nutrient management in the Ethiopian sub-humid highlands. Front Sustain Food Syst 8:1373698

Jaramillo DM et al (2021) Enhancing the sustainability of temperate pasture systems through more diverse swards. Agronomy 11(10):1912

Jeuffroy MH, Baranger E, Carrouée B, De Chezelles ELSDC, Gosme M, Hénault C, Schneider A, Cellier P (2013) Nitrous oxide emissions from crop rotations including wheat, oilseed rape and dry peas. Biogeosciences 10(3):1787–1797

Johnston MR, Balster NJ (2020) Aboveground traits define differences in biomass allocation among 16 prairie species. Agrosyst Geosci Environ 3(1):e20059

Joshi BK, Bhattarai M, Sharma S, Niru KC, Ghimire KH (2023) Millet genetic resources and diversity in Nepal. National Agriculture Genetic Resources Center (Genebank), NARC, Khumaltar, Khatmandu

Khan A, Munsif F, Akhtar K, Afridi MZ, Ahmad Z, Fahad S, Ullah R, Khan FA, Din M (2014) Response of fodder maize to various levels of nitrogen and phosphorus. Am J Plant Sci 5:2323–2329
Kijowska-Oberc J et al (2020) Adaptation of forest trees to rapidly changing climate. Forests 11(2):123
Kremen C (2020) Ecological intensification and diversification approaches to maintain biodiversity, ecosystem services and food production in a changing world. Emerg Top Life Sci 4(2):229–240
Kremsa VŠ (2021) Sustainable management of agricultural resources (agricultural crops and animals). In: Sustainable resource management. Elsevier, pp 99–145. https://doi.org/10.1016/B978-0-12-820643-6.00005-4
Kubkomawa H (2019) Nutrient requirements of livestock for sustainable productivity in tropical Africa: a review. J Emerg Trends Eng Appl Sci 10(5):247–272
Kumar D, Yadav MR, Arif M, Mahala D, Kumar D, Ghasal P, Verma R, Yadav KC (2020) Multiple agrosystem services of forage legumes towards agricultural sustainability: an overview. Indian J Agric Sci 90(8):1367–1377. https://doi.org/10.56093/ijas.v90i8.105882
Landsman AP et al (2020) Invasive grass (Microstegium vimineum) indirectly benefits spider community by subsidizing available prey. Ecol Evol 10(20):11133–11143
Lazali M, Drevon JJ (2023) Legume ecosystemic services. Communications 13:013–018
Lemaire G et al (2023) Domestic herbivores, the crucial trophic level for sustainable agriculture: avenues for reconnecting livestock to cropping systems. Agronomy 13(4):982
Liliane TN, Charles MS (2020) Factors affecting yield of crops. In: Agronomy-climate change & food security, p 9. https://doi.org/10.5772/intechopen.90672
Mackiewicz-Walec E et al (2024) Smooth brome (Bromus inermis L.)—a versatile grass: a review. Agriculture 14(6):854
Manoj KN, Shekara BG (2020) Agronomic management practices for enhancing quality and productivity of fodder: a review. Ind J Pure App Biosci 8(5):403–410
Maphosa Y, Jideani VA (2017) The role of legumes in human nutrition. In: Functional food-improve health through adequate food, vol 1, p 13. https://doi.org/10.5772/intechopen.69127
Marchão RL et al (2024) Integrated crop–livestock–forestry systems for improved soil health, environmental benefits, and sustainable production. In: Soil health series: volume 3 soil health and sustainable agriculture in Brazil, pp 19–61
McElroy M et al (2016) Interactions between cultivars of legume species (*Trifolium pratense* L., *Medicago sativa* L.) and grasses (*Phleum pratense* L., *Lolium perenne* L.) under different nitrogen levels. Can J Plant Sci 97(2):214–225
Mengistu S, Nurfeta A, Tolera A, Bezabih M, Adie A, Wolde-Meskel E, Zenebe M (2021) Livestock production challenges and improved forage production efforts in the Damot Gale District of Wolaita Zone, Ethiopia. Adv Agric (1):5553659
Michalk DL et al (2019) Sustainability and future food security—a global perspective for livestock production. Land Degrad Dev 30(5):561–573
Mijena D (2022) Improved forage production in Ethiopia: efforts done, success achieved, challenges, and future opportunities. Int J Food Sci Agric 6(3):260–266
Mitchell RB, Nelson CJ (2017) Structure and morphology of legumes and other forbs. In: Forages, an introduction to grassland agriculture, vol 1, pp 35–50
Moorthy PK et al (2022) Pests and their management in cruciferous vegetables: (cabbage, cauliflower, knol khol, broccoli, radish, turnip, beet root). In: Trends in horticultural entomology. Springer, Berlin/Heidelberg, pp 997–1011
Mundt CC (2014) Durable resistance: a key to sustainable management of pathogens and pests. Infect Genet Evol 27:446–455
Mwendia SW et al (2017) Assessment of Napier grass accessions in lowland and highland tropical environments in East Africa: productivity and forage quality. Exp Agric 53(1):27–43
Nadeem MA, Baloch FS (2023) Genome-wide association studies revealed DArTseq loci associated with seed traits in Turkish common bean germplasm. Turk J Agric For 47(4):479–496

Nazir MJ et al (2024) Harnessing soil carbon sequestration to address climate change challenges in agriculture. Soil Tillage Res 237:105959

Neres M et al (2021) Expansion of hay production and marketing in Brazil. Heliyon 7(4):e06787

Nosirov B, Sobirova F (2022) Effect of production of quality products on productivity in livestock farms. Proc Digit Econ Financ Res 1:18–24

Obrycki JF, Karlen DL (2020) Forages for conservation and improved soil quality. In: Forages: the science of grassland agriculture, vol 2, pp 227–247

Ogbu, OC, and Ilo, SU (2021) Pasture and Forage Crops Production. In: Agricultural Technology for Colleges_Revised Edition. Page 181–192. ISBN-13: 979-859-742-521-4

Ojeda JJ et al (2016) Evaluation of the agricultural production systems simulator simulating Lucerne and annual ryegrass dry matter yield in the Argentine Pampas and South-Eastern Australia. Agric Syst 143:61–75

Perić K et al (2024) The role of crop wild relatives and landraces of forage legumes in pre-breeding as a response to climate change. Agronomy 14(7):1385

Phocas F et al (2016) Towards the agroecological management of ruminants, pigs and poultry through the development of sustainable breeding programmes. II. Breeding strategies. Animal 10(11):1760–1769

Powell LS (2020) Evaluating the impact of forage brassicas in a cool season mix on stocker cattle performance and soil health conditions. University of Georgia

Putnam DH (2021) Factors influencing yield and quality in alfalfa. In: The alfalfa genome. Springer International Publishing, Cham, pp 13–27

Ranck EJ, LA Holden, JA Dillon, CA Rotz, KJ Soder (2020) Economic and environmental effects of double cropping winter annuals and corn using the Integrated Farm System Model, 103:(4), Journal of Dairy Science 3804–3815

Rao IM et al (2015) LivestockPlus: the sustainable intensification of forage-based agricultural systems to improve livelihoods and ecosystem services in the tropics. Trop Grassl Forra Trop 3(2):59–82

Raza A et al (2020) Nitrogen fixation of legumes: biology and physiology. In: The plant family Fabaceae: biology and physiological responses to environmental stresses. Springer, Singapore, pp 43–74

Reheul D et al (2017) Sustainable intensification in the production of grass and forage crops in the low countries of north-west Europe. Grass Forage Sci 72(3):369–381

Roberts DP, Mattoo AK (2018) Sustainable agriculture—enhancing environmental benefits, food nutritional quality and building crop resilience to abiotic and biotic stresses. Agriculture 8(1):8

Sarıkaya MF, Tatar M (2023) Global climate change and crop production. Uluslararası Sivas Bilim ve Teknoloji Üniversitesi Dergisi 2(2):65–78

Sarıkaya MF, İleri O, Erkovan Ş, Erkovan Hİ, Koç A (2023) Growing forage pea (Pisum arvense L.) for hay: different sowing dates and plant densities in Central Anatolia. Atatürk Üniversitesi Ziraat Fakültesi Dergisi 54(2):75–80

Shah F, Wu W (2019) Soil and crop management strategies to ensure higher crop productivity within sustainable environments. Sustainability 11(5):1485

Shankar S et al (2019) Brassicaceae-a classical review on its pharmacological activities. Int J Pharm Sci Rev Res 55(1):107–113

Singh P, Sumeriya HK (2012) Effect of nitrogen on yield, economics and quality of fodder sorghum genotypes. Ann Plant Soil Res 14(2):133–135

Singh V, Singh G, Pandey VK, Kumar M, Singh A (2019) Performance of chickpea-mustard intercropping on yield and economics of chickpea and mustard crop under different fertility management and various row combinations. Int J Curr Microbiol App Sci 8(1):236–249

Solomon JK (2022) Legumes for animal nutrition and dietary energy. In: Advances in legumes for sustainable intensification. Elsevier, London, pp 227–244

Spaeth KE Jr, Spaeth KE Jr (2020) The ecological connection in farming, ranching, and gardening. In: Soil health on the farm, ranch, and in the garden. Springer, Cham, pp 27–83

Stagnari F, Maggio A, Galieni A, Pisante M (2017) Multiple benefits of legumes for agriculture sustainability: an overview. Chem Biol Technol Agric 4:1–13

Subedi U, Burton Hughes K, Chen G, Hannoufa A, Singer SD (2023) Eliciting targeted mutations in Medicago sativa using CRISPR/Cas9-mediated genome editing: a potential tool for the improvement of disease resistance. In: Plant-pathogen interactions. Springer US, New York, pp 219–239

Takawale P, Kauthale V, Kale R (2022) Production performance and economics of forage cropping systems under irrigated condition of Western Maharashtra. Agric Sci Dig Res J. https://doi.org/10.18805/ag.D-5505

Taran T (2019) Nutritious feed for farm animals during lean period: silage and hay-a review. Forage Res 45(1):10–22

Trivedi B (2010) Grasses and legumes for tropical pasture. Indian Agricultural Research Institute, New Delhi

Tulu D, Gadissa S, Hundessa F, Kebede E (2023) Contribution of climate-smart forage and fodder production for sustainable livestock production and environment: lessons and challenges from Ethiopia. Adv Agric 2023(1):8067776

Vanzant ES, Cochran RC, Coblentz WK (2020) Animal methods for evaluating forage quality. In: Forages: the science of grassland agriculture, vol 2, pp 673–685

Wang SL, Ball E, Nehring R, Williams R, Chau T (2018) Impacts of climate change and extreme weather on US agricultural productivity: evidence and projection. In: Agricultural productivity and producer behavior. University of Chicago Press, Chicago, pp 41–75

Zhang L et al (2023) Asymmetric gene expression and cell-type-specific regulatory networks in the root of bread wheat revealed by single-cell multiomics analysis. Genome Biol 24(1):65

Breeding Success in Forage Crops

Muhammad Naeem, Ali Raza, Abdul Rehman,
Muhammad Nosherwan, Muhammad Saqib,
and Zeeshan Ali

Abstract

The breeding of forage crops is highly susceptible to the risks of lack of attention due to the numerous challenges and dilemmas associated with the vast array of species that can be bred, desired traits, target environments, crop uses, breeding procedures, and seed markets. Additionally, contrasting demands from farmers, the seed industry, and the society further exacerbate the situation. Although there have been many successful outcomes in fodder crop breeding, it is common to encounter challenges due to a lack of proper attention. They have incorporated several forms of inconsistency between the chosen environments and the desired outcomes, neglecting significant characteristics such as seed yield or employing poor methods of selecting multiple features, and not fully utilizing the available genetic diversity, among other issues. This chapter examines conventional breeding strategies in addition to modern biotechnological tools, including genetic engineering and marker-assisted selection, emphasizing their role in improving pasture crop varieties. An essential factor in maximizing the effectiveness of grassland systems is to maximize the protein/energy ratio of forage and assign it a comparable value to other types of animal feed. Grasses contain a high amount of energy in the form of both structural and nonstructural carbohydrates, whereas forage legumes are abundant in protein. The selection of a species with one or more evident phenotypic restrictions or deficits, such as weak seedling vigor, production of harmful alkaloids, or high vulnerability to a significant pest, often determines the breeding aims in advance. For species that do not have these

---

M. Naeem (✉) · A. Raza · A. Rehman · M. Nosherwan · Z. Ali
Department of Plant Breeding and Genetics, Faculty of Agriculture and Environment,
The Islamia University of Bahawalpur, Bahawalpur, Pakistan
e-mail: mniub@iub.edu.pk; abdul_rehman@iub.edu.pk

M. Saqib
Department of Agronomy, Faculty of Agriculture and Environment,
The Islamia University of Bahawalpur, Bahawalpur, Pakistan
e-mail: muhammad.saqib@iub.edu.pk

F. S. Baloch et al. (eds.), *Forage Crops in the Omics Era*,
https://doi.org/10.1007/978-981-95-9249-4_2

restrictions, breeders have the advantage of establishing breeding goals that are less strict and more adaptable.

Keywords

Forage crop breeding • Genetic improvement • Climate resilience • Forage productivity • Livestock feed improvement

1 Introduction

Forage crops play a vital role in sustainable livestock production, impacting animal health, productivity, and end-product quality. Therefore, breeding for forage crops for yield gains, disease resistance, and quality traits is crucial (Reddy et al. 2023). Climate change poses challenges to forage production, emphasizing the need for abiotic stress-tolerant varieties and disease management strategies. Forage crops adapted to harsh environments have certain unique features such as a perennially, deep root system, high resource-use efficiency (light, nutrients, and water), and low production of methane and N_2O, making them suitable for future use under climate change. The impact of climate change on forage crops underscores the importance of developing climate-smart forage crops through advanced breeding and biotechnological tools (Indu et al. 2023). In 2022–23, in Pakistan, the production of sorghum and barley declined to 23.4% and 2.6%, respectively (Finance Division 2023).

Nowadays, the tools that are essential for the breeding of forage crops are genomic selection, QTLs, GWASs, and gene editing. Genomic selection has shown promise in enhancing genetic gain for traits in forage grass species, despite challenges related to diversity and relatedness within populations (Barre et al. 2022). Genetic transformation and biotechnology enhance breeding efforts for forage crops (Zhang et al. 2015). Significant progress has been made in the molecular breeding of forage crops through the identification of various quantitative trait loci (QTLs) associated with key agronomic traits, including yield, quality, disease resistance, and tolerance to environmental stresses. For instance, a comprehensive genetic map for rye (*Secale cereale* L.) was recently developed, leading to the identification of a prominent QTL related to spike length, a key factor in determining yield potential (Che et al. 2023).

Moreover, the incorporation of genome-wide association studies (GWASs) into forage crop breeding programs has greatly enhanced our understanding of the genetic basis underlying essential traits such as productivity, quality, and resilience to stresses. GWAS enables researchers to identify associations between genetic variants across the genome and complex phenotypic traits by analyzing diverse germplasm collections. This approach provides a robust method for pinpointing genetic markers useful in breeding programs (Chen 2024). For example, in a study focused on fodder yield, GWAS proved instrumental in identifying genetic markers closely linked to biomass accumulation in perennial ryegrass (*Lolium perenne*), a crucial grass species widely used in temperate agroecosystems (Jaškūnė et al. 2020).

In addition to QTL mapping and GWAS, genome-editing technologies, particularly CRISPR/Cas9, have emerged as powerful tools for enhancing disease resistance and other desirable traits in forage crops. Research on alfalfa (*Medicago sativa*) demonstrated the potential of CRISPR/Cas9-mediated genome editing for making precise and efficient genetic modifications. By inducing a single nonhomologous end-joining-derived insertion/deletion (indel) at a specific target site, researchers achieved accurate genetic alterations to improve disease resistance (Subedi et al. 2023).

Similarly, in red clover (*Trifolium pratense*), CRISPR/Cas9 was used to knock out the isoflavone synthase gene, leading to a significant reduction in isoflavone concentrations without disrupting the plant's ability to form root nodules. This finding indicates that CRISPR/Cas9 could play a role in protecting the rhizosphere while preserving symbiotic relationships with rhizobia (Dinkins et al. 2021).

Together, these advancements highlight the transformative potential of molecular breeding techniques in optimizing forage crops for yield, quality, and resilience to biotic and abiotic stresses. By integrating QTL mapping, GWAS, and genome-editing technologies, breeders can accelerate the development of forage crops better suited to modern agricultural challenges.

The purpose of this review paper on the breeding success of forage crops is to comprehensively summarize and assess the present status of research on the genetic improvement of forage plants. The objective is to evaluate the effectiveness of selective breeding programs in improving characteristics such as yield, quality, and resilience in forage crops. The review should analyze the historical context of forage plant breeding and estimate the gradual increase in crop production over time. Furthermore, it should address the economic significance of features that are the focus of breeding programs, as well as the future possibilities and difficulties encountered in maximizing breeding methods.

2 Yield and Productivity Enhancement

Forage crops are primarily cultivated due to their rapid and alacritous biomass production for they are either grazed directly by the animals or are cut down. The nutritional quality also depends on the biomass production rate. According to another investigation, plant height positively correlates with the biomass productivity for crops and this aspect provides a solid base for assessing yield criteria. Maintaining the optimal water and nutrient supply is crucial for biomass production in forage crops (Freeman et al. 2007). Another study demonstrated that N-supply is very important for regrowth after pruning and cutting grass (Dawson et al. 2004).

As biomass production is important for improvement of the forage crops, new cultivars with rapid growth can be developed. A study in which genome-wide association studies were carried out to identify several markers associated with the heading date (Fè et al. 2015). Skøt et al. (2011) studied the flowering time in perennial ryegrass using the genome prediction (GP) model and QTL mapping as described by Nuñez and Yamada (2017). Biomass production can also be increased by manipulating genes associated with delayed senescence and thus, by introducing the 5′ flanking region of *SEE1* gene (maize cysteine protease gene) in ryegrass (*Lolium multiflorum*) increases the lifespan of the leaf by 8–16 days (Li et al. 2004). Another study was carried out by Calderini et al. (2007), using *SAG12* promotor (Arabidopsis Senescence-Associated Gene12), that also delayed the senescence in *Madia sativa* and increased the chlorophyll content as well as yield significantly. Another example is the expression of two transcription factors of the NAC family isolated from *Panicum virgatum* to recover the wild phenotypes in the *Arabidopsis-mutant* lines with defective leaf senescence (Yang et al. 2015).

Also, the fast growth of the plants must be combined with the regrowth ability of the plant in response to the cutting and pruning. This is why, recovery traits have been a point of interest in breeding trials for several decades in *Medicago* spp. and *Lolium* spp. (Wilman et al. 1977; Vance et al. 1979). Another study demonstrated that management practices are very important for soil health by performing cutting experiments by using dissolved ^{15}C and ^{13}C as well as nitrogen (Schmitt et al. 2013). Therefore, management practices are important for forage yields such as time for cutting or grazing for the subsequent regrowth ability of plants (Bumb et al. 2016).

Mixed cropping systems are also commonly used for growing forages along with legumes as the legumes are capable of synthesizing soil nitrogen through N-fixation with rhizobium, the value ranging from 32 to 115 kg/ha (Iannetta et al. 2016), requiring less fertilizer application by approximately 23–31 kg N/ha (Preissel et al. 2015) though it may also have disadvantages, i.e., competition for resources, disease incidence, and so on. Forage breeding is focused more on monoculture selection criteria as there is evidence of better mixed species crops that can planted in trials for new varieties. Unlike other crops that consider mono-cropping to be the most productive, forage crops possess the advantage that they can be grown (without lowering yield) in combination as shown by multiple trials (Tilman et al. 2006).

Dry matter (DM) refers to the portion of feed that remains after all water content has been removed. In contrast, moisture content represents the water present in the feed, which contributes to the overall weight but does not offer any nutritional value to livestock. Only the dry matter contains the essential nutrients needed by animals for various physiological functions such as maintenance, growth, reproduction, and lactation. Since livestock production is a significant part of Pakistan's agricultural sector, contributing 11.53% to the national GDP (Economic Survey of Pakistan, 2012), understanding and managing dry matter in feeds is critical for efficient livestock farming.

Several factors influence dry matter and the potential quality losses of forage. These factors include field and pre-ensiling conditions, respiration and temperature during ensiling, the fermentation process, the techniques used for covering and weighing down silage, and how aerobic deterioration is managed. Although it is impossible to eliminate all DM and quality losses, implementing effective management practices can significantly mitigate these losses and ensure that animals receive high-quality forage suitable for their nutritional needs (Rotz 2003). Enhancing these factors helps minimize losses in both the quantity and quality of dry matter.

One key practice for minimizing DM losses is ensuring rapid wilting of forage in the field. Quick drying can be achieved by conditioning the forage and spreading it evenly immediately after cutting. This approach accelerates the drying process, helping to retain the nutritional value of the forage (Wilkinson et al. 2003). Studies have shown that dry matter losses, particularly those involving leaf material, are closely linked to the DM content of the forage at the time of processing and the intensity of conditioning applied.

After harvesting, before the active fermentation phase can begin, oxygen trapped within the packed forage facilitates biological and chemical reactions that degrade nutrients and reduce energy availability. These reactions produce by-products such as water, carbon dioxide, heat, and free ammonia (McAllister and Hristov 2000). The generation of heat increases the silage temperature, leading to further nutrient losses and diminished forage quality (Holmes 2006). These losses during the fermentation process are primarily due to carbon dioxide production, which directly impacts the overall dry matter content of the silage.

To minimize DM losses during storage, it is essential to properly seal silos and silage piles. Effective sealing reduces oxygen exposure and prevents the entry of air during the storage period, thereby limiting aerobic deterioration. This practice helps preserve the integrity and nutritional quality of the silage by maintaining a low-oxygen environment conducive to efficient fermentation.

In conclusion, ensuring optimal conditions during the harvesting, ensiling, and storage phases can significantly reduce dry matter and quality losses in forage. By adopting strategies such as rapid wilting, proper conditioning, and airtight sealing, farmers can produce high-quality feed that supports the health and productivity of their livestock. This is particularly important for countries such as Pakistan, where the livestock sector plays a crucial role in the agricultural economy and food security.

Plant breeders have relied on mainly three approaches for the forage seed yield, i.e., recurrent, post-synthesis, and indirect selection (Casler et al. 1996), which likely improved the forage seed yield. Forage yield is seen to be significantly improved by indirect selection of the morphophysiological traits, i.e., leaf length, leaf area, specific leaf weight, mesophylls size. Casiler and Pederson (1996) showed the effect of management practices under defined systems, i.e., frequent clipping vs. infrequent clipping, to be effective for seed yields of forage. The selection for large leaf size and high point of stolon emergence has long been restricted due to negating correlations between these two traits; however, intensive breeding has proved to be successful in reducing these correlations in emerging cultivars (Caradus et al. 1995).

The development of a simple in vitro, two-stage technique (Tilley and Terry 1963), which is repeatable and inexpensive, is the only most important criterion for the selection of cool-season forages for measuring in vitro digestibility. Evidence of high genetic variability for the IVDMD (in vitro dry matter digestibility) is present among all forage species. However, the genetic progress for digestibility is only reported in *Medicago*, *Dactylis*, *Lolium*, *Bromus*, and *Phleum* spp. (Casler and Vogel 1999). Near-infrared reflectance spectrophotometry that is cost- and time-effective helps increase the selection efficiency (Reich and Casler 1985). Neutral detergent fiber (NDF) is a measure of total insoluble fiber that includes hemicellulose, cellulose, and lignin and it increases as the forage matures while decreasing the digestibility.

Forage yield and nutritive value are negatively correlated (Bogaert 1977); however, the improvements in forage digestibility do not cause a decline in the overall yield (Casler and Jung 1999). These quality traits (IVDMD and NDF) of forage crops are very insensitive to the Genotype × Environment (G × E) interactions. However, some traits are far more sensitive to the G × E interactions such as lignin concentration and need several environmental replications to be assessed (Casler and Jung 1999).

3 Breeding Successes in Forage Crops in Pest and Disease Resistance

Currently, pasture/forage crops exhibit more environmental compatibility and sustainability compared to other crops, and they are expected to become the foundation of future agricultural systems. Pasture and pasture crops play a crucial role in the agricultural sector of the USA and are probably the most significant crops cultivated in the country now (Leath et al. 1996). The extent of production decline caused by diseases affecting forage and pasture crops is based upon the genetic makeup of the host, the particular kind of pathogen, and the final use of the crops. To develop effective policies and markets, conduct problem-solving research, and make informed decisions about management and production investments, it is necessary to estimate the losses and inefficiencies in both forage/pasture crops and livestock enterprises (Nutter Jr and Gaunt 1996).

In Pakistan, fodder crops are susceptible to various kinds of diseases that can have major effects on both the quantity and quality of the crops. Rust is a prevalent fungal infection that impacts forage crops such as wheat, barley, and oats. The presence of orange-brown pustules on the leaves and stems is observed, resulting in a decrease in photosynthesis and ultimately causing a loss in crop output (Ijaz et al. 2021). Fungal infections are responsible for powdery mildew and downy mildew, which can impact several forage crops including alfalfa and clover. The presence of a white powdery growth on the leaves is observed, which leads to a decrease in photosynthesis and a weakening of the plants (Gupta and Singh 2010). Fusarium wilt impacts several forage crops, such as alfalfa and clover. The disease is due to soil-borne fungi and results in the withering, yellowing, and demise of plants. Anthracnose results in the formation of black lesions on the leaves, stems, and pods, which ultimately leads to a decrease in plant health and productivity (Pakdaman Sardrood and Mohammadi Goltapeh 2018). Besides affecting cereal grains, rusts also target many other grasses, causing a decrease in their forage quality. Cattle that consume rusty grasses often find them less appetizing and may develop allergies as a result of consuming contaminated feed (Lucas et al. 1992).

Aphids are sap-sucking insects that can damage forage crops and spread viral pathogens. These factors can result in inhibited development, leaf discoloration, and decreased vigor in plants (Singh and Singh 2021). Aphids also harm the crop indirectly by transferring numerous plant viruses. To preserve the crops, it is necessary to regulate the number of these organisms due to their significant economic impact. This contribution provides a detailed description of various aspects of aphid systematic and biology, including endemism, host–plant association, diversity, morphology, feeding behavior, life history, polymorphism, factors influencing it, migration, defense mechanisms, aphid-ant association, endosymbiosis, economic importance, and population management (Perring et al. 2018). Whiteflies are little insects that suck sap from plants by feeding on the undersides of leaves. Their effects include leaf yellowing, decreased photosynthesis, and the spread of viral diseases (Soumia et al. 2021). In addition to causing direct damage through eating, they also serve as vectors for economically significant plant viruses such as *Begomovirus* and *Geminivirus*. They transmit viral particles through persistent and semipermanent modes of transmission. Armyworms consume foliage, stems, and other components of plants, resulting in the loss of leaves and decreased crop production. Cutworms are the immature forms of moths that consume the stems of forage crops by severing them at the lower part. Termites can cause harm to forage crops by consuming the roots, stems, and other components of the plants (Mahendra et al. 2023). Pathogens can diminish the strength of plants, rendering them more vulnerable to other illnesses and environmental pressures.

In integrated pest management (IPM), the development and application of pest-resistant crop varieties typically follow a structured process consisting of four key stages: screening, categorization, breeding, and implementation (Stout 2014; Mortazavi et al. 2025; Ali et al. 2025). Each stage plays a crucial role in ensuring the successful identification and integration of resistant traits to minimize crop damage caused by pests. The first step involves evaluating a wide range of crop germplasm to identify genotypes that display resistance to specific pests. This process may include controlled laboratory tests, greenhouse trials, and field assessments to observe how different plant varieties respond to pest infestations. Through rigorous screening, researchers can pinpoint those plant lines that exhibit strong resistance or reduced susceptibility to pests, forming the foundation for further development. Once resistant genotypes are identified, the next step is to classify the type of resistance they exhibit. Resistance can be categorized into three primary mechanisms: This form of resistance affects the biology or behavior of the pest, often reducing its survival, growth, reproduction, or overall fitness. For example, certain plants may produce compounds that are toxic to pests. Also known as non-preference, this mechanism deters pests from feeding, laying eggs, or seeking shelter on the plant. Physical traits, such as hairy leaves or waxy surfaces, may discourage pest activity. Unlike antibiosis and antixenosis, tolerance refers to the plant's ability to endure pest damage and continue growing and producing yield without significant loss. Tolerant plants can recover from pest attacks better than susceptible ones. After identifying and categorizing the sources of resistance, the next stage involves incorporating these resistance genes into commercially viable crop varieties. This is achieved through traditional breeding techniques, such as crossing resistant genotypes with high-yielding or otherwise desirable cultivars, or through modern molecular breeding and genetic engineering. The goal is to develop new varieties that not only retain their resistance to pests but also exhibit favorable agricultural traits such as high yield, good quality, and adaptability to local growing conditions. The final stage focuses on integrating the newly developed resistant varieties into practical pest management strategies. This involves working with farmers to adopt these varieties as part of a broader IPM approach. Resistant varieties can be used alongside other pest control measures, such as crop rotation, biological control agents, and targeted pesticide applications, to create a more sustainable and effective pest management system. Education, outreach, and support for farmers are critical during this stage to ensure proper utilization and maximize the benefits of resistant varieties. By systematically progressing through these four stages—screening, categorization, breeding, and implementation—agricultural scientists and practitioners can develop robust pest-resistant crops that enhance productivity, reduce reliance on chemical pesticides, and promote more sustainable agricultural practices. Breeding tactics designed to control insects and pests in crops have undergone major advancements to tackle the problems presented by changing pest populations and climatic conditions. Transgenic methods involve the incorporation of genes from different organisms into crop plants to provide them with the ability to resist pests. As an example, crops can be genetically modified to produce insecticidal proteins obtained from *Bacillus thuringiensis* (Bt), which harm specific insect pests. The process of breeding for host–plant resistance entails the careful selection and cultivation of crop types that possess natural resistance or tolerance to particular insect pests (Mookiah et al. 2021). The resistance can be attributed to different processes, such as antixenosis (lack of preference), antibiosis (negative impact on insect growth), and tolerance (capacity to endure injury). Pyramiding is the process of combining numerous resistance genes in a single plant to create a wide-ranging resistance against multiple pests. The study investigated three distinct gene pyramiding methods, one utilizing a cascade pedigree and the other two depending on the sequence of crosses between the founding parents. The objective was to assess the transmission probabilities of the target genes and determine the required population size for each scheme (Holland 2004). Breeders can boost the resilience of crops against evolving insect populations by combining genes with various mechanisms of action. Gene editing tools such as CRISPR-Cas9 allow for accurate modification of the plant genome to introduce or fortify genes that make the plant more resistant to pests (Rato et al. 2021). This method enables breeders to carefully manipulate plants to enhance their resistance against specific pests while limiting any unintentional alterations in their genetic makeup. The sorghum shoot fly, scientifically known as *Atherigona soccata* (Rondani), is a highly destructive insect that has a significant impact on sorghum production worldwide. Four quantitative trait loci (QTLs) were discovered, with SBI-05 being the primary locus associated with non-preference for oviposition, whereas SBI-01, SBI-07, and SBI-10 contributed to resistance to shoot fly (Kiranmayee et al. 2016). A study found that marker CS132–2 was located in the same position for both leaf toughness and stem tunneling qualities in two specific QTLs of *Busseolafusca* and *Chilo partellus*. This suggests that these two traits can be enhanced using the same linked marker (Muturi et al. 2021).

Marker-assisted selection (MAS) allows the identification and selection of plants based on unique genetic markers that are linked to resistance against diseases. This enables breeders to effectively evaluate a substantial quantity of plant samples and identify those containing desirable resistance genes, hence speeding the breeding process (Ben-Ari and Lavi 2012). A gene pyramiding system that uses a cascading pedigree, implemented using a Marker-assisted selection, is cost-effective due to its span of five generations and the shortest required cumulated population size compared to other schemes (Boopathi and Boopathi 2020). The integration of bioinformatics tools and big data analysis allows breeders to leverage vast genomic datasets to identify candidate genes, predict trait performance, and optimize breeding strategies for disease resistance (Figs. 1, 2 and 3) (Hu et al. 2018). Genome assembly,

Fig. 1 Three major breeding strategies in forage crops transgenic breeding

Fig. 2 Marker-assisted selection

sequence alignment, and variant calling are common bioinformatics activities for interpreting sequenced data. However, these tasks need complex algorithms, and there are multiple computational approaches with distinct biases that compete with each other (Hwang et al. 2015).

4 Strategies and Success in Breeding Forage Crops against Abiotic Stress Conditions

The strategy for breeding forage crops is essential to improving their ability to withstand and tolerate drought conditions (Liu et al. 2023). Drought is a major environmental factor that adversely impacts crop yield, particularly in areas characterized by wetland conditions (Žaltauskaitė et al. 2021). There is a rising demand for drought-resistant cultivars because of the increasing severity and frequency of

Fig. 3 Traditional breeding

droughts caused by climate change (Sustek-Sánchez et al. 2023). Breeding success of forage plants in drought is challenging due to genetic complexity. Strategies such as masala and recurrent phenotypic selection are used, but progeny tests are ineffective for low inheritability traits (Acuña et al. 2012). The conventional and molecular approaches are combined to improve the forage crop, mainly focusing on root morphology, proline estimates, leaf rolling, and physiological activity (Rauf et al. 2016). Forage diverse plant species can enhance drought resistance by incorporating drought-resistant traits and promoting positive species interactions, reducing vulnerability and increasing resilience to drought stress (Lüscher et al. 2022). Sorghum plants can enhance drought tolerance, impacting traits such as plant height, biomass, and nutritional content. Genotypes with specific genetic compositions, such as Spidfid, have the potential to increase crop production in conditions with limited water availability (Fardin et al. 2023). The process of breeding for drought resistance in forage crops such as *Vicia faba* L. includes selecting variants with xeromorphic traits, considering the impact of temperature and precipitation on productivity, and identifying varieties with high levels of drought resistance (Kurkina et al. 2021). Maize crops can enhance drought resistance, impacting forage yield, quality, and energy productivity under stressful conditions, aiding in livestock production sustainability (Rad et al. 2021). Forage crops can become more resilient to abiotic stresses such as drought, salt, heat, and cold by using gene editing techniques such as CRISPR-Cas. This improvement allows for better adaptation to changing climates (Sustek-Sánchez et al. 2023). One notable breakthrough in enhancing forage crops against drought stress involves the use of gene editing techniques. Recent successes have permitted the alteration of genetic profiles in forage crops to offer better tolerance to environmental challenges such as drought. These genetic alterations have successfully improved biomass yield and enhanced the nutritional content of forage crops under stress circumstances. This strategy not only aids in preserving productivity in severe climatic conditions but also increases the overall resilience of the crops (Capstaff and Miller 2018).

Breeding forage crops for heat stress resistance involves selecting varieties with greater tolerance to high temperatures, which can increase productivity and sustain nutritional value in challenging environments (Nour et al. 2023). Breeding forage crops for tolerance to acute stress such as heat shock is more beneficial than focusing on long-term climate changes, as per the study on forage grass growth under future climate scenarios (Hart et al. 2022). Improving forage crops for temperate–tropical transition zones enhances adaptability to changing climates, including heat stress, by selecting traits such as genetic photoperiod responses and resilience to extreme temperatures (Quesenberry et al. 2022). Higher yield and heat resilience can improve oats' resistance to terminal heat stress, resulting in improved sustainability and productivity in challenging environments (Niazi et al. 2020). Breeding forage crops involves selecting genotypes with traits such as membrane thermostability, photosynthetic efficiency, and heat shock proteins to enhance heat tolerance, aiding in developing resilient varieties (Chaudhary et al. 2020). Through genetic modification, transgenic forage crops containing heat shock proteins (HSPs) exhibit improved heat stress tolerance. It works by protecting photosystem II and minimizing oxidative damage (Alam et al. 2013). Alfalfa for miR156 overexpression and SPL13 RNAi knockdown improves heat stress tolerance by modulating water potential, antioxidants, anthocyanin, and chlorophyll levels, as well as stress-related gene expression (Matthews et al. 2019).

Breeding forage sorghum with specific root traits such as main root length and angle can enhance salt stress resistance, positively impacting yield and performance under challenging conditions (Amombo et al. 2022). Breeding can enhance the tolerance of forage crops to salt stress by enhancing key physiological processes such as photosynthesis, antioxidant gene expression, and glyoxalase system activity. Bermuda grass has demonstrated this response to combined stress conditions (Zhou et al. 2022). A method of breeding fodder crops involves the combination of several plant varieties through hybridization and the incorporation of stress tolerance genes, including those that enhance resistance to salt stress. Intergeneric hybrids, such as *Lolium* and Festuca, have the potential to enhance genetic diversity (Singh and

Chahal 2020). Selecting genotypes with high-stress tolerance traits, such as enhanced catalase activity and mechanisms that exclude Na^+ ions, allows alfalfa to adapt to salt stress. This enhances resistance to salt stress and leads to improved yields (Benabderrahim et al. 2020).

5 Climate Change Adaptation

5.1 Breeding Cultivars of Forage Crops Resilient to Changing Environmental Conditions

Forage crops, especially legumes, play a pivotal role in maintaining productivity in organic and low-input mixed farming systems. These crops contribute significantly to soil fertility by fixing atmospheric nitrogen, improving soil structure, and supporting crop rotations. In addition, forage legumes enhance the sustainability of farming practices by reducing the need for synthetic fertilizers and increasing overall system resilience. The cultivation of maize for silage is expected to grow in importance in both northern and southern regions, driven by its high yield potential and versatility. Additionally, the use of catch crops planted in the autumn is likely to expand, serving dual purposes: enhancing environmental protection by reducing nutrient leaching and extending the grazing season by providing supplemental forage (Wilkins and Kirilov 2003). Some successes, such as biomass, nutritive value, and stress resilience improvements in forage crops, are listed in Table 1.

5.2 Impact of Climate Change on Forage Production

Climate change poses significant challenges to the reliability and quality of forage production. Its effects are projected to influence not only the quantity of forage available but also the quality, water requirements for cultivation, and large-scale rangeland vegetation dynamics. The primary impact of climate change will be seen in the overall productivity of forage crops and rangelands. Developing countries are particularly vulnerable due to their heavy reliance on agriculture, higher baseline temperatures, and limited capacity to adopt advanced technologies (Giridhar and Samireddypalle 2015). As climate change progresses, forage crops will face challenges from elevated temperatures, increased carbon dioxide concentrations, and unpredictable water availability caused by shifting precipitation patterns.

5.2.1 Adapting Forage Systems to Climate Variability

Forage crops exhibit a range of physiological and morphological adaptations that enable them to cope with moderate climatic variability. However, significant climate changes will necessitate adjustments in forage management practices and may alter the selection of forage species for different regions (Baron and Bélanger 2020). As population growth, resource scarcity, and climate change continue to shape global agriculture, developing resilient agroecosystems is essential for sustaining

Table 1 Biomass, nutritive value, and stress resilience improvements in forage crop

Forage crop	Traits	Improvements	References
Alfalfa	Biomass yield	50% yield increment has been estimated for 3 years of mass selection	Annicchiarico and Pecetti (2021)
	Nutritional quality	High protein content (21.89%) and low neutral detergent fiber content (NDF), and acid detergent fiber content (ADF)	Tucak et al. (2021)
	Stress resilience	Resistance against biotic factors including diseases (anthracnose, verticillium wilt, fusarium wilt, powdery mildew) and insects (thrip, pea aphid, beet leaf moth) has been achieved. Resistance for abiotic factors drought and salinity has also much improvement	Yang et al. (2022), Vozhehova et al. (2021), Yu et al. (2021)
Ryegrass	Biomass yield	30–40% increase in dry matter yield and optimizing photosynthesis efficiency, and using hybrid breeding techniques	Gebremedhin et al. (2020)
	Nutritional quality	Increasing crude protein content, improving digestibility by reducing lignin, enhancing water-soluble carbohydrates for better energy value	Tubritt et al. (2020)
	Stress resilience	Resistance for rust, leaf spot, and crown rust as well as pests such as the Argentine stem weevil and grass grub Lower than −10 °C temperatures with slim snow cover determined low damages of the perennial ryegrass	Miao et al. (2022), Kemesyte et al. (2017)
Maize	Biomass yield	Improvements have been reported to range from 15% to 50%, depending on factors such as genetic advancements, planting density adjustments, and climate resource utilization	Lorenz et al. (2010)
	Nutritional quality	Increasing protein content, enhancing digestibility by reducing fiber (especially lignin), and optimizing the balance of essential amino acids	Katoch et al. (2017)
	Stress resilience	Resistance to pests such as the corn earworm and diseases such as gray leaf spot Improvements in drought, heat, and soil salinity tolerance	Yang et al. (2023), Gadag et al. (2021)
Bermuda grass	Biomass yield	Exhibits high biomass yield, especially in warm climates, with potential increases of up to 30%	
	Nutritional quality	Have good digestibility and protein content	Anderson et al. (2010)
	Stress resilience	Highly drought and heat-tolerant, making it well-suited for arid regions and challenging climates	Bela (2023)

(continued)

Table 1 (continued)

Forage crop	Traits	Improvements	References
Timothy grass	Biomass yield	Producing high biomass yield per hectare under optimal conditions and being used as hay, silage, and fodder	Katoch et al. (2012)
	Nutritional quality	Highly digestible and favored for feeding small animals due to its low fiber content and high nutrient availability	Chand et al. (2022)
	Stress resilience	Moderate tolerance to cold temperatures, but it is less drought and heat-tolerant compared to other forage grasses	
Sorghum	Biomass yield	Improvement in sorghum biomass yield was achieved by combining bmr-6 allele lines and selecting hybrids with superior traits using the FAI-BLUP index and partial diallel analysis	da Silva et al. (2018)
	Nutritional quality	Dry matter digestibility of stem increased from 2% to 19% and of leaves by 2.8%	Contreras-Govea et al. (2010)
	Stress resilience	Developing drought-tolerant, heat-resistant, and pest-resistant varieties using advanced breeding techniques	Somegowda et al. (2023)

food production and protecting the environment. Resilience in this context refers to the ability of agricultural systems to absorb disturbances and reorganize while maintaining essential functions, structure, and productivity (Walker et al. 2004).

Agronomists are increasingly focusing on resilience in crop production, particularly in response to the impacts of extreme weather events linked to climate change (Shiu et al. 2012). The ability of forage crops to recover from adverse conditions is closely linked to their species-specific characteristics and their interactions within ecosystems. Diversified forage systems offer several benefits, such as improved carbon assimilation, nitrogen fixation, and nutrient cycling, all of which enhance soil microbial activity and promote resilience against drought and poor soil conditions (Islam and Ashilenje 2018). In managed systems, implementing strategies that support multiple ecological functions can boost productivity despite environmental challenges.

5.2.2 Climate-Resilient Crop Varieties and Biotechnology Advances

To ensure global food security under changing climatic conditions, developing crop varieties with enhanced tolerance to multiple abiotic stresses is critical. Recent advancements in molecular biology, particularly whole-genome sequencing and functional genomics, have paved the way for sophisticated breeding techniques. These include marker-assisted backcrossing, gene pyramiding, recurrent selection, genomic selection, and genome editing, all of which offer new opportunities to improve crop resilience (Saini et al. 2024). When plants detect water deficiency, they activate early-response genes that regulate downstream genes, leading to physiological and biochemical adaptations that increase drought survival rates.

Transcription factors are particularly important in orchestrating these drought responses (Xiong et al. 2010).

5.3 Carbohydrate Content in Forage Crops

Forage crops generally contain 50–80% dry matter (DM) in the form of carbohydrates. If carbohydrate levels are insufficient, grain supplements can be added to balance the nutritional content. Forage carbohydrates are categorized into structural carbohydrates (such as cellulose and hemicellulose) and storage carbohydrates (such as starch and water-soluble polymers such as fructans). These complex carbohydrates are broken down into simple sugars either by the animal's digestive enzymes (in nonruminants and ruminants) or by microbial fermentation in the rumen (for ruminants). The specific ratio of carbohydrates in forage crops influences digestibility, particularly when cell wall structures limit microbial access to nutrients (Capstaff and Miller 2018).

5.3.1 Challenges and Strategies for Resilient Forage–Livestock Systems

If current trends persist, climate change will continue to affect both biotic factors (such as pests and pathogens) and abiotic factors (such as temperature, water availability, and solar radiation). These challenges pose risks to sustainable agricultural production worldwide, including forage–livestock systems in the United States. Developing resilient forage–livestock systems is critical for maintaining economic viability in this sector. The goal is to adopt simple and cost-effective strategies that producers can readily implement to preserve ecosystem functions without compromising their livelihoods (Tracy et al. 2018).

5.3.2 Complementary Forage Systems for Resilience

Integrating different functional groups of forage plants can provide complementary benefits in grazing systems. For example, C3 grasses thrive during cooler seasons (spring and fall), while C4 grasses are better adapted to hot, dry summer conditions (Shay and Kubien 2013). Strategic use of both types of grasses can extend the grazing season and enhance resilience to weather variability (Tracy et al. 2010). The Noble Research Institute in Ardmore, Oklahoma, has focused on combining cool-season grasses and legumes with warm-season grasses to develop resilient forage systems. For instance, combining alfalfa (*Medicago sativa* L.) with tall fescue (*Festuca arundinacea* Schreb.) in alternating rows or a checkerboard pattern improves establishment and persistence of both species (Butler et al. 2011).

Maintaining proper soil pH, nutrient levels, and organic matter is fundamental to resilient forage systems. While limited research in the United States has specifically examined the connection between soil fertility and resilience, grasslands often grow on nutrient-poor, acidic, or highly weathered soils. Applying fertilizers to these soils can significantly enhance productivity, especially when efficient nutrient-extraction capabilities are combined with high-yielding forage varieties (Hanna et al. 2004).

6 Future Direction and Challenges

In conclusion, the breeding of forage crops holds immense potential for addressing current and future challenges in agriculture, yet it also faces significant hurdles that must be navigated. The future direction of forage crop breeding is shaped by the need to enhance traits such as resilience, nutritional value, and sustainability while tackling complex genetic landscapes, resource limitations, regulatory frameworks, and environmental pressures. One of the primary future directions in forage crop breeding is the shift toward trait-focused approaches. Breeding programs will increasingly prioritize specific traits such as drought tolerance, pest resistance, and improved nutritional content to meet the diverse needs of livestock and address challenges associated with climate change. This necessitates the integration of advanced molecular techniques such as CRISPR/Cas9 for precision breeding, allowing for faster and more targeted genetic modifications to enhance desired traits. A further important aspect of growing future fodder crops is climate resistance. Cultivars that can survive in unstable environments, such as high heat, drought, and floods, are becoming more and more necessary as climate change contributes to environmental stresses. The goal of breeding for resilience is to maintain the production and stability of forage crop systems in the face of climatic change by locating and implementing genetic features that impart resistance to these challenges. However, there are difficulties in the way of accomplishing these objectives. The genetic diversity of fodder crops makes it challenging to recognize and efficiently control favorable features. Breeding initiatives face obstacles due to a lack of money, infrastructure, and trained labor, especially in poorer nations where the need for better fodder crops is greatest. Guidelines related to genetically modified organisms (GMOs) and new breeding methods, for example, might make it more difficult to implement creative approaches and prevent access to new cultivars.

Global forage crop supply is also threatened by continuous pest and disease pressures, which calls for ongoing efforts to create resistant cultivars using both conventional breeding techniques and modern biotechnological technologies. The acceptability of genetically modified fodder crops and innovative breeding methods by consumers is still an issue that affects adoption rates and commercial acceptance. This is especially true in areas where agricultural practices are heavily influenced by consumer preferences. Despite these challenges, multidisciplinary cooperation and creativity provide opportunities for fodder crop development to advance. Breeding programs can benefit from a variety of experts by establishing collaborations among breeders, geneticists, agronomists, and other specialists.

Author Contributions

Muhammad Naeem: Conceptualizing the chapter, defining the scope, and identifying key themes.
Ali Raza: Conducting an extensive literature review and summarizing relevant studies.
Abdul Rehman: Organizing and synthesizing the collected literature into a coherent structure.
Muhammad Nosherwan: Writing the main sections of the review, including critical analysis and comparisons.
Muhammad Saqib: Editing, proofreading, and ensuring proper citations and referencing.
Zeeshan Ali: Formatting the chapter as per publisher guidelines and finalizing the submission.

References

Acuña H, Inostroza L, Tapia G (2012) Strategies for selecting drought tolerant germplasm in forage legume species. In: Water stress. In Tech, Croatia, pp 277–300
Alam I, Kim Y, Kim K, Ahn N, Lee B (2013) Generation of transgenic tall fescue plants with enhanced abiotic stress tolerance. In: Proceedings of the 22nd international Grassland congress, pp 159–160
Ali A, Ölmez F, Tatar M, Mortazavi P, Altaf MT, Turgay EB, Aktaş H, Nadeem MA, Javed J, Gou J-Y, Aasim M, Dababat AA, Baloch FS (2025) Molecular screening of septoria-resistant genes in historical Turkish bread wheat germplasm using the validated gene-specific SSR markers. Turk J Agric For 49:89–109. https://doi.org/10.55730/1300-011X.32513251
Amombo E, Ashilenje DS, Lazaar K, Hirich A, Kouisni L, Devkota KP, Oukarroum A, Cherki G, El Gharous M, Nilahyane A (2022) Will roots play a decisive role in forage sorghum production under salt stress? Environ Sci Proc 16:61
Anderson WF, Dien BS, Jung H-JG, Vogel KP, Weimer PJJBR (2010) Effects of forage quality and cell wall constituents of Bermuda grass on biochemical conversion to ethanol. Bioenergy Res 3:225–237
Annicchiarico P, Pecetti L (2021) Comparison among nine alfalfa breeding schemes based on actual biomass yield gains. Crop Sci 61:2355–2371
Baron VS, Belanger G (2020) Climate, climate-change, and forage adaptation. Forages: The science of grassland agriculture 2:151–186
Barre P, Asp T, Byrne S, Casler M, Faville M, Rognli OA, Roldan-Ruiz I, Skøt L, Ghesquière M (2022) Genomic prediction of complex traits in forage plants species: perennial grasses case. In: Genomic prediction of complex traits: methods and protocols. Springer, pp 521–541
Bela KJA (2023) Crop tolerance under biotic and abiotic stresses. Agronomy 13(12):3024
Benabderrahim MA, Guiza M, Haddad M (2020) Genetic diversity of salt tolerance in tetraploid alfalfa (Medicago sativa L.). Acta Physiol Plant 42:5
Ben-Ari G, Lavi U (2012) Marker-assisted selection in plant breeding. In: Plant biotechnology and agriculture. Elsevier, pp 163–184
Boopathi NM, Boopathi NM (2020) Marker-assisted selection (MAS). In: Genetic mapping and marker assisted selection: basics, practice and benefits. Springer, Singapore, pp 343–388
Bumb I, Garnier E, Bastianelli D, Richarte J, Bonnal L, Kazakou E (2016) Influence of management regime and harvest date on the forage quality of rangelands plants: the importance of dry matter content. AoB Plants 8:plw045
Butler TJ, Stein JD, Interrante SM, Malinowski DP (2011) Novel approaches to establishing alfalfa-tall fescue pastures in the southern great plains. Forage & Grazinglands 9:1–8
Calderini O, Bovone T, Scotti C, Pupilli F, Piano E, Arcioni S (2007) Delay of leaf senescence in Medicago sativa transformed with the ipt gene controlled by the senescence-specific promoter SAG12. Plant Cell Reports 26(5):611–615
Che Y, Yang Y, Yang Y, Wei L, Guo J, Yang X, Li X, Liu W, Li L (2023) Construction of a high-density genetic map and mapping of a spike length locus for rye. PloS one 18:e0293604
Chen S (2024) Advances in molecular breeding of forage crops: Technologies, applications and prospects. Agriculture 14:279
Capstaff NM, Miller AJ (2018) Improving the yield and nutritional quality of forage crops. Front Plant Sci 9:338501
Caradus J, Hay R, Woodfield D (1995) The positioning of white clover cultivars in New Zealand. NZGA Res Prac Ser 6:45–49
Casiler M, Pederson G (1996) Host resistance/tolerance and its deployment. In: Pasture and forage crop pathology. American Society of Agronomy, Madison, pp 473–507
Casler MD, Jung HJG (1999) Selection and evaluation of smooth bromegrass clones with divergent lignin or etherified ferulic acid concentration. Crop Sci 39:1866–1873
Casler M, Vogel KP (1999) Accomplishments and impact from breeding for increased forage nutritional value. Crop Sci 39:12–20

Casler M, Pedersen JF, Eizenga G, Stratton S (1996) Germplasm and cultivar development. In: Cool-season forage grasses, vol 34. American Society of Agronomy, Madison, pp 413–469

Chand S, Indu, Singhal RK, Govindasamy P (2022) Agronomical and breeding approaches to improve the nutritional status of forage crops for better livestock productivity. Grass Forage Sci 77:11–32

Chaudhary S, Devi P, Bhardwaj A, Jha UC, Sharma KD, Prasad PV, Siddique KH, Bindumadhava H, Kumar S, Nayyar H (2020) Identification and characterization of contrasting genotypes/cultivars for developing heat tolerance in agricultural crops: current status and prospects. Front Plant Sci 11:587264

Contreras-Govea FE, Marsalis MA, Lauriault LM, Bean BW (2010) Forage sorghum nutritive value: a review. Forage Grazingl 8:1–6

da Silva MJ, Carneiro PCS, de Souza Carneiro JE, Damasceno CMB, Parrella NNLD, Pastina MM, Simeone MLF, Schaffert RE, da Costa Parrella RAJIC (2018) Evaluation of the potential of lines and hybrids of biomass sorghum. Ind Crops Prod 125:379–385

Dawson L, Thornton B, Pratt S, Paterson E (2004) Morphological and topological responses of roots to defoliation and nitrogen supply in Lolium perenne and Festuca ovina. New Phytol 161:811–818

Dinkins RD, Hancock J, Coe BL, May JB, Goodman JP, Bass WT, Liu J, Fan Y, Zheng Q, Zhu H (2021) Isoflavone levels, nodulation, and gene expression profiles of a CRISPR/Cas9 deletion mutant in the isoflavone synthase gene of red clover. Plant Cell Reports 40:517–528

Fardin F, Sani B, Moaveni P, Afsharmanesh G, Mozafari H (2023) Nutritional value and agronomic traits of forage sorghum under drought stress. Biocatal Agric Biotechnol 48:102624

Fè D, Cericola F, Byrne S, Lenk I, Ashraf BH, Pedersen MG, Roulund N, Asp T, Janss L, Jensen CS (2015) Genomic dissection and prediction of heading date in perennial ryegrass. BMC Genomics 16:1–15

Finance Division (2023) Pakistan economic survey 2022–23. Finance Division, Government of Pakistan

Freeman KW, Girma K, Arnall DB, Mullen RW, Martin KL, Teal RK, Raun WR (2007) By-plant prediction of corn forage biomass and nitrogen uptake at various growth stages using remote sensing and plant height. Agron J 99:530–536

Gadag R, Bhat JS, Mukri G, Gogoi R, Suby S, Das AK, Yadav S, Yadava P, Nithyashree M, Gopalakrishna Naidu (2021) Resistance to biotic stress: theory and applications in maize breeding. In: Genomic designing for biotic stress resistant cereal crops. Springer, Cham, pp 129–175

Gebremedhin A, Badenhorst P, Wang J, Shi F, Breen E, Giri K, Spangenberg GC, Smith K (2020) Development and validation of a phenotyping computational workflow to predict the biomass yield of a large perennial ryegrass breeding field trial. Front Plant Sci 11:689

Gupta P, Singh J (2010) Disease management in forage legumes. In: Forage legumes, pp 138–141

Giridhar K, Samireddypalle A (2015) Impact of climate change on forage availability for livestock. Climate change impact on livestock: adaptation and mitigation 97–112

Hanna WW, Chaparro CJ, Mathews BW, Burns JC, Sollenberger LE, Carpenter JR (2004) Perennial pennisetums. Warm-season (C4) grasses 45:503–535

Holmes B (2006) Density in silage storage. Silage for Dairy Farms: Growing, Harvesting, Storing and Feeding Conference Proceedings (NRAES-181). Natural Resource, Agriculture and Engineering Service, Ithaca, New York

Hart EH, Christofides SR, Davies TE, Rees Stevens P, Creevey CJ, Müller CT, Rogers HJ, Kingston-Smith AH (2022) Forage grass growth under future climate change scenarios affects fermentation and ruminant efficiency. Sci Rep 12:4454

Holland JB (2004) Implementation of molecular markers for quantitative traits in breeding programs-challenges and opportunities. In: New directions for a diverse planet. Crop Science Society of America

Hu H, Scheben A, Edwards D (2018) Advances in integrating genomics and bioinformatics in the plant breeding pipeline. Agriculture 8:75

Hwang S, Kim E, Lee I, Marcotte EM (2015) Systematic comparison of variant calling pipelines using gold standard personal exome variants. Sci Rep 5:17875

Iannetta PP, Young M, Bachinger J, Bergkvist G, Doltra J, Lopez-Bellido RJ, Monti M, Pappa VA, Reckling M, Topp CF (2016) A comparative nitrogen balance and productivity analysis of legume and non-legume supported cropping systems: the potential role of biological nitrogen fixation. Front Plant Sci 7:1700

Ijaz S, Haq IU, Habib Z, Mukhtar S, Nasir B (2021) Biotechnological applications for developing resistance against biotic and abiotic stresses and other quality traits in fodder crops. In: Sustainable winter fodder. CRC Press, pp 31–80

Indu I, Mehta BK, Shashikumara P, Gupta G, Dikshit N, Chand S, Yadav PK, Ahmed S, Singhal RK (2023) Corrigendum to: forage crops: a repository of functional trait diversity for current and future climate adaptation. Crop Past Sci 74:1103–1103

Islam MA, DS Ashilenje (2018) Diversified forage cropping systems and their implications on resilience and productivity. Sustainability 10:3920

Jaškūnė K, Aleliūnas A, Statkevičiūtė G, Kemešytė V, Studer B, Yates S (2020) Genome-wide association study to identify candidate loci for biomass formation under water deficit in perennial ryegrass. Frontiers in Plant Science 11:570204

Katoch R, Thakur M, Kumar N, Bandhari JJ (2012) Golden Timothy–present status and future perspectives in North-West Himalayas. Range Manag Agrofor 33:1–7

Katoch R, Apoorva TA, Sood SJFR (2017) Improving nutritive value and digestibility of maize stover—a review. Forage Res 43:174–180

Kemesyte V, Statkeviciute G, Brazauskas G (2017) Perennial ryegrass yield performance under abiotic stress. Crop Sci 57:1935–1940

Kiranmayee KU, Kishor PK, Hash CT, Deshpande SP (2016) Evaluation of QTLs for shoot fly (Atherigona soccata) resistance component traits of seedling leaf blade glossiness and trichome density on sorghum (Sorghum bicolor) chromosome SBI-10L. Trop Plant Biol 9:12–28

Kurkina Y, Batlutskaya I, Klueva V, Degtyareva K, Mackanina O (2021) Breeding of Vicia faba L. in relation to drought resistance. IOP Conf Ser Earth Environ Sci 775:012015. IOP Publishing

Leath K, Welty R, Pratt R, Sonoda R (1996) Pasture/forage crops and diseases in the United States. In: Pasture and forage crop pathology. American Society of Agronomy, Madison, pp 33–58

Li Q, Robson PR, Bettany AJ, Donnison IS, Thomas H, Scott IM (2004) Modification of senescence in ryegrass transformed with IPT under the control of a monocot senescence-enhanced promoter. Plant Cell Rep 22:816–821

Liu S, Wang H, Qin F (2023) Genetic dissection of drought resistance for trait improvement in crops. Crop J 11(4):975–985

Lorenz A, Gustafson T, Coors J, De Leon NJCS (2010) Breeding maize for a bioeconomy: a literature survey examining harvest index and stover yield and their relationship to grain yield. Crop Sci 50:1–12

Lucas GB, Campbell CL, Lucas LT, Lucas GB, Campbell CL, Lucas LT (1992) Diseases caused by airborne fungi. In: Introduction to plant diseases: identification and management. Van Nostrand Reinhold, New York, pp 192–242

Lüscher A, Barkaoui K, Finn JA, Suter D, Suter M, Volaire F (2022) Using plant diversity to reduce vulnerability and increase drought resilience of permanent and sown productive grasslands. Grass Forage Sci 77:235–246

Mahendra K, Anokhe A, Thakur S (2023) A brief overview of nutri-millet's insect pests and their management. Indian Farm 73:17–20

Matthews C, Arshad M, Hannoufa A (2019) Alfalfa response to heat stress is modulated by microRNA156. Physiol Plant 165:830–842

Miao C, Zhang Y, Bai X, Qin TJL (2022) Insights into the response of perennial ryegrass to abiotic stress: underlying survival strategies and adaptation mechanisms. Life 12:860

Mookiah S, Sivasubramaniam B, Thangaraj T, Govindaraj S (2021) Host plant resistance. In: Molecular approaches for sustainable insect pest management. Springer, Singapore, pp 1–56

Mortazavi P, Ali A, Tatar M, Ölmez F, Altaf MT, Nadeem MA et al (2025) Molecular screening of diverse tomato germplasm for root-knot nematode resistance using the Mi23 marker. Physiol Mol Plant Pathol 136:102607

Muturi PW, Mgonja M, Rubaihayo P, Mwololo JK (2021) QTL mapping of traits associated with dual resistance to the African stem borer (Busseola fusca) and spotted stem borer (Chilo partellus) in sorghum (Sorghum bicolor). Int J Genomics 2021:7016712

McAllister T, Hristov A (2000) The fundamentals of making good quality silage. Adv Dairy Technol 12:381–399

Niazi IAK, Akhtar S, Kohli S, Naveed A, Rauf S, Shehzad M (2020) Oat (Avena sativa L.) advanced lines outperform existing cultivars for forage yield and its components under terminal heat stress. Pak J Agric Sci 57:327–331. https://doi.org/10.21162/PAKJAS/2020.8557

Nour R, Sanchez-Ballesta MT, Mounawer B (2023) Influence of drought and heat stresses in fodder legumes productivity and nutritive value. J Oasis Agric Sustain Dev 5:18

Nuñez FD, Yamada T (2017) Molecular regulation of flowering time in grasses. Agronomy 7:17

Nutter F Jr, Gaunt R (1996) Recent developments in methods for assessing disease losses in forage/pasture crops. In: Pasture and forage crop pathology. American Society of Agronomy, Madison, pp 93–118

Pakdaman Sardrood B, Mohammadi Goltapeh E (2018) Weeds, herbicides and plant disease management. In: Sustainable agriculture reviews 31: biocontrol. Springer, Cham, pp 41–178

Perring TM, Battaglia D, Walling LL, Toma I, Fanti P (2018) Aphids: biology, ecology, and management. In: Sustainable management of arthropod pests of tomato. Elsevier, pp 15–48

Preissel S, Reckling M, Schläfke N, Zander P (2015) Magnitude and farm-economic value of grain legume pre-crop benefits in Europe: a review. Field Crop Res 175:64–79

Quesenberry KH, Rios EF, Kenworthy KE, Blount AR, Reith PE (2022) Breeding forages with climate resiliency in temperate/tropical transition zones. Grass Forage Sci 77:124–130

Rad RD, Sharifabad HH, Torabi M, Azizinejad R, Salemi H, Heidari Soltanabadi M (2021) Impact of drought stress on biochemical responses, energy, and water productivity on maize forage (Zea mays L.). SN Appl Sci 3:1–11

Rato C, Carvalho MF, Azevedo C, Oblessuc PR (2021) Genome editing for resistance against plant pests and pathogens. Transgenic Res 30:427–459

Rauf S, Al-Khayri JM, Zaharieva M, Monneveux P, Khalil F (2016) Breeding strategies to enhance drought tolerance in crops. In: Advances in plant breeding strategies: agronomic, abiotic and biotic stress traits. Springer, Cham, pp 397–445

Reddy NK, Rakhonde G, Purushotham P, Patel PS, Ahale S (2023) Impact of climate change on forage crop production with special emphasis on diseases and mitigation strategies through breeding and molecular approaches. In: Molecular interventions for developing climate-smart crops: a forage perspective. Springer, Singapore, pp 75–97

Reich J, Casler M (1985) Effect of maturity and alfalfa competition on expected selection response for smooth bromegrass forage quality traits 1. Crop Sci 25:635–640

Rotz CA (2003) How to maintain forage quality during harvest and storage. Advances in Dairy Tecnology 15:227–239

Saini DK, Kumar S, Kaur R (2024) Applying Genomics Resources to Accelerate the Development of Climate Resilient Crops. Pages 43–120 Adapting to Climate Change in Agriculture-Theories and Practices: Approaches for Adapting to Climate Change in Agriculture in India. Springer

Shay PE, Kubien DS (2013) Field analysis of photoprotection in co-occurring cool climate C 3 and C 4 grasses. Physiologia Plantarum 147:316–328

Shiu CJ, Liu SC, Fu C, Dai A, Sun Y (2012) How much do precipitation extremes change in a warming climate? Geophysical Research Letters 39

Stout MJ (2014) Host-plant resistance in pest management. Pages 1–21 Integrated pest management. Elsevier

Subedi U, Burton Hughes K, Chen G, Hannoufa A, Singer SD (2023) Eliciting targeted mutations in Medicago sativa using CRISPR/Cas9-mediated genome editing: A potential tool for the improvement of disease resistance. Pages 219–239 Plant-Pathogen Interactions. Springer

Schmitt A, Pausch J, Kuzyakov Y (2013) C and N allocation in soil under ryegrass and alfalfa estimated by 13 C and 15 N labelling. Plant Soil 368:581–590

Singh A, Chahal HS (2020) Management of abiotic stress in forage crops. In: Abiotic stress in plants, vol 13. IntechOpen, London

Singh R, Singh G (2021) Aphids. In: Polyphagous pests of crops. Springer, Singapore, pp 105–182
Skøt L, Sanderson R, Thomas A, Skøt K, Thorogood D, Latypova G, Asp T, Armstead I (2011) Allelic variation in the perennial ryegrass FLOWERING LOCUS T gene is associated with changes in flowering time across a range of populations. Plant Physiol 155:1013–1022
Somegowda VK, Reddy SD, Gaddameedi A, Kiranmayee KU, Naravula J, Kishor PK, Penna S (2023) Genomics breeding approaches for developing Sorghum bicolor lines with stress resilience and other agronomic traits. Curr Plant Biol 37:100314
Soumia P, Guru Pirasanna Pandi G, Krishna R, Ansari WA, Jaiswal DK, Verma JP, Singh M (2021) Whitefly-transmitted plant viruses and their management. In: Emerging trends in plant pathology. Springer, Singapore, pp 175–195
Sustek-Sánchez F, Rognli OA, Rostoks N, Sõmera M, Jaškūnė K, Kovi MR, Statkevičiūtė G, Sarmiento C (2023) Improving abiotic stress tolerance of forage grasses–prospects of using genome editing. Front Plant Sci 14:1127532
Tracy B, Foster J, Butler T, Islam M, Toledo D, Vendramini J (2018) Resilience in forage and grazinglands. Crop Science 58:31–42
Tracy BF, Maughan M, Post N, Faulkner DB (2010) Integrating annual and perennial warm-season grasses in a temperate grazing system. Crop Science 50:2171–2177
Tilley J, Terry R (1963) A two-stage technique for the in vitro digestion of forage crops. Grass Forage Sci 18:104–111
Tilman D, Reich PB, Knops JM (2006) Biodiversity and ecosystem stability in a decade-long grassland experiment. Nature 441:629–632
Tubritt T, Delaby L, Gilliland T, O'Donovan M (2020) The relationship between the grazing efficiency and the production, morphology and nutritional traits of perennial ryegrass varieties. J Agric Sci 158:583–593
Tucak M, Ravlić M, Horvat D, Čupić TJA (2021) Improvement of forage nutritive quality of alfalfa and red clover through plant breeding. Agronomy 11:2176
Van Bogaert G (1977) Scope for improving the yield of grasses and legumes by breeding and selection. In: Proceedings of the international meeting on animal production from temperate grasslands. Animal Production Association, Agricultural Institute, Belclare, Tuam, Co. Galway, Ireland, pp. 29–35
Vance CP, Heichel GH, Barnes DK, Bryan JW, Johnson LE (1979) Nitrogen fixation, nodule development, and vegetative regrowth of alfalfa (Medicago sativa L.) following harvest. Plant Physiol 64:1–8
Vozhehova R, Tyshchenko A, Tyshchenko O, Dymov O, Piliarska O, Lykhovyd P (2021) Evaluation of breeding indices for drought tolerance in alfalfa (Medicago) genotypes. Sci Pap Ser A Agron 64:435–444
Wilman D, Droushiotis D, Mzamane MN, Shim J (1977) The effect of interval between harvests and nitrogen application on initiation, emergence and longevity of leaves, longevity of tillers and dimensions and weights of leaves and 'stems' in Lolium. J Agric Sci 89:65–79
Walker B, Holling CS, Carpenter SR, Kinzig A (2004) Resilience, adaptability and transformability in social–ecological systems. Ecology and society 9
Wilkins R, Kirilov A (2003) Role of forage crops in animal production systems
Wilkinson J, Bolsen K, Lin C, Buxton D, Muck RE, Harrison J (2003) Silage science and technology. Agronomy Monograph 42:1–30
Xiong X, James VA, Zhang H, Altpeter F (2010) Constitutive expression of the barley HvWRKY38 transcription factor enhances drought tolerance in turf and forage grass (Paspalum notatum Flugge). Molecular Breeding 25:419–432
Yang J, Worley E, Torres-Jerez I, Miller R, Wang M, Fu C, Wang Z-Y, Tang Y, Udvardi M (2015) PvNAC1 and PvNAC2 are associated with leaf senescence and nitrogen use efficiency in switchgrass. Bioenergy Res 8:868–880
Yang B, Zhao Y, Guo ZJP (2022) Research progress and prospect of alfalfa resistance to pathogens and pests. Plants 11:2008
Yang Z, Cao Y, Shi Y, Qin F, Jiang C, Yang SJMP (2023) Genetic and molecular exploration of maize environmental stress resilience: toward sustainable agriculture. Mole Plant 16:1496–1517

Yu R, Wang G, Yu X, Li L, Li C, Song Y, Xu Z, Zhang J, Guan CJPB (2021) Assessing alfalfa (Medicago sativa L) tolerance to salinity at seedling stage and screening of the salinity tolerance traits. Plant Biol 23:664–674

Žaltauskaitė J, Dikšaitytė A, Miškelytė D, Sujetovienė G, Januškaitienė I, Kacienė G (2021) Forage crops nitrogen use efficiency changes under drought conditions. In: CEST 2021: proceedings of the 17th international conference on environmental science and technology, Athens, Greece, 1 to 4 September 2021. Global Network of Environmental Science and Technology, Athens, pp 1–5

Zhang HD, Chen P, Zhang J, Gao JD, Li JL, Zhang YM (2015) Breeding forage crop meets the needs of a sustainable agriculture. Adv Mater Res 1073:1104–1107

Zhou X, Yin Y, Wang G, Amombo E, Li X, Xue Y, Fu J (2022) Mitigation of salt stress on low temperature in bermudagrass: resistance and forage quality. Front Plant Sci 13:1042855

Genome Editing and Forage Crops

Zemran Mustafa

Abstract

Forage crops are vital for livestock nutrition and agricultural sustainability. However, they face many challenges such as diseases, pests, climate change, and low yields. Genome editing technology has emerged as an important tool that promises fixing these challenges by increasing tolerance to abiotic stress and diseases and improving nutritional quality and yield. This revolutionary technology can be applied to improve the genetic characteristics of plants, including forage crops.

This chapter examines the progress, applications, and difficulties in genome editing technologies. Significant accomplishments in genome editing such as an increase in protein content, reduction of anti-nutritional elements, and enhancing resistance to drought and salinity will be discussed thoroughly. Genome editing together with genomic selection and synthetic biology has accelerated generation of climate-resilient forage crops. Even though there is a great potential in adoption of genome editing technology issues such as off-target mutations and regulatory obstacles, public acceptance slows down the process. There is an immediate necessity for international cooperation and technology-friendly legislation for this technology to progress and achieve its full potential.

This chapter defends the idea of improving and implementing genome editing in forage crops and in agriculture as a whole in order to preserve the environment and provide global security.

Z. Mustafa (✉)
Faculty of Agricultural Sciences and Technology, Sivas University of Science and Technology, Sivas, Turkey

F. S. Baloch et al. (eds.), *Forage Crops in the Omics Era*,
https://doi.org/10.1007/978-981-95-9249-4_3

Keywords

Forage · CRISPR/Cas9 · Abiotic and biotic stress tolerance · Quality and quantity improvement · Climate change and adaptation · Regulatory frameworks

1 Introduction

1.1 Genome Editing

In the past few years, genome editing in plants has advanced a lot, which has altered agricultural biotechnology and plant breeding. The origins of genome editing can be traced back to the early days of genetic engineering when common methods such as mutagenesis and somatic hybridization were used to make mutations in genes. In spite of this, these methods were not always precise and required a lot of work. Thus, more advanced tools were in great demand to make position specific changes in plant genomes.

The invention of zinc-finger nucleases (ZFNs) and transcription activator-like effector nucleases (TALENs), which are sequence-specific nucleases, allowed scientists to make precise changes in specific parts of the genomes (Kamburova et al. 2017; Sedeek et al. 2019). These early genome editing methods laid the groundwork for CRISPR/Cas9 system, which is easier to manipulate, as well as implement in genome editing (Mao et al. 2019; Chen et al. 2019). The discovery of the CRISPR/Cas9 system paved the way to uncover gene functions to improve crops by designing new varieties that are resistant to disease and drought and have better nutritional values (Sedeek et al. 2019). Using this technology, it was possible to edit multiple genes at once increasing the pace of advance in plant breeding, making it easier to deal with challenges such as climate change and food security (Chen et al. 2019; Kaul et al. 2020).

Besides CRISPR/Cas9, other CRISPR-related systems, such as CRISPR/Cpf1, were explored for the potential of genome editing (Kim et al. 2017). One of the advantageous feature of these technologies was usage of purified CRISPR/Cas9 ribonucleoproteins (RNPs), which gives plant biotechnology researchers opportunities to edit genomes without integrating a foreign DNA in the genome (Malnoy et al. 2016). These new delivery systems such as RNP complexes have boosted the effectiveness of genome editing in notoriously recalcitrant plants (Yanagawa et al. 2023; Odahara et al. 2021). Genome editing in plants using a CRISPR/Cas system involves choosing the plant of interest, finding the specific genes related to the desired trait, designing one or more guide RNAs (gRNA) to modify the gene and choose the delivery system. Selecting a codon-optimized Cas gene for the crop in question will enhance the success rate. Following the transformation of cells with the CRISPR/Cas system, the entire plant is cultivated from the transformed totipotent cells, and the edits are verified through PCR, sequencing and confirming the desired traits (Fig. 1).

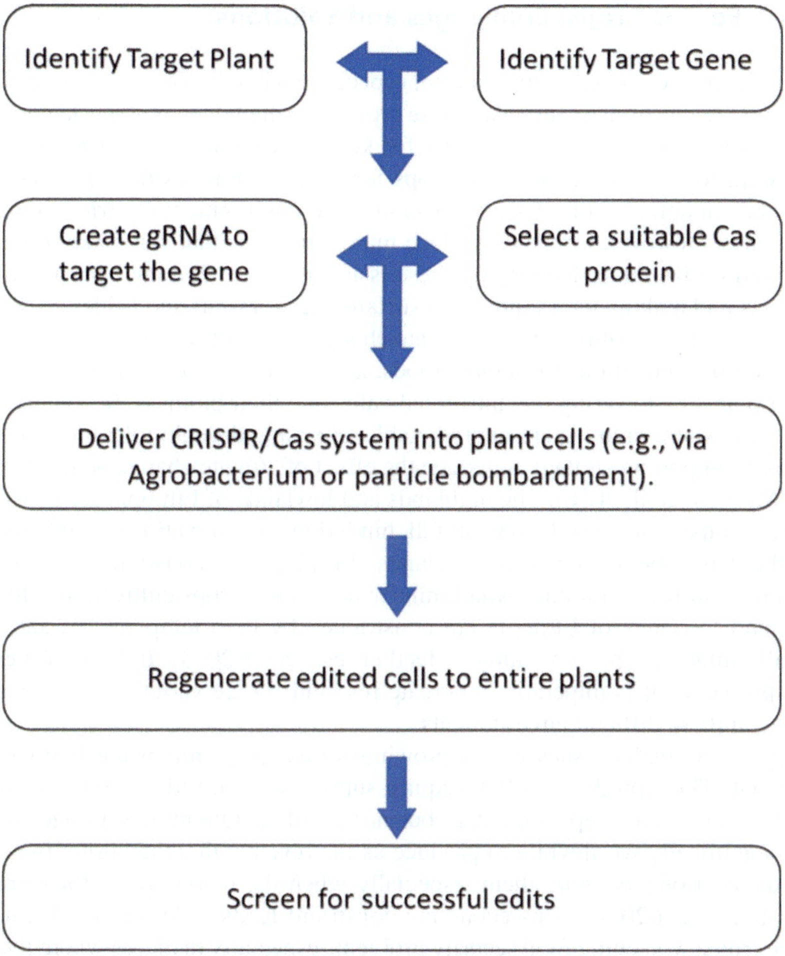

Fig. 1 Flowchart of the genome editing process in plants using CRISPR/Cas system

Even though genome editing technology has come a long way, it still faces a lot of challenges, especially when it comes to regulatory systems and public acceptance. In numerous instances, calling genome-edited plants genetically modified organisms (GMOs) changes how they are viewed, sold, and used (Berg et al. 2021). As technology improves, it is important for everyone involved, including scientists, lawmakers, and the public, to engage in a constructive conversation about ethical issues and make sure that the benefits of genome editing are used in a responsible way (Helliwell et al. 2019).

1.2 Forage Crops: Challenges and Solutions

Forage crops are crucial to agriculture, especially when in livestock production, as they provide nutrition to animals. These crops, which include grasses, legumes, and other herbaceous plants, are essential for keeping ruminant livestock healthy and producing fodder. The growing of crops for fodder improves the nutrition of animals and supports healthy soil, biodiversity, and environmentally friendly farming techniques. However, forage crops face many problems that can make it harder to grow and utilize them. Identifying these issues is key to coming up with effective solutions and making their cultivation sustainable (Capstaff and Miller 2018).

As it is for most of the plants, climate change is one of the biggest challenge for forage crops cultivation. Droughts, floods, and other extreme weather events harm fodder plants by lowering the nutritional value and their biomass. In countries such as Jordan, where rainfalls are unpredictable, choosing drought tolerant forage varieties is an important strategy to reduce the effect of climate change on their production (Massimi et al. 2018). The highlands and lowlands of Ethiopia are constrained by high temperatures and heavy rainfall, hindering the cultivation of feed crops. On the other hand, the highlands and lowlands of Ethiopia are constrained by high temperatures and heavy rainfall, which hinder the forage crops cultivation. The highlands and lowlands of Ethiopia are constrained by high temperatures and heavy rainfall, hindering the cultivation of feed crops (Desta 2023). In light of these climate problems, it is imperative to create resilient forage varieties that are able to survive in these difficult environments.

One of the major issues is that growing forage crops might not be financially beneficial. Thus, producers often require support and subsidies from government agencies in order to keep their forage businesses afloat. Oftentimes, forage crops are not economically worthwhile to produce as the revenue that they bring in is lower than the cost of producing them, especially when the market price for them goes down (Kadıoğlu 2024). This economic constraint leads to less land allocated for forage crops, worsening food security problems especially in places where livestock production is important. Hence, improving the economic success of forage crop farmers through better management and easier access to markets will encourage their long-term production (Leal et al. 2023).

Controlling weeds is a significant challenge in forage crops growing areas. They need to be managed effectively to prevent quality and quantity losses. Glyphosate resistant forage varieties have been used intensively as a way to control weeds, making herbicide use an effective tool. However, continuous use of these pesticides impacts the environment negatively and causes spread of the pesticide-resistant weeds. Thus, using a mix of cultural, mechanical, and chemical methods to control weeds is indispensable (Wilson and Burgener 2009). The genetic improvement of forage crops has been lacking far behind that of main food crops such as corn and soybeans. Over the last 20 years, while significant improvements in yield have been observed in main crops, it is hard to say the same for the forage crops. Multiple factors including insufficient funding for research, a lack of genetic resources, and the difficulty of breeding fodder crops has caused this discrepancy (Ruckle et al. 2017).

To address this problem, research funds for forage crops must be increased and modern breeding techniques such as genomic selection must be implemented in order to enhance the yield and quality of these crops (Capstaff and Miller 2018). Soil health should be taken into account when producing forage crops, as the repeated cultivation of these crops without efficient soil management measures might cause nutrient depletion and soil degradation (Xu 2024). Adding forage crops to crop rotations and using cover cropping systems can help sustainability by increasing the amount of organic matter and nutrients in the soil (Blanco-Canqui et al. 2013; Nielsen et al. 2015). Providing the necessary tools and advising the farmers about these practices would reduce the misuse of the land.

2 Applications of Genome Editing in Forage Crops

2.1 Enhancing Nutritional Quality

To tackle the problems of food security and nutritional deficiencies, genome editing represents an important advancement in agricultural biotechnology for the improvement of nutritional quality of forage crops. Forage crops provide necessary nutrients that improve animal health and productivity and are essential for livestock production in the agricultural ecosystem. Improvements in nutritional quality, yield, and stress resilience can be addressed through genome editing technologies, particularly CRISPR/Cas9 by performing accurate modifications of the genetic structure (Capstaff and Miller 2018; Nagamine and Ezura 2022).

Recent developments in molecular breeding techniques allowed for the identification of specific genes associated with corresponding traits, facilitating the application of genome editing in forage crops. For example, genome-wide association studies (GWAS) have the ability to identify single nucleotide polymorphisms (SNPs) linked to increased protein levels and essential nutrients in forage crops (Capstaff and Miller 2018; Chen 2024). Researchers can improve the nutritional quality of forage crops by targeting these SNPs through genome editing techniques, enhancing the overall health of livestock that depend on these feeds. Moreover, forage varieties that maintain high nutritional value even under adverse climatic conditions, such as drought and salinity, can be developed by altering the associated genes (Sustek-Sánchez et al. 2023; Singh and Chahal 2020). Genome editing techniques can improve the nutritional quality of forage crops by enhancing their digestibility and palatability. Enhanced digestibility is crucial for maximizing nutrient uptake in livestock, improving growth rates and overall production of livestock (Kulkarni et al. 2018; Talla et al. 2023).

Furthermore, modification of particular genes can produce forage crops with superior nutritional value, optimizing feed conversion ratios. These interventions can reduce the environmental impact of animal agriculture, enabling sustainable livestock production (Roy 2024; Bao et al. 2022).

The integration of genome editing and traditional breeding techniques will accelerate the development of new forage varieties that can meet the ever-increasing demands of the agricultural sector. Utilizing the existing genetic diversity of forage crops together with accuracy of genome editing enables breeders to develop specialized cultivars adapted to specific regional conditions while meeting high standards of nutritional requirements (Wei et al. 2023; Ku and Ha 2020). This method allows the increment of nutritional quality and resilience to stress conditions in the midst of climate change and fluctuating market demands (Zenda et al. 2021; Salgotra 2024).

The regulatory framework for genome-edited crops is evolving, leading many countries to adopt more favorable regulations for genome-edited crops categorizing them as non-transgenic. These regulatory revisions are crucial for the widespread application of genome editing technologies to improve forage crops, accelerating commercialization of improved varieties bypassing the lengthy approval procedures associated with traditional genetically modified organisms (GMOs) (Tripathi et al. 2022; Awais 2024; Gudeta 2019).

Public acceptance of genome-edited crops will lead farmers to increase their production of these crops, improving livestock productivity and nutritional outcomes.

2.2 Increasing Yield

One of the benefits of genome editing is its ability to perform precise genomic modifications to increase the yield significantly. Studies have shown that modifying genes linked to plant architecture can increase height and biomass in plants by improving growth patterns (Wolabu et al. 2023). This is particularly relevant for forage crops where enhancing overall yield can be achieved by optimizing vegetative development. Traits like drought tolerance and disease resistance, which are essential for maintaining high yields even in harsh environments, can be enhanced through genome editing (Dogar et al. 2022). These traits allow sustainable forage crops production, ensuring a consistent supply of high-quality feed for livestock.

Editing and modifying flowering time genes can extend the vegetative growth phase of forage crops, ensuring better biomass accumulation before flowering (Wolabu et al. 2023). An example is alfalfa plant that had improved biomass yield and quality, which is attributed to delayed flowering time.

For yield increment in forage crops, genome editing can be applied to enhance nutrient uptake efficiency, utilizing more of the nutrients found in soil (Bao et al. 2022; Chen 2024). The economic importance of improving fodder crop yields is substantial, improving profitability for livestock production by reducing feed costs and increasing its productivity. Development and growing of high-yielding forage varieties will reduce the need for additional land conversion for forage crops in the light of increasing global population and thus the demand for animal products (Fuglie et al. 2021).

2.3 Abiotic Stress Tolerance

The increasing prevalence of abiotic stress conditions such as drought, salinity, and extreme temperatures create major obstacles to the productivity of feed crops. Genome editing technologies can be exploited to enhance abiotic stress tolerance of crops by modifying genes related to stress responses. In this way, forage crops' resilience toward adverse environmental conditions can be improved, contributing to agricultural sustainability (Hazzouri et al. 2020; Sustek-Sánchez et al. 2023). Transcription factors that regulate stress response pathways can be targets of genome editing to increase abiotic stress tolerance. An important example is addressing of the GmAITR gene family in soybean using CRISPR/Cas9. The deletion of the specific gene improved salinity tolerance without compromising plant fitness (Wang et al. 2021; Chen et al. 2021). This study highlights the effectiveness of targeting genes that are members of specific pathways to enhance the abiotic stress resilience and ensuring high yield despite unfavorable conditions.

Another method is the application of microRNAs to modulate stress responses in plants. Studies have demonstrated that expression of microRNA156 in alfalfa improved its resistance to heat stress (Matthews et al. 2018). Researchers can employ genome editing techniques to integrate and express microRNAs that can boost resilience to extreme temperatures of plethora of forage crops.

The study of fodder crops' wild relatives has unearthed important genetic resources that can be used for improvement of abiotic stress tolerance. Research performed on *Medicago ruthenica*, a wild legume species that exhibits significant resilience to various abiotic stresses, unveiled molecular mechanisms underlying its special attributes. Homologs of these genes can be targeted in fodder crops such as alfalfa to achieve similar performances (Wang et al. 2021).

Omics technology combined with genome editing has make possible identifying stress tolerance pathways. Transcriptomics, proteomics, and metabolomics allow to further pinpoint the genes associated with abiotic stress responses, pointing out the targets that researchers can work upon. Precise modulation of multiple genes associated with stress tolerance facilitates the development of forage crops that can withstand extreme environmental conditions (Sustek-Sánchez et al. 2023; Kamali 2023).

The alteration of ion transporter genes that are related to salt tolerance has shown promise in enhancing the resilience of fodder crops to saline conditions (Huang et al. 2020). Forage crops that are resilient to environmental shocks lead to higher yields, reduced costs, and enhanced animal productivity, mitigating the economic risks associated with climatic change for farmers (Singh and Chahal 2020).

2.4 Disease Resistance

Biotic stresses such as fungal, bacterial, and viral infections can also be tackled through genome editing, providing development of new fodder crop varieties that can withstand them (Su et al. 2023).

Major application of CRISPR/Cas9 technology to enhance disease resistance in plants includes knocking out susceptibility (S) genes that are needed for pathogen entry or infection. Resistance toward diseases such as powdery mildew and rust has been introduced to various plants by targeting their corresponding S genes (Schenke and Cai 2020). In species such as alfalfa and clover, knocking out of S genes has resulted in notable improvements in resistance to common pathogens, resulting in reduction of reliance on chemicals (Vaghela et al. 2023; Capstaff and Miller 2018).

Genome editing can be utilized in modification of the resistance (R) genes, which recognize and initiate a response when encountering the pathogen. In legumes, R gene alterations have shown great promise in combating root-knot nematodes and various soil-borne diseases (Erdoğan et al. 2023). Precise genome editing to increase activity or expression of existing R genes enables researchers to develop forage crops with robust resistance to various diseases (Mushtaq et al. 2019).

Combining the existing genetic variability of forage crops with genome editing accuracy allows breeders to generate cultivars resistant to prevalent diseases in their respective ecosystems (Su et al. 2023). This integrative method increases disease resistance and strengthens the overall resilience of fodder crops stressors to biotic stress under climate change constraints. Improved disease resistance translates to reduced costs associated with pest and disease management and increased fodder yields, easing economical barriers in cultivating them (Paul et al. 2021; Yıldırım 2023). In this way, farmers can mitigate the risks of crop losses due to disease outbreaks, ensuring a steady supply of high-quality fodder and guaranteeing their livelihoods (Ghimire 2017).

2.5 Improving Ecosystem Services

The benefits derived from natural ecosystems, which include supplying goods (such as food and fiber), regulating services (like climate regulation and water purification), supporting services (such as nutrient cycling), and cultural services (offering recreational and aesthetic fields), are referred as ecosystem services (Ashilenje et al. 2022; Ashworth et al. 2022). Ecosystem services can be significantly improved using genome editing to cultivate fodder crops that enhance soil health. Forage crops such as legumes play a crucial role in atmospheric nitrogen fixation, improving soil fertility and reducing dependence on synthetic fertilizers (Capstaff and Miller 2018). Nitrogen fixation and its utilization efficiency can be improved by promoting symbiotic nitrogen-fixing bacteria in rhizosphere through means of genome editing.

Genes associated with root nodule development can be altered for improvement of nitrogen fixation efficiency. In return, healthier soils with reduced environmental impacts due to fertilizers and healthier ecosystems will be promoted in local agricultural systems (Acharya et al. 2023).

The growth of perennial crops such as Kernza and Silphium, which have extensive root systems, can be optimized through genome editing to enhance water quality and decrease soil erosion (Ashworth et al. 2022).

Modifying genes related to root architecture and biomass production in perennial fodder crops will help with soil erosion mitigation and water infiltration. For ecosystem resilience, preserving soil structure and moisture content is essential, particularly in regions prone to flooding or drought (Yang 2023). Genome-edited forage crops have the potential to enhance soil health, improve water quality, and foster biodiversity in agricultural systems.

Maintaining ecological balance and productivity can be achieved by planting multiple forage varieties as a means of diverse cropping systems, which can support a wider range of beneficial insects and pollinators (Yıldırım 2023).

Genome editing creates the opportunity to improve biodiversity in agricultural systems by providing habitats for beneficial insects and increase pollinator health (Zaeem et al. 2019).

The introduction of flowering forage crops into agricultural systems provides continuous forage resources for pollinators throughout the growing season. This increases in pollinator population and the ecological benefits they provide.

Improved soil health, water quality, and biodiversity reduce input costs for farmers by lowering the necessity of fertilizer and pesticide usage, giving them a high economic leverage (Pilati and Prestamburgo 2016).

Integrating various forage crops in rotational grazing systems enhances nutrient cycling and reduces dependence on external inputs to the soil, promoting sustainable agricultural practices (Jeavons et al. 2023). The economic viability and ecological benefits coming together with forage crop adaptations underscore the importance of genome editing technologies in paving the road for crop improvement.

3 Challenges in Genome Editing for Forage Crops

Genome editing technologies, particularly CRISPR/Cas9, present a substantial opportunity for enhancing agricultural production and sustainability in forage crops. Numerous challenges must be addressed to fully realize the potential of these technologies in improving fodder crops. The issues encompass technical, regulatory, and sociological factors that may restrict the effective implementation of genome editing in agriculture.

The primary technological challenge in genome editing for forage crops is the effectiveness of transformation and regeneration processes. Many forage species exhibit recalcitrance to transformation, hindering the application of preferred genetic modifications (Razzaq et al. 2019; Nivya 2023). Legumes play a crucial role in improving soil health through nitrogen fixation, yet they often present difficulties in achieving consistent transformation and regeneration of modified plants. This obstinacy can significantly hinder the breeding process and limit the application of genome editing technologies in these essential crops (Nivya 2023). The complexity of polyploid genomes in certain forage species presents challenges, as multiple gene copies can complicate the editing process and lead to unpredictable alterations (Ahmad et al. 2023; Nadakuduti et al. 2018).

A significant challenge is the potential for off-target effects associated with CRISPR/Cas9 genome editing. The method allows for precise adjustments; however, unintended modifications may occur at similar genomic sites, leading to undesirable traits or the inactivation of critical genes (Ahmad et al. 2021; Zhao and Wolt 2017). This risk necessitates the development of advanced editing techniques, such as base editing and prime editing, aimed at reducing off-target effects while maintaining editing efficiency (Hassan et al. 2020; Azameti and Palnam 2021). The requirement for rigorous validation of modified lines to ensure specificity and safety adds to the complexity and time required for the development of genome-edited fodder crops.

Regulatory barriers pose significant challenges to the application of genome editing in forage crops. The legal framework governing genome-edited crops varies considerably across countries, with some jurisdictions implementing stringent regulations akin to those applied to genetically modified organisms (GMOs) (Ishii and Araki 2016; Lassoued et al. 2019). This discrepancy may create ambiguity for researchers and developers, thereby hindering investment in genome editing technologies for fodder crops. In specific regions, genome-edited crops may encounter extended regulatory processes that can delay their market entry, thereby reducing the incentive for breeders to pursue these innovations (Lassoued et al. 2019; Awais 2024). The public perception of genome editing and GMOs influences regulatory decisions, requiring stakeholders to communicate clearly regarding the benefits and safety of genome-edited forage crops (Araki and Ishii 2015; Menz et al. 2020).

The economic viability of genome editing in forage crops is contingent upon the development of market-oriented attributes that meet the needs of both farmers and consumers. Genome editing can enhance traits such as disease resistance, drought tolerance, and nutritional quality; however, the economic benefits must be clearly demonstrated to encourage adoption (Maaß et al. 2019; Kalaitzandonakes et al. 2022). Farmers often hesitate to invest in new technologies due to an unclear understanding of potential returns on investment, especially in regions where traditional breeding methods dominate (Evanega et al. 2024). Thus, comprehensive economic evaluations and demonstration projects are crucial to demonstrate the advantages of genome-edited fodder crops in real-world agricultural settings.

Ultimately, these recent methodologies present both opportunities and challenges. The integration of these methodologies can accelerate the development of improved forage varieties, requiring a thorough understanding of the genetic basis of traits and the interactions between modified genes and the complete genome (Chen 2024; Razzaq et al. 2021). Breeders must have the necessary skills and knowledge to effectively utilize genome editing alongside traditional methods, which may require additional training and resources (Tsakirpaloglou 2023). The integration of these technologies into breeding programs is crucial for maximizing their benefits in fodder crop improvement.

4 Future Perspectives

Future prospects for forage crops utilizing genome editing appear promising, particularly with advancements in CRISPR/Cas9 technology and other genome editing methodologies. The potential of these technologies to enhance forage crop traits, such as yield, nutritional quality, disease resistance, and tolerance to abiotic stress, is substantial. The increasing global demand for sustainable agricultural practices necessitates the application of genome editing in forage crops to address challenges related to food security, climate change, and environmental sustainability. Focusing on specific genetic mechanisms, such as those that regulate tillering and branching, researchers could develop forage varieties that enhance biomass accumulation, thereby providing greater feed for livestock. This is particularly important due to the increasing demands for animal production, as improved yields can result in more efficient use of land and resources. Modifying genes linked to protein content and digestibility can lead to the development of fodder varieties that improve livestock nutrition, thereby promoting animal health and productivity (Assem 2023; Ku and Ha 2020). The evolution of customer preferences toward sustainably sourced animal products underscores the growing importance of enhancing the nutritional quality of fodder crops. The development of disease-resistant forage crops through precise gene editing can reduce reliance on chemical pesticides, thereby promoting more sustainable agricultural practices (Joo et al. 2021). Enhancing the inherent defenses of fodder crops against diseases enables researchers to maintain consistent production levels, even in the face of increasing disease prevalence driven by climate change and globalization (Li et al. 2022; Yang et al. 2022). Climate change leads to increased drought frequency, higher salinity, and extreme temperatures, necessitating the development of forage varieties that can withstand these conditions to maintain productivity (Sustek-Sánchez et al. 2023; Yadav et al. 2023). Genome editing technology can modify genes associated with stress responses, enabling the creation of resilient fodder crops that can thrive in challenging environmental conditions. This resilience supports farmers and improves the overall stability of agricultural systems in the context of climate uncertainty. However, the future of genome editing in forage crops raises significant concerns. Technical constraints, such as transformation efficiency and off-target effects, must be addressed to ensure the effective implementation of these technologies. The regulatory framework for genome-edited crops is complex and varies by region, which may impact the commercialization of improved cultivars. Engaging with policymakers and the public to enhance understanding and acceptance of genome-edited crops is crucial for overcoming these challenges.

References

Acharya P, Ghimire R, Lehnhoff E, Marsalis M (2023) Cover crop forage potential and subsequent sorghum silage yield and nutritive value. Agron J 115(4):1723–1734. https://doi.org/10.1002/agj2.21334

Ahmad A, Munawar N, Khan Z, Qusmani A, Khan S, Jamil A et al (2021) An outlook on global regulatory landscape for genome-edited crops. Int J Mol Sci 22(21):11753. https://doi.org/10.3390/ijms222111753
Ahmad N, Fatima S, Mehmood M, Zaman Q, Atif R, Zhou W et al (2023) Targeted genome editing in polyploids: lessons from brassica. Front Plant Sci 14. https://doi.org/10.3389/fpls.2023.1152468
Araki M, Ishii T (2015) Towards social acceptance of plant breeding by genome editing. Trends Plant Sci 20(3):145–149. https://doi.org/10.1016/j.tplants.2015.01.010
Ashilenje D, Amombo E, Hirich A, Kouisni L, Devkota K, Mouttaqi A et al (2022) Crop species mechanisms and ecosystem services for sustainable forage cropping systems in salt-affected arid regions. Front Plant Sci 13:899926. https://doi.org/10.3389/fpls.2022.899926
Ashworth A, Katuwal S, Moore P, Adams T, Anderson K, Owens P (2022) Perenniality drives multifunctional forage–biomass filter strips' ability to improve water quality. Crop Sci 63(1):336–348. https://doi.org/10.1002/csc2.20878
Assem S (2023) Development of an in vitro regeneration system from immature inflorescences and CRISPR/Cas9-mediated gene editing in sudangrass. J Genet Eng Biotechnol 21(1):58. https://doi.org/10.1186/s43141-023-00517-6
Awais M (2024) The regulatory landscape of genetically engineered crops in Pakistan. Food Energy Secur 13(1). https://doi.org/10.1002/fes3.494
Azameti M, Palnam W (2021) Base editing in plants: applications, challenges, and future prospects. Front Plant Sci 12:664997. https://doi.org/10.3389/fpls.2021.664997
Bao Q, Wolabu T, Zhang Q, Zhang T, Liu Z, Sun J et al (2022) Application of CRISPR/Cas9 technology in forages. Grassl Res 1(4):244–251. https://doi.org/10.1002/glr2.12036
Berg J, Bouwman L, Battaglia E, Kleter G (2021) Future-proofing EU legislation for genome-edited plants: Dutch stakeholders' views on possible ways forward. Agronomy 11(7):1331. https://doi.org/10.3390/agronomy11071331
Blanco-Canqui H, Holman J, Schlegel A, Tatarko J, Shaver T (2013) Replacing fallow with cover crops in a semiarid soil: effects on soil properties. Soil Sci Soc Am J 77(3):1026–1034. https://doi.org/10.2136/sssaj2013.01.0006
Capstaff NM, Miller AJ (2018) Improving the yield and nutritional quality of forage crops. Front Plant Sci 9:535
Chen S (2024) Advances in molecular breeding of forage crops: technologies, applications and prospects. Agriculture 14(2):279. https://doi.org/10.3390/agriculture14020279
Chen K, Wang Y, Zhang R, Zhang H, Gao C (2019) CRISPR/Cas genome editing and precision plant breeding in agriculture. Annu Rev Plant Biol 70(1):667–697. https://doi.org/10.1146/annurev-arplant-050718-100049
Chen S, Zhang N, Zhou G, Hussain S, Ahmed S, Tian H et al (2021) Knockout of the entire family of AITR genes in Arabidopsis leads to enhanced drought and salinity tolerance without fitness costs. BMC Plant Biol 21(1). https://doi.org/10.1186/s12870-021-02907-9
Desta AG (2023) Challenges of the forage crops in northwestern Ethiopia, 30 March 2023, PREPRINT (Version 1). Available at Research Square. https://doi.org/10.21203/rs.3.rs-2745885/v1
Dogar A, Ali M, Riaz Z, Ali Q, Ahmad S, Javed M (2022) Role of CRISPR to improve stress tolerance in plants. Biol Clin Sci Res J 2022(1). https://doi.org/10.54112/bcsrj.v2022i1.97
Erdoğan İ, Cevher-Keskin B, Bilir Ö, Hong Y, Tör M (2023) Recent developments in CRISPR/Cas9 genome-editing technology related to plant disease resistance and abiotic stress tolerance. Biology 12(7):1037. https://doi.org/10.3390/biology12071037
Evanega S, Brown Z, Bubeck D, Falck-Zepeda JB, Chou FL, Graham N et al (2024) Applications, benefits, and challenges of genome edited crops. CAST issue paper 74. Council for Agricultural Science and Technology (CAST), Ames. https://doi.org/10.62300/hjvu4083
Fuglie K, Peters M, Burkart S (2021) The extent and economic significance of cultivated forage crops in developing countries. Front Sustain Food Syst 5:712136. https://doi.org/10.3389/fsufs.2021.712136

Ghimire B (2017) Use of CRISPR/Cas9 for development of disease resistant cultivars in plant breeding. Int J Appl Sci Biotechnol 5(4):403–409. https://doi.org/10.3126/ijasbt.v5i4.18523

Gudeta D (2019) Genome editing: tools and application in plants. Open Access J Microbiol Biotechnol 4(1). https://doi.org/10.23880/oajmb-16000135

Hassan M, Yuan G, Chen J, Tuskan G, Yang X (2020) Prime editing technology and its prospects for future applications in plant biology research. Biodes Res 2020. https://doi.org/10.34133/2020/9350905

Hazzouri K, Flowers J, Nelson D, Lemansour A, Masmoudi K, Amiri K (2020) Prospects for the study and improvement of abiotic stress tolerance in date palms in the post-genomics era. Front Plant Sci 11:293. https://doi.org/10.3389/fpls.2020.00293

Helliwell R, Hartley S, Pearce W (2019) Ngo perspectives on the social and ethical dimensions of plant genome-editing. Agric Hum Values 36(4):779–791. https://doi.org/10.1007/s10460-019-09956-9

Huang L, Wu D, Zhang G (2020) Advances in studies on ion transporters involved in salt tolerance and breeding crop cultivars with high salt tolerance. J Zhejiang Univ Sci B 21(6):426–441. https://doi.org/10.1631/jzus.b1900510

Ishii T, Araki M (2016) A future scenario of the global regulatory landscape regarding genome-edited crops. Gm Crops Food 8(1):44–56. https://doi.org/10.1080/21645698.2016.1261787

Jeavons E, Lann C, Baaren J (2023) Interactions between natural enemies and pollinators: combining ecological theory with agroecological management. Entomol Gen 43(2):243–259

Joo K, Muszynski M, Kantar M, Wang M, He X, Valle-Echevarria A (2021) Utilizing CRISPR-Cas in tropical crop improvement: a decision process for fitting genome engineering to your species. Front Genet 12:786140. https://doi.org/10.3389/fgene.2021.786140

Kadıoğlu S (2024) Factors effective on farmer satisfaction for the forage crops supports and the tendency to continue forage crop production (tra-1). Turk J Range Forage Sci 5(1):36–45. https://doi.org/10.51801/turkjrfs.1453514

Kalaitzandonakes N, Willig C, Zahringer K (2022) The economics and policy of genome editing in crop improvement. Plant Genome 16(2). https://doi.org/10.1002/tpg2.20248

Kamali S (2023) Genomic and transcriptomic approaches to developing abiotic stress-resilient crops. Agronomy 13(12):2903. https://doi.org/10.3390/agronomy13122903

Kamburova VS, Nikitina EV, Shermatov SE, Buriev ZT, Kumpatla SP, Emani C, Abdurakhmonov IY (2017) Genome editing in plants: an overview of tools and applications. Int J Agron 2017(1):7315351. https://doi.org/10.1155/2017/7315351

Kaul T, Sony SK, Verma R, Motelb KFA, Prakash AT, Eswaran M et al (2020) Revisiting CRISPR/Cas-mediated crop improvement: special focus on nutrition. J Biosci 45:1–37. https://doi.org/10.1007/s12038-020-00094-7

Kim H, Kim ST, Ryu J, Kang BC, Kim JS, Kim SG (2017) CRISPR/Cpf1-mediated DNA-free plant genome editing. Nat Commun 8(1):14406. https://doi.org/10.1038/ncomms14406

Ku H, Ha S (2020) Improving nutritional and functional quality by genome editing of crops: status and perspectives. Front Plant Sci 11:577313. https://doi.org/10.3389/fpls.2020.577313

Kulkarni K, Tayade R, Asekova S, Song J, Shannon J, Lee J (2018) Harnessing the potential of forage legumes, alfalfa, soybean, and cowpea for sustainable agriculture and global food security. Front Plant Sci 9:1314. https://doi.org/10.3389/fpls.2018.01314

Lassoued R, Phillips P, Smyth S, Hesseln H (2019) Estimating the cost of regulating genome edited crops: expert judgment and overconfidence. GM Crops Food 10(1):44–62. https://doi.org/10.1080/21645698.2019.1612689

Leal VN, Santos DDC, Paim TDP, Santos LPD, Alves EM, Claudio FL et al (2023) Economic results of forage species choice in crop–Livestock integrated systems. Agriculture 13(3):637

Li J, Yu X, Zhang C, Li N, Zhao J (2022) The application of CRISPR/Cas technologies to brassica crops: current progress and future perspectives. aBIOTECH 3(2):146–161. https://doi.org/10.1007/s42994-022-00076-3

Maaß O, Consmuller N, Kehlenbeck H (2019) Socioeconomic impact of genome editing on agricultural value chains: the case of fungal-resistant and coeliac-safe wheat. Sustainability 11(22):6421. https://doi.org/10.3390/su11226421

Malnoy M, Viola R, Jung MH, Koo OJ, Kim S, Kim JS et al (2016) DNA-free genetically edited grapevine and apple protoplast using CRISPR/Cas9 ribonucleoproteins. Front Plant Sci 7:1904

Mao Y, Botella JR, Liu Y, Zhu JK (2019) Gene editing in plants: progress and challenges. Natl Sci Rev 6(3):421–437. https://doi.org/10.1093/nsr/nwz005

Massimi M, Bader N, Khamish K, Al-Suod A (2018) Economic analysis for forage agronomic crops grown using treated wastewater in Kherbeh Als-Samra region, Jordan. Int J Plant Soil Sci 22(4):1–9. https://doi.org/10.9734/ijpss/2018/40960

Matthews C, Arshad M, Hannoufa A (2018) Alfalfa response to heat stress is modulated by microrna156. Physiol Plant 165(4):830–842. https://doi.org/10.1111/ppl.12787

Menz J, Modrzejewski D, Hartung F, Wilhelm R, Sprink T (2020) Genome edited crops touch the market: a view on the global development and regulatory environment. Front Plant Sci 11:586027. https://doi.org/10.3389/fpls.2020.586027

Mushtaq M, Sakina A, Wani S, Shikari A, Tripathi P, Zaid A et al (2019) Harnessing genome editing techniques to engineer disease resistance in plants. Front Plant Sci 10:550. https://doi.org/10.3389/fpls.2019.00550

Nadakuduti S, Buell C, Voytas D, Starker C, Douches D (2018) Genome editing for crop improvement—applications in clonally propagated polyploids with a focus on potato (solanum tuberosum l.). Front Plant Sci 9:1607. https://doi.org/10.3389/fpls.2018.01607

Nagamine A, Ezura H (2022) Genome editing for improving crop nutrition. Front Genome Ed 4:850104. https://doi.org/10.3389/fgeed.2022.850104

Nielsen D, Lyon D, Hergert G, Higgins R, Holman J (2015) Cover crop biomass production and water use in the central great plains. Agron J 107(6):2047–2058. https://doi.org/10.2134/agronj15.0186

Nivya V (2023) Recalcitrance to transformation, a hindrance for genome editing of legumes. Front Genome Ed 5:1247815. https://doi.org/10.3389/fgeed.2023.1247815

Odahara M, Watanabe K, Kawasaki R, Tsuchiya K, Tateishi A, Motoda Y et al (2021) Nanoscale polyion complex vesicles for delivery of cargo proteins and Cas9 ribonucleoprotein complexes to plant cells. ACS Appl Nano Mater 4(6):5630–5635. https://doi.org/10.1021/acsanm.1c00695

Paul N, Park S, Liu H, Choi S, Ma J, MacCready J et al (2021) Plant and fungal genome editing to enhance plant disease resistance using the CRISPR/Cas9 system. Front Plant Sci 12:70092. https://doi.org/10.3389/fpls.2021.700925

Pilati L, Prestamburgo M (2016) Sequential relationship between profitability and sustainability: the case of migratory beekeeping. Sustainability 8(1):94. https://doi.org/10.3390/su8010094

Razzaq A, Saleem F, Kanwal M, Mustafa G, Yousaf S, Arshad H et al (2019) Modern trends in plant genome editing: an inclusive review of the CRISPR/Cas9 toolbox. Int J Mol Sci 20(16):4045. https://doi.org/10.3390/ijms20164045

Razzaq A, Kaur P, Akhter N, Wani S, Saleem F (2021) Next-generation breeding strategies for climate-ready crops. Front Plant Sci 12:620420. https://doi.org/10.3389/fpls.2021.620420

Roy A (2024) Advancements in genetic enhancement: CRISPR/Cas-mediated genome editing in leguminous crops. J Adv Biol Biotechnol 27(6):670–681. https://doi.org/10.9734/jabb/2024/v27i6927

Ruckle M, Meier M, Frey L, Eicke S, Kölliker R, Zeeman S et al (2017) Diurnal leaf starch content: an orphan trait in forage legumes. Agronomy 7(1):16. https://doi.org/10.3390/agronomy7010016

Salgotra R (2024) Dynamics of genome edited products in the world market for future food security: a review. Plant Arch 24(1):189–207. https://doi.org/10.51470/plantarchives.2024.v24.no.1.029

Schenke D, Cai D (2020) Applications of CRISPR/Cas to improve crop disease resistance: beyond inactivation of susceptibility factors. IScience 23(9):101478. https://doi.org/10.1016/j.isci.2020.101478

Sedeek KE, Mahas A, Mahfouz M (2019) Plant genome engineering for targeted improvement of crop traits. Front Plant Sci 10:114. https://doi.org/10.3389/fpls.2019.00114

Singh A, Chahal HS (2020) Management of abiotic stress in forage crops. In: Abiotic stress in plants, vol 13. IntechOpen, London

Su H, Wang Y, Xu J, Omar AA, Grosser JW, Calovic M et al (2023) Efficient generation of transgene-free canker-resistant Citrus sinensis using Cas12a/crRNA ribonucleoprotein. bioRxiv. https://doi.org/10.1101/2023.02.12.528187

Sustek-Sánchez F, Rognli O, Rostoks N, Sõmera M, Jaškūnė K, Kovi M et al (2023) Improving abiotic stress tolerance of forage grasses—prospects of using genome editing. Front Plant Sci 14. https://doi.org/10.3389/fpls.2023.1127532

Talla V, Sandal S, Walia P, Vidya A (2023) Quality enhancement in forage crops of cereals and legumes. Int J Plant Soil Sci 35(16):166–179. https://doi.org/10.9734/ijpss/2023/v35i163142

Tripathi L, Dhugga KS, Ntui VO, Runo S, Syombua ED, Muiruri S et al (2022) Genome editing for sustainable agriculture in Africa. Front Genome Ed 4:876697. https://doi.org/10.3389/fgeed.2022.876697

Tsakirpaloglou N (2023) Guidelines for performing CRISPR/Cas9 genome editing for gene validation and trait improvement in crops. Plants 12(20):3564. https://doi.org/10.3390/plants12203564

Vaghela U, Sonagara M, Pratibha et al (2023) A review on revolutionary strategy for crop improvement: genome editing. Int J Environ Clim Change 13(8):2005–2018. https://doi.org/10.9734/ijecc/2023/v13i82158

Wang T, Ren L, Li C, Zhang D, Zhang X, Zhou G et al (2021) The genome of a wild Medicago species provides insights into the tolerant mechanisms of legume forage to environmental stress. BMC Biol 19(1):96. https://doi.org/10.1186/s12915-021-01033-0

Wei X, Long Y, Yi C, Pu A, Hou Q, Liu C et al (2023) Bibliometric analysis of functional crops and nutritional quality: identification of gene resources to improve crop nutritional quality through gene editing technology. Nutrients 15(2):373. https://doi.org/10.3390/nu15020373

Wilson R, Burgener P (2009) Evaluation of glyphosate-tolerant and conventional alfalfa weed control systems during the first year of establishment. Weed Technol 23(2):257–263. https://doi.org/10.1614/wt-08-082.1

Wolabu T, Mahmood K, Jerez I, Cong L, Yun J, Udvardi M et al (2023) Multiplex CRISPR/Cas9-mediated mutagenesis of alfalfa flowering locus ta1 (msfta1) leads to delayed flowering time with improved forage biomass yield and quality. Plant Biotechnol J 21(7):1383–1392. https://doi.org/10.1111/pbi.14042

Xu L (2024) Yield and nutrient composition of forage crops and their effects on soil characteristics of winter fallow paddy in South China. Front Plant Sci 14:1292114. https://doi.org/10.3389/fpls.2023.1292114

Yadav R, Tripathi M, Tiwari S, Tripathi N, Asati R, Chauhan S et al (2023) Genome editing and improvement of abiotic stress tolerance in crop plants. Life 13(7):1456. https://doi.org/10.3390/life13071456

Yanagawa Y, Suenaga Y, Iijima Y, Endo M, Sanada N, Katoh E et al (2023) Genome editing by introduction of Cas9/sgRNA into plant cells using temperature-controlled atmospheric pressure plasma. PLoS One 18(2):e0281767. https://doi.org/10.1371/journal.pone.0281767

Yang X (2023) Optimizing cropping systems to close the gap between economic profitability and environmental health. New Phytol 240(6):2498–2512. https://doi.org/10.1111/nph.19310

Yang Y, Xu C, Shen Z, Yan C (2022) Crop quality improvement through genome editing strategy. Front Genome Ed 3:819687. https://doi.org/10.3389/fgeed.2021.819687

Yildirim K (2023) Genome editing for healthy crops: traits, tools and impacts. Front Plant Sci 14. https://doi.org/10.3389/fpls.2023.1231013

Zaeem M, Nadeem M, Pham T, Ashiq W, Ali W, Gilani S et al (2019) The potential of corn-soybean intercropping to improve the soil health status and biomass production in cool climate boreal ecosystems. Sci Rep 9(1). https://doi.org/10.1038/s41598-019-49558-3

Zenda T, Liu S, Dong A, Duan H (2021) Advances in cereal crop genomics for resilience under climate change. Life 11(6):502. https://doi.org/10.3390/life11060502

Zhao H, Wolt J (2017) Risk associated with off-target plant genome editing and methods for its limitation. Emerg Top Life Sci 1(2):231–240. https://doi.org/10.1042/etls20170037

Transgenic Forage Crops

Aslı Küçükrecep and Dilek Tekdal

Abstract

Forage crops, essential for livestock feed and ecosystem services, are constantly under pressure to meet the demands of a growing global population. Biotechnology, specifically genetic engineering, offers powerful tools to enhance forage quality, yield, and resilience against biotic and abiotic stresses. Transgenic forage crops are promising to enhance livestock production, improve environmental sustainability, and address global food security challenges. This chapter will explore the emerging prospects in this field and critically analyze the ethical considerations accompanying these advancements. Also, this chapter provides an overview of the principles and applications of transgenesis in forage crops, highlighting key achievements and future directions.

Keywords

Food security · Forage crops · Global population · Stress · Transgenesis · Yield

1 Introduction

The transformation of Turkiye's agricultural landscape hinges critically on advancing animal husbandry, mainly through developing high-value fodder plant production. In the Turkish context, where traditional feeding practices often rely on straw, the challenge extends beyond mere production increases—it encompasses a fundamental shift in how we approach animal nutrition and agricultural sustainability. This challenge is particularly evident in regions where low-quality feeds continue to

A. Küçükrecep · D. Tekdal (✉)
Department of Biotechnology, Faculty of Science, Mersin University, Mersin, Turkey
e-mail: dilektekdal@mersin.edu.tr

F. S. Baloch et al. (eds.), *Forage Crops in the Omics Era*,
https://doi.org/10.1007/978-981-95-9249-4_4

limit animal productivity, creating a cycle of reduced efficiency in the livestock sector.

The multifaceted benefits of forage crops extend well beyond their immediate nutritional value. As Açıkgöz et al. (2005) demonstrated in their comprehensive study of Turkish agricultural systems, these crops serve dual purposes: providing nutrient-rich livestock feed while enriching soil quality for subsequent plantings. This symbiotic relationship is particularly relevant in Turkiye's diverse agricultural zones, where soil enhancement through forage crop rotation can significantly impact agricultural productivity.

Turkiye's agricultural sector faces mounting pressure against global demographic shifts. While the UN (2019) projects global population expansion from 7.6 billion in 2021 to 9.8 billion by 2050 and ultimately reaching 11.2 billion by 2100, these figures hold particular significance for Turkiye's position as a key agricultural producer. The food security challenge becomes especially acute in Turkey, where agricultural innovation must keep pace with population growth and changing dietary patterns. Approximately 60% of the world's population is undernourished, so the need for cereals, staple foods, and animal products is paramount. Due to the continuous increase in population and per capita food consumption, there is also an increase in the demand for animal products, especially in developing countries (Nardone 2002). Within Turkiye's agricultural landscape, the livestock sector has emerged as a crucial economic driver, reflecting global trends where it accounts for 40% of agricultural production value (FAO 2019). This sector's significance extends beyond mere statistics—it represents a vital source of livelihood for millions of Turkish farmers and agricultural workers, mirroring the global pattern where approximately 1.3 billion people depend on livestock-related activities. The sector's rapid growth in Turkiye presents opportunities and challenges for sustainable forage crop development. Feeding animals with rations that will meet their needs sufficiently and balanced is essential for increasing production. In addition, feed is a large part of the costs and inputs in the animal husbandry sector. The economic dynamics of Turkish animal husbandry reveal a critical cost structure where roughage dominates, comprising 78% of feed costs compared to 22% for mixed feed (Harmanşah 2018). This distribution adds significance when considered alongside global patterns, where grain roughage accounts for 13% of animal husbandry's dry matter intake (CGIAR 2019). These figures underscore the strategic importance of optimizing forage crop production within Turkiye's agricultural framework. FAO (2017) stated that animals need an average of 3 kg of grain to produce 1 kg of meat globally. Global grain production in 2023 is approximately 1,772,295.22 million tons. Approximately 52% of grain production is corn, 36% is wheat, 7% is barley, 2% is sorghum, 1% is oats, and 2% is other grains (rye, etc.).

Forty-four percent of grains are used in the feed sector, 33% in the food sector, 8% in biofuel production, and 16% in other industrial purposes (FAOSTAT 2021). Many global grains comprise corn (52%) and wheat (36%). Fifty-seven percent of corn produced worldwide is used in the feed sector, 7% in the food sector, 33% in biofuels, and 3% in other industrial purposes. In Turkiye, 79% of total corn is used in the feed sector, 16% in the food sector, 1% in biofuels, and 4% in other industrial purposes. Of the wheat produced in the world, 32% is used in the feed sector, 56% in the food sector, 3% in biofuels, and 9% in other industrial purposes; 8% of the wheat produced in Turkiye is used in the feed sector, 78% in the food sector, 1% in biofuels, and 4% in other industrial purposes (FAOSTAT 2021).

Forage crops constitute an important part of agriculture in terms of meeting the feed needs of animals and the protein that animals provide to humans. In addition, forage crops are the most crucial input for animal feeding in terms of containing the necessary substances for the stomach flora of animals, being rich in minerals and vitamins, increasing the reproductive power of animals, and providing quality products (Yolcu and Tan 2008). During animal production, 55–80% of the costs are feed costs, which increases the importance of accelerating studies on the production of feed plants (IFCN 2014).

2 Effects of Climate Change on Forage Crops

Climate change is a global concern that affects every living organism and every sector, with plant and animal production feeling its impacts more intensely each day. It leads to significant adverse outcomes worldwide. Defined as any change that modifies the composition of the global atmosphere—directly or indirectly—over comparable periods, climate change arises from natural climate variability and human activities (UN 1992). It is now regarded as one of the most pressing issues of our time (Türkeş 2014; Koç 2018; Sarıkaya and Tatar 2023).

Climate change produces substantial economic losses and adverse effects on agricultural productivity by intensifying and increasing the frequency of climate-related disasters such as floods, droughts, and storms (Başoğlu and Telatar 2013). Reduced feed crop yields due to climate change and price hikes driven by supply shortages diminish livestock production and limit public access to animal-based products (ÇŞB 2012). Furthermore, lower yield and quality in forage crops, declining biodiversity, and shifting cropping patterns are among the key consequences of climate change.

It will lead to a severe decrease in plant productivity and the extinction of some crops. The increase in carbon dioxide levels and temperature in the environment will primarily affect the amount and yield of forage crops (Chapman et al. 2012).

A study examining the effects of high carbon dioxide on the environment concluded that the quality of forage crops may decrease as carbon dioxide concentration increases (Thornton et al. 2015). Craine et al. (2010) suggested that forage quality measures of crude protein and digestible organic matter decrease with increasing temperature and increase with increasing precipitation. In a study investigating the effects of climate change on the productivity of plants such as wheat, corn, sunflower, and barley in Turkiye in 2050, it was emphasized that there would be a loss of productivity in agricultural products. It was claimed that wheat and barley yields would decrease by 7.6%, corn yield by 10%, and sunflower yield by 6.5%. In addition, researchers stated that due to climate change, there would be changes in the cultivation areas of wheat, barley, corn, and sunflower products, depending on the region. They predicted that cultivation areas would decrease by 2.24% for barley, 9.11% for corn, and 8.18% for wheat (Dellal et al. 2011). It was determined that global wheat production decreased by 88.2 kg/ha, corn production by 90.3 kg/ha, and barley production by 144.9 kg/ha due to climatic changes from 1981 to 2002 (Lobell and Field 2007).

3 Principles of Transgenic Forage Crop Development

Developing transgenic forage crops requires the stable insertion of foreign DNA into the plant genome. One of the most widespread techniques employed in this process is *Agrobacterium tumefaciens*-mediated transformation (Gelvin 2003). Under natural conditions, this soil bacterium transfers a portion of the Ti plasmid (T-DNA) to plant cells, allowing it to integrate into the host genome. Scientists can incorporate genes of interest into the T-DNA to introduce desired traits. For species that are more difficult to transform, biolistic transformation (particle bombardment) and protoplast fusion can also be used (Altpeter et al. 2016).

Efficient selection and regeneration protocols play a critical role in achieving successful transformation. Selectable marker genes—often those providing resistance to antibiotics or herbicides—aid in identifying cells that have transformed. Regenerating entire plants from these modified cells then depends on well-optimized tissue culture conditions, which can vary significantly among forage crop species (Wang and Sticklen 1994).

4 Applications of Transgenic Forage Crops

Transgenic approaches have been employed to address several key challenges in forage production, which are discussed in the following sections.

4.1 Enhanced Nutritional Value

Elevating the digestibility and nutritional quality of forages is a priority area of research. For instance, transgenic alfalfa expressing fungal cellulase enzymes has improved fiber digestibility (Reddy et al. 2005). Similar genetic interventions in ryegrass can potentially boost essential amino acids like lysine and methionine, enhancing animal performance (El-Shemy et al. 2010). Many forage species also contain anti-nutritional compounds such as tannins and saponins that adversely impact animal health and productivity. RNA interference (RNAi) technology has been used successfully to silence genes integral to the production of these compounds (Wang et al. 2012).

Furthermore, transgenic forage crops can be developed to contain higher concentrations of proteins, vitamins, and minerals, improving both animal health and livestock productivity. Adjusting forage composition to lower methanogenic potential may reduce ruminant methane emissions (Muetzel et al. 2016). Similarly, engineering forage crops to enhance livestock phosphorus utilization can reduce phosphorus excretion, minimizing environmental pollution.

4.2 Improved Abiotic Stress Tolerance

Forage crops frequently encounter environmental stresses such as drought, salinity, and extreme temperatures. Transgenic strategies have been utilized to strengthen tolerance to these conditions. For example, overexpressing genes that encode osmoprotectants or stress-related transcription factors has been shown to bolster drought tolerance in alfalfa and other forages (Zhang et al. 2005). Introducing genes that produce osmoprotectants (e.g., proline and glycine betaine) or modifying genes in stress response pathways can help maintain crop productivity under challenging environmental conditions (Shinozaki and Yamaguchi-Shinozaki 2007).

4.3 Herbicide Resistance

Developing herbicide-resistant forages enables more effective weed control, boosts yields, and lowers resource competition. Glyphosate-resistant alfalfa is a prominent example (Padgette et al. 1995).

4.4 Reduced Lignin Content

Lignin, a complex polymer in plant cell walls, negatively impacts forage digestibility. Downregulating key enzymes in the lignin biosynthesis pathway through genetic engineering has reduced lignin content and improved digestibility in forages such as alfalfa and switchgrass (Jung and Casler 2001).

4.5 Disease Resistance

In sustainable forage production, managing fungal and viral diseases is paramount. Transgenic approaches have successfully introduced resistance to alfalfa mosaic virus and anthracnose in alfalfa (Hill et al. 2006).

4.6 Enhanced Protein Content

Many forage crops lack sufficient levels of essential amino acids, prompting the need for dietary protein supplements in livestock. Strategies to raise protein levels involve overexpressing genes responsible for amino acid synthesis or introducing genes encoding high-quality storage proteins, such as zeins (Huang et al. 2004).

4.7 Future Directions

More precise gene editing tools such as CRISPR-Cas9 will likely reshape transgenic forage crop development, allowing targeted gene modifications with greater accuracy (Jaganathan et al. 2018). Moreover, creating multi-transgenic forages that combine multiple beneficial traits, known as "stacked traits," promises to address various challenges simultaneously. Ongoing research that delves into the interactions between transgenes and the plant's inherent genetic makeup remains essential for maximizing the advantages of genetic engineering in forage crops.

5 Genetic Engineering Techniques for Forage Crop Improvement

Forage crops are essential in sustaining livestock production and contribute substantially to global food security. Enhancing their nutritional content, productivity, and resilience to environmental challenges is paramount. Genetic engineering furnishes robust methods for pursuing these objectives, helping to overcome certain limitations of traditional plant breeding.

5.1 Agrobacterium-Mediated Transformation

The soilborne plant pathogen *Agrobacterium tumefaciens* has been pivotal in genetic engineering due to its innate ability to transfer T-DNA (transfer DNA) into plant genomes. Researchers insert the gene of interest (flanked by T-DNA border sequences) into *Agrobacterium* through binary vectors, and the bacterium subsequently infects plant tissues (e.g., leaf discs or embryogenic calli). The T-DNA then integrates randomly into the plant genome, facilitating stable expression of the transgene (Gelvin 2003).

This method is still the most frequently employed, taking advantage of *A. tumefaciens* and its T-DNA transfer system (Gelvin 2003). It has been successfully applied to a range of forage crops, including alfalfa (*Medicago sativa*) and ryegrass (*Lolium perenne*), to improve traits such as drought resistance and nitrogen use efficiency (Bouton 2007).

5.2 Direct Gene Transfer

Several techniques bypass the requirement for *Agrobacterium*. Among these methods are:

Particle bombardment (biolistics): Gold or tungsten particles coated with DNA are shot into plant cells via a gene gun. This approach is often preferred for species that respond poorly to *Agrobacterium*-based methods (Altpeter et al. 2016). For example, particle bombardment has introduced genes into maize (*Zea mays*)—used as a forage crop—to enhance its digestibility.

Protoplast transformation: Protoplasts which are plant cells stripped of their cell walls, are exposed to DNA in the presence of polyethylene glycol (PEG) or subjected to electroporation, promoting DNA uptake. Although obtaining fully regenerated plants from protoplasts can be challenging, this technique holds value for certain forage species (Potrykus 1991).

5.3 Genome Editing Technologies

Recent advances in genome editing—specifically CRISPR-Cas systems—have brought about a paradigm shift in genetic engineering. CRISPR-Cas9 allows targeted modifications of specific regions in the plant genome, enabling gene knockouts, insertions, or replacements at precise loci (Cong et al. 2013). This precision minimizes unwanted alterations and provides greater control over gene functions.

In forage crops, CRISPR-Cas9 can improve nutritional quality by adjusting the lignin biosynthesis pathway or boosting disease resistance by modifying genes that confer susceptibility (Jaganathan et al. 2018). Such technologies significantly enhance our capability to steer transgene integration and expression (Jaganathan et al. 2018).

Aside from conventional gene insertion techniques, genome editing tools such as CRISPR-Cas9 are transforming the possibilities for forage crop enhancement. They allow highly targeted gene editing that can be harnessed to create forage varieties with traits such as higher biomass production, better digestibility, enhanced nutritional value (e.g., increased protein or altered fatty acid composition), and improved tolerance to both biotic and abiotic stresses (including drought, salinity, pests, and diseases). For instance, CRISPR-Cas9-mediated gene editing in alfalfa (*Medicago sativa*) has successfully modified lignin biosynthesis, lowering lignin levels and increasing ruminant digestibility (Li et al. 2019). Likewise, editing genes connected with osmotic stress responses in ryegrass (*Lolium perenne*) has improved drought tolerance (Yu et al. 2021).

Moreover, the development of multigene stacking strategies offers promising avenues for engineering forage crops that carry multiple desirable traits. In this approach, various genes are introduced and expressed simultaneously, combining traits such as enhanced digestibility and improved disease resistance.

5.4 Challenges and Future Directions

Despite considerable advancements, certain hurdles persist in engineering forage crops, including:

Genotype dependence: Different genotypes frequently demand tailored transformation and regeneration protocols.

Public acceptance: Societal concerns over genetically modified organisms (GMOs) can hinder the widespread adoption of genetically modified forage crops.

Complex trait engineering: Many key traits, such as yield and stress resilience, involve multiple genes and require more advanced engineering approaches. Future research should focus on:

(a) Developing efficient transformation protocols for a broader range of forage species.
(b) Advanced gene editing tools such as base and prime editors are used for precise gene modification.
(c) Implementing multigene engineering strategies to address complex traits.
(d) Addressing public concerns through transparent communication and robust risk assessment.

By overcoming these challenges and harnessing the power of genetic engineering, we can develop improved forage crops that contribute to sustainable and efficient livestock production, ultimately enhancing global food security.

6 Trait Development in Transgenic Forage Crops: Enhancing Yield and Quality

As transgenic forage crops are fundamental to livestock farming, they are a significant focus for interventions designed to enhance their nutritional composition, digestibility, tolerance to environmental stresses, and overall biomass production. Here, principal trait targets, key engineering approaches, and the hurdles of creating and disseminating these optimized varieties are explored.

6.1 Targets for Trait Improvement

Efforts to raise forage quality and yield generally involve addressing multiple interconnected aspects. Central objectives include:

Enhanced digestibility: High lignin levels in plant cell walls hinder ruminants' ability to break down carbohydrates. Lowering lignin through the downregulation of specific enzymes—such as cinnamyl alcohol dehydrogenase (CAD) and caffeoyl-CoA O-methyltransferase (COMT)—has been shown to raise digestibility and promote greater forage intake (Baucher et al. 1998a; Reddy et al. 2005).

Elevated protein content: Many forages lack sufficient essential amino acids, thus requiring livestock diets to be supplemented with additional protein. Genetic modifications that bolster amino acid biosynthesis or introduce high-quality storage proteins (e.g., zeins) can help meet this shortfall (Huang et al. 2004).

Improved abiotic stress tolerance: Drought, high salinity, and temperature extremes can significantly reduce forage yield. Introducing genes that encode osmoprotectants (e.g., proline, glycine betaine) or modulate stress–signal pathways can safeguard productivity when circumstances are harsh (Shinozaki and Yamaguchi-Shinozaki 2007).

Reduction of anti-nutritional compounds: Certain forage species contain tannins, saponins, and other substances that can undermine animal health and performance. RNA interference (RNAi) has been successfully applied to suppress genes key to producing these compounds (Wang et al. 2012).

7 Environmental and Economic Impacts of Transgenic Forage Crops

Forage crops form a substantial portion of agricultural production worldwide and are vital for sustaining livestock. Incorporating transgenic or genetically modified (GM) varieties engineered to possess traits such as enhanced nutritional value, resistance to pests, or herbicide tolerance presents both opportunities and complexities. This section examines how these modifications influence the environment and the economy, acknowledging the numerous factors that shape their overall effects.

7.1 Environmental Impacts: Herbicide Tolerance and Pesticide Use

Herbicide tolerance, often granted by introducing genes that confer resistance to glyphosate or glufosinate, has been a significant driver behind transgenic forage crop development. In theory, such tolerance can streamline weed control and reduce reliance on plowing, yet its impact on herbicide usage is not straightforward. Some initial data showed a decrease in herbicide application (Somers 1996), whereas other studies highlight the prospect of overreliance on particular herbicides, potentially causing the emergence of resistant weed species (Heap 2014). Should resistant weeds proliferate, farmers might then need more herbicide applications or alternative control measures, eroding the environmental gains initially anticipated. Potential herbicide runoff and its repercussions for nontarget organisms must be carefully monitored (Giesy et al. 2000).

7.2 Impact on Biodiversity

Gene flow from transgenic forage crops to nearby wild relatives can raise concerns over biodiversity preservation. If transgenes enter wild populations, it may alter their fitness or influence broader ecological processes. The risk depends on the nature of the transgene, the crop in question, and the presence of compatible wild relatives (Ellstrand 2003). In alfalfa (*Medicago sativa*), gene flow to its wild counterpart, *Medicago falcata*, has been documented (Lu et al. 2009). While further evaluation is needed to ascertain the long-term consequences, the potential for unintentional impacts calls for risk assessments and measures such as buffer zones or sterile plant lines.

7.3 Effects on Soil Health

Transgenic forage crops can reshape soil conditions through changed nutrient cycles and shifts in microbial populations. Improvements in nitrogen use efficiency in GM forages, for example, may reduce the need for nitrogen fertilizers and reduce nitrogen runoff and greenhouse gas emissions (Taiz and Zeiger 2010). Nonetheless, the broader effects on soil microbial ecology, including species diversity or function changes, require additional research.

7.4 Economic Impacts: Increased Productivity and Reduced Input Costs

Forage varieties with traits such as higher biomass, enhanced digestibility, and pest resistance can support higher levels of livestock production and decrease certain input costs (Wang et al. 2018). Varieties resistant to specific herbicides, for example, simplify weed management, potentially cutting labor and fuel expenditures. However, the actual economic payoff depends on various factors, including the traits introduced, the particular farm system, and prevailing market prices.

7.5 Market Acceptance and Regulatory Costs

Public attitudes toward GM crops and the regulatory framework for their approval play a key role in whether transgenic forages become economically viable. Stringent oversight can elevate development expenses and delay market entry. Additionally, consumer skepticism and the need for segregated markets may limit GM forages' uptake, dampening their potential economic benefits.

Although transgenic forage crops present considerable promise for improving livestock production and tackling environmental challenges, their adoption should be guided by a balanced evaluation of environmental and economic outcomes. Ongoing research into gene flow, herbicide resistance, and long-term soil health effects will be pivotal. Open communication with stakeholders, consumers, policymakers, and others is vital to fostering trust and ensuring these technologies are deployed responsibly.

8 Transgenic Forage Crops Species

Transgenesis provides a powerful avenue for enhancing various traits in forage crop species, as illustrated by the following examples drawn from the materials provided:

Lignin biosynthesis modulation: Reducing lignin levels and altering lignin composition by downregulating enzymes, such as caffeic acid O-methyltransferase (COMT), cinnamyl alcohol dehydrogenase (CAD), or 4-coumarate:CoA ligase (4CL), has shown promising results in boosting the dry matter digestibility of transgenic forages such as tobacco and alfalfa (Baucher et al. 1998a, 1999). Such approaches can diminish lignin content without markedly affecting plant growth (Chabannes et al. 2000).

Fructan metabolism: Introducing microbial fucosyltransferase genes (like SacB) that promote fructan accumulation can potentially elevate soluble carbohydrate levels and enhance drought resistance (Pilon-Smits et al. 1995; LePage et al. 2000).

Rumen by-pass proteins: Including genes coding for high-value proteins, such as sunflower seed albumin (SFA8), in forage legumes may raise the availability of essential amino acids for ruminants (Rafiqul et al. 1996; Wang et al. 2000). However, more work is needed to achieve sufficiently high protein expression levels to attain a meaningful nutritional impact.

Antifungal proteins (AFPs): Introducing AFPs into legumes (e.g., subterranean clover) can improve resistance against fungal pathogens such as *Phytophthora clandestina* (Aldao et al. 2000).

Viral resistance: Using coat protein-mediated resistance or modified viral movement proteins has effectively conferred resistance or immunity to viruses, including alfalfa mosaic virus (AMV) and white clover mosaic virus (WCMV) in transgenic forage legumes (Chu et al. 2000; Kalla et al. 2000; Voisey et al. 2001).

Insect resistance: Incorporating *Bacillus thuringiensis* (Bt) crystal proteins or proteinase inhibitors (PIs) in forage legumes shows potential in boosting resistance to insect pests (Strizhov et al. 1996; Voisey et al. 2001).

Delayed leaf senescence: Deploying senescence-specific promoters to regulate the isopentenyl transferase (*ipt*) gene can elevate cytokinin levels, which extends leaf lifespan and enhances forage quality (Gan and Amasino 1995; Ludlow et al. 2000).

Flowering control: Manipulating genes that oversee floral meristem initiation, such as the LEAFY or APETALA1 equivalents, allows for the possibility of limiting or abolishing flowering in transgenic forage crops, with potential advantages such as enhanced forage quality and reduced pollen allergens (Wu et al. 1997; Colasanti et al. 1998).

Overall, these approaches showcase the scope of genetic engineering in improving the productivity, nutritional profile, and resilience of forage species in grassland agriculture (Table 1). Nevertheless, achieving high and consistent transgene expression, addressing biosafety concerns, including gene flow, and complying with intricate regulatory requirements for field release and commercialization remain key hurdles.

9 Conclusion

Over the past six decades in Turkiye, large cattle numbers have climbed by around 4.5 million, while small ruminants have declined by about 13 million. This shift in livestock dynamics underscores the vital need for policies that support forage crop production—an essential pillar of sustainable animal husbandry. Expanded

Table 1 Transgenic forage plant varieties and applications

Crop	Trait targeted	Gene source/method	Application/ benefit	Reference
Alfalfa	Increased dry matter digestibility	Downregulation of CAD	Improved forage quality for ruminants	Baucher et al. (1998b)
Alfalfa	Increased phosphorus acquisition	Expression of phytase and acid phosphatase genes	Enhanced nutrient uptake, reduced need for phosphate fertilizers	Ma et al. (2012)
Alfalfa	Delayed leaf senescence	Expression of *ipt* gene under senescence-specific promoter	Extended period of optimal grazing, improved forage quality	Calderini et al. (2007)
Alfalfa	Resistance to *Spodoptera*	Expression of synthetic *cryIC* gene	Protection against insect pest damage	Strizhov et al. (1996)
Alfalfa	Resistance to alfalfa mosaic virus (AMV)	Expression of AMV coat protein gene	Field immunity to AMV	Hill et al. (2006)
Bahiagrass	Improved turf quality	Constitutive expression of *ATHB16* gene	Enhanced turf density and reduced vertical growth	Zhang et al. (2007)
Creeping Bentgrass	Resistance to dollar spot	Overexpression of rice *TLPD34*	Reduced disease incidence and severity	Fu et al. (2005)
Perennial Ryegrass	Increased dry matter digestibility	Downregulation of COMT	Improved forage quality for ruminants	Chen et al. (2003)
Perennial Ryegrass	Downregulation of pollen allergens (Lol p1, Lol p2)	Antisense gene expression under pollen-specific promoter	Reduced allergenicity, potential for developing hypoallergenic cultivars	Wu et al. (1997)
Perennial Ryegrass	Enhanced salt tolerance	Expression of wheat fructosyltransferase genes	Increased tolerance to salinity stress	Hisano et al. (2004)
Red Clover	Resistance to AMV and WCMV	Expression of AMV and WCMV coat protein genes	Virus resistance in a dual-purpose forage legume	Kalla et al. (2000)
Subterranean Clover	Resistance to fungal diseases	Expression of antifungal protein (AFP) genes	Protection against root rot and other fungal pathogens	Aldao et al. (2000)

(continued)

Table 1 (continued)

Crop	Trait targeted	Gene source/method	Application/ benefit	Reference
Subterranean Clover	Accumulation of sulfur-rich protein	Expression of sunflower seed albumin gene	Improved nutritional value for ruminants	Rafiqul et al. (1996)
Switchgrass	Increased saccharification efficiency	Downregulation of CAD	Improved biofuel production	Fu et al. (2011)
Switchgrass	Improved biomass yield	Overexpression of miR156	Altered morphology leading to increased biomass	Fu et al. (2012)
Tall Fescue	Improved dry matter digestibility	Downregulation of COMT	Improved forage quality for ruminants	Chen et al. (2004)
Tall Fescue	Accumulation of sulfur-rich protein	Expression of sunflower seed albumin gene	Improved nutritional value for ruminants	Wang et al. (2000)
Tall Fescue	Resistance to gray leaf spot and brown patch	Expression of bacteriophage T4 lysozyme gene	Reduced disease incidence and severity	Dong et al. (2008)
White Clover	Resistance to white clover mosaic virus (WCMV)	Expression of mutated WCMV movement protein or replicase gene	High levels of resistance or immunity to WCMV	Voisey et al. (2001)
White Clover	Resistance to clover yellow vein virus (CYVV)	Expression of CYVV coat protein gene	Resistance to CYVV	Chu et al. (2000)

financial incentives for forage cultivation, calibrated to match the growing number of farmers and inflationary pressures, will help maintain planting momentum without eroding actual returns. Addressing the root causes of high seed costs—particularly by investing in local seed programs—can strengthen production values, reduce reliance on costly imports, and encourage more widespread adoption of improved forage varieties. This way, strategic support for forage crops can bolster the livestock sector, safeguarding economic viability and long-term food security.

References

Açıkgöz E, Hatipoğlu R, Altınok S, Sancak C, Tan A, Uraz D (2005) Forage crops production and problems. In: Turkish agricultural engineering 6th technical congress, pp 503–518

Aldao G, Drayton M, Kalla R, Cammue B, Spangenberg G (2000) Development of transgenic subterranean clover expressing different chimeric AFP genes for enhanced resistance to fungal disease. In: 2nd International symposium molecular breeding of forage crops, p 110

Altpeter F, Springer NM, Bartley LE, Blechl AE, Brutnell TP, Citovsky V, Voytas DF (2016) Advancing crop transformation in the era of genome editing. Plant Cell 28(7):1510–1520

Başoğlu A, Telatar OM (2013) Impacts of climate change: an econometric application on the agriculture sector. Karadeniz Tech Univ J Soc Sci 6:7–25

Baucher M, Monties B, Van Montagu M, Boerjan W (1998a) Biosynthesis and genetic engineering of lignin. Crit Rev Plant Sci 17(2):125–197

Baucher M et al (1998b) Down-regulation of cinnamyl alcohol dehydrogenase in transgenic alfalfa (*Medicago sativa* L.) and the effect on lignin composition and digestibility. Plant Mol Biol 39(4):437–447

Baucher M, Bernard-Vailhé MA, Chabbert B, Besle JM, Opsomer C, Van Montagu M, Botterman J (1999) Down-regulation of cinnamyl alcohol dehydrogenase in transgenic alfalfa (*Medicago sativa* L.) and the effect on lignin composition and digestibility. Plant Mol Biol 39:437–447

Bouton JH (2007) Molecular breeding of switchgrass for use as a biofuel crop. Curr Opin Plant Biol 10(2):186–192

Calderini O et al (2007) Delay of leaf senescence in *Medicago sativa* transformed with the ipt gene controlled by the senescence-specific promoter SAG12. Plant Cell Rep 26(5):611–615

CGIAR (2019) FAO sets the record straight–86% of livestock feed is inedible by humans. https://cgiar.org/news-events/news/fao-sets-the-record-straight-86-of-livestock-feed-is-inedible-by-humans/. Accessed: 4.12.2024

Chabannes M, Yoshinaga A, Ruel K, Boudet, AM (2000) Simultaneous down-regulation of CAD and CCR genes induces differences in amount, nature and topology of lignins in the cell wall. In: 6th International congress of plant molecular biology

Chapman SC, Chakraborty S, Dreccer MF, Howden SM (2012) Plant adaptation to climate change—opportunities and priorities in breeding. Crop Past Sci 63(3):251–268

Chen L, Auh CK, Dowling P, Bell J, Chen F, Hopkins A, Wang ZY (2003) Improved forage digestibility of tall fescue (Festuca arundinacea) by transgenic down-regulation of cinnamyl alcohol dehydrogenase. Plant Biotechnol J 1(6):437–449

Chen L, Auh CK, Dowling P, Bell J, Lehmann D, Wang ZY (2004) Transgenic down-regulation of caffeic acid O-methyltransferase (COMT) led to improved digestibility in tall fescue (*Festuca arundinacea*). Funct Plant Biol 31(3):235–245

Chu P, Holloway B, Venables I, Lane L, Gibson J, Larkin P, Higgins TJ (2000) Development of transgenic white clover (*Trifolium repens*) with resistance to clover yellow vein virus. In: 2nd International symposium molecular breeding of forage crops

Colasanti J, Yuan Z, Sundaresan V (1998) The indeterminate gene encodes a zinc finger protein and regulates a leaf-generated signal required for the transition to flowering in maize. Cell 93(4):593–603

Cong L, Ran FA, Cox D, Lin S, Barretto R, Habib N, Zhang F (2013) Multiplex genome engineering using CRISPR/Cas systems. Science 339(6121):819–823

Craine JM, Elmore AJ, Olson KC, Tolleson D (2010) Climate change and cattle nutritional stress. Glob Chang Biol 16(10):2901–2911

ÇŞB (2012) Effects of climate change on agriculture and food security in Turkey. In: Publication of the preparation of Turkey's 2nd national communication on climate change project, ministry of environment and urbanization of the Republic of Turkey, 34 p

Dellal I, McCarl BA, Butt T (2011) The economic assessment of climate change on Turkish agriculture. J Environ Prot Ecol 12(1):376–385

Dong S, Shew HD, Tredway LP, Lu J, Sivamani E, Miller ES, Qu R (2008) Expression of the bacteriophage T4 lysozyme gene in tall fescue confers resistance to gray leaf spot and brown patch diseases. Transgenic Res 17:47–57

Ellstrand NC (2003) Dangerous liaisons? When cultivated plants mate with their wild relatives. Johns Hopkins University Press, Baltimore

El-Shemy HA, Aboul-Soud MAM, Khalafalla MM (2010) Field evaluation of transgenic perennial ryegrass over-expressing the HMW-glutenin subunit 1Ax1 gene. J Appl Genet 51(1):63–74

FAO (2017) The future of food and agriculture: trends and challenges. https://www.fao.org/publications/en. Accessed: 1.11.2024

FAO (2019) Agricultural output growth to keep food prices low over the coming decade, but many uncertainties are ahead. https://www.fao.org/newsroom/en. Accessed: 1.11.2024

FAOSTAT (2021) Food and Agricultural Organization of the United Nations. https://www.fao.org/faostat/en/#data. Accessed: 1.11.2024

Fu D, Tisserat NA, Xiao Y, Settle D, Muthukrishnan S, Liang GH (2005) Overexpression of rice TLPD34 enhances dollar-spot resistance in transgenic bentgrass. Plant Sci 168(3):671–680

Fu C, Xiao X, Xi Y, Ge Y, Chen F, Bouton J, Wang ZY (2011) Downregulation of cinnamyl alcohol dehydrogenase (CAD) leads to improved saccharification efficiency in switchgrass. Bioenergy Res 4:153–164

Fu C, Sunkar R, Zhou C, Shen H, Zhang JY, Matts J, Wang ZY (2012) Overexpression of miR156 in switchgrass (*Panicum virgatum* L.) results in various morphological alterations and leads to improved biomass production. Plant Biotechnol J 10(4):443–452

Gan S, Amasino RM (1995) Inhibition of leaf senescence by autoregulated production of cytokinin. Science 270(5241):1986–1988

Gelvin SB (2003) *Agrobacterium*-mediated plant transformation: the biology behind the "gene-jockeying" tool. Microbiol Mol Biol Rev 67(1):16–37

Giesy JP, Dobson S, Solomon KR (2000) Ecotoxicological risk assessment for roundup herbicide. Rev Environ Contam Toxicol 167:35–120

Harmanşah F (2018) Quality forage production in Turkey, problems and suggestions. TÜRKTOB J 25:9–13

Heap I (2014) Global perspective of herbicide-resistant weeds. Pest Manag Sci 70(9):1306–1315

Hill KK, Bancroft JB, Chenault KD, Jarausch W (2006) Evaluation of transgenic alfalfa for resistance to alfalfa mosaic virus. Plant Dis 90(3):363–367

Hisano H, Kanazawa A, Kawakami A, Yoshida M, Shimamoto Y, Yamada T (2004) Transgenic perennial ryegrass plants expressing wheat fructosyltransferase genes accumulate increased amounts of fructan and acquire increased tolerance on a cellular level to freezing. Plant Sci 167(4):861–868

Huang S, Adams WR, Zhou Q, Malloy KP, Voyles DA, Anthony J, Larkins BA (2004) Improving nutritional quality of maize proteins by expressing sense and antisense zein genes. J Agric Food Chem 52(7):1958–1964

IFCN (2014) Dairy report (2014). International Farm Comparison Network Dairy Research Center, Kiel

Jaganathan D, Ramasamy K, Sridhar J, Krishna S, Janakiraman V, Shanthi S (2018) CRISPR for crop improvement: an update review. Front Plant Sci 9:985

Jung HG, Casler MD (2001) Relationship of lignin and esterified phenolics to in vitro dry matter digestibility in sudangrass. Crop Sci 41(4):1166–1171

Kalla R, Chu P, Spangenberg G (2000) Molecular breeding of forage legumes for virus resistance (Chapter 13). In: Spangenberg G (ed) Molecular breeding of forage crops. Kluwer Academic Publishers, Dordrecht

Koç G (2018) Analysis of the effects of climate change on dairy farms in terms of economic and agricultural policies; the example of Thrace region, Ege University, Institute of Science, Department of Agricultural Economics, Master's thesis, 157 p

LePage C, Mackin L, Lidgett A, Spangenberg G (2000) Development of transgenic white clover expressing chimeric bacterial levansucrase genes for enhanced tolerance to drought stress. In: 2nd International symposium molecular breeding of forage crops, p 80

Li X, Su L, Li C, Shu Q (2019) Targeted mutagenesis of lignin biosynthesis genes in alfalfa (*Medicago sativa* L.) using CRISPR/Cas9 system. Plant Biotechnol J 17(1):127–139

Lobell D, Field CB (2007) Global scale climate-crop yield relationships and the impacts of recent warming. Environ Res Lett 2:1–7

Lu BR, Rong J, Arnold ML, Clements DR, Coupland RT, Maurice DC (2009) Hybridization between transgenic glyphosate-resistant alfalfa and its wild relative *Medicago falcata*. Theor Appl Genet 118(6):1089–1096

Ludlow E, Lin YH, Chalmers J, Kalla R, Pallaghy C, Spangenberg G (2000) Development of transgenic white clover with delayed leaf senescence. In: 2nd International symposium molecular breeding of forage crops, p 84

Ma XF, Tudor S, Butler T, Ge Y, Xi Y, Bouton J, Wang ZY (2012) Transgenic expression of phytase and acid phosphatase genes in alfalfa (Medicago sativa) leads to improved phosphate uptake in natural soils. Mol Breed 30:377–391

Muetzel S, Baldinger S, Goopy JP, Rea SM, McMahon KD, Yu Z, Duval SM (2016) Reducing methanogenesis in ruminants by genetic manipulation of rumen methanogens. J Anim Sci Biotechnol 7(1):1–11

Nardone A (2002) Evolution of livestock production and quality of animal products. In: Proceedings of 39th annual meeting of the Brazilian Society of Animal Science, Brazil, pp 486–513

Padgette SR, Kolacz KH, Delannay X, Re DB, LaVallee BJ, Tinius CN, Kishore GM (1995) Development, identification, and characterization of a glyphosate-tolerant soybean line. Crop Sci 35(5):1451–1461

Pilon-Smits EAH, Ebskamp MJM, Paul MJ, Jeuken MJW, Weisbeek PJ, Smeekens SCM (1995) Improved performance of transgenic fructan-accumulating tobacco under drought stress. Plant Physiol 107(1):125–130

Potrykus I (1991) Gene transfer to plants: assessment of published approaches and results. Annu Rev Plant Physiol Plant Mol Biol 42(1):205–225

Rafiqul M, Khan I, Ceriotti A, Tabe L, Aryan A, McNabb W, Higgins TJ (1996) Accumulation of a sulphur-rich seed albumin from sunflower in the leaves of transgenic subterranean clover (*Trifolium subterraneum* L.). Transgenic Res 5:179–185

Reddy MSS, Chen F, Shadle G, Jackson L, Aljoe H, Dixon RA (2005) Targeted down-regulation of caffeoyl-CoA O-methyltransferase in alfalfa impacts lignin monomer composition, reduction in lignin content, and improved forage quality. Planta 221(1):133–142

Sarıkaya MF, Tatar M (2023) Global climate change and crop production. USBTU 2(2): 65-78.

Shinozaki K, Yamaguchi-Shinozaki K (2007) Gene networks involved in drought stress response and tolerance. J Exp Bot 58(2):221–227

Somers DA (1996) Aryloxyphenoxypropionate- and cyclohexanedione- resistant crops. In: Duke SO (ed) Herbicide-resistant crops: agricultural, environmental, economic, regulatory, and technical aspects. CRC Press and Lewis, Boca Raton, pp 175–188

Strizhov N et al (1996) A synthetic cryIC gene, encoding a *Bacillus thuringiensis* delta-endotoxin, confers Spodoptera resistance in alfalfa and tobacco. Proc Natl Acad Sci 93(29):15012–15017

Taiz L, Zeiger E (2010) Plant physiology, 5th edn. Sinauer Associates

Thornton PK, Boone RB, Ramirez-Villegas J (2015) Climate change impacts on livestock. CCAFS working paper no. 120. CGIAR Research Program on Climate Change, Agriculture and Food Security CCAFS, Copenhagen, 19 p

Türkeş M (2014) The effects of climate change on agricultural food security. Turk J Agric Food Sci Technol 22:7185

UN (1992) United Nations framework convention on climate change. United Nations, New York. https://unfccc.int/resource/docs/convkp/conveng.pdf. Accessed: 1.11.2024

UN (2019) Population. United Nations, New York. https://www.un.org/en/global-issues/population. Accessed: 1.11.2024

Voisey CR, Dudas B, Biggs R, Burgess EPJ, Wigley PJ, McGregor PG, White DWR (2001) Transgenic pest and disease resistant white clover plants. In: Molecular breeding of forage crops: proceedings of the 2nd international symposium, molecular breeding of forage crops, Lorne and Hamilton, Victoria, Australia, November 19–24, 2000. Springer Netherlands, Dordrecht, pp 239–250

Wang ZY, Sticklen MB (1994) Plant regeneration from protoplasts of alfalfa (*Medicago sativa* L.) via somatic embryogenesis. Plant Cell Rep 13(9):618–623

Wang Z-Y, Ye X-D, Nagel J, Potrykus I, Spangenberg G (2000) Expression of a Sulphur-rich sunflower albumin gene in transgenic tall fescue (*Festuca arundinacea* Schreb.) plants. Plant Cell Rep 20(2):213–219

Wang Y, Wang Z, Sun Y, Li Q, Zhang Y, Wang Y, Liu C (2012) Metabolic engineering of flavonoids in alfalfa (*Medicago sativa* L.) by down-regulating chalcone synthase using RNA interference. Plant Biotechnol Rep 6(4):355–364

Wang ZY, Dixon RA, Zhong R (2018) Transgenic forage crops: development and impact. Crit Rev Biotechnol 38(2):240–251

Wu XL, Ye XD, Wang ZY, Potrykus I, Spangenberg G (1997) Gene transfer to ryegrasses: down-regulation of major pollen allergens in transgenic plants. In: XVIII International grassland congress

Yolcu H, Tan M (2008) A general view of forage crops agriculture in our country. Ankara University Faculty of Agriculture. J Agric Econ 14:303–312

Yu X, Wang C, Zhang X, Wang Z, Pan X (2021) CRISPR/Cas9-mediated genome editing for drought tolerance improvement in perennial ryegrass (*Lolium perenne* L.). Front Plant Sci 12:667973

Zhang L, Oh DH, Zhang H, Reymond M, Lee C, Zhang S (2005) An improved *Agrobacterium*-mediated transformation and regeneration system for switchgrass (*Panicum virgatum* L.). Plant Sci 168(4):995–1002

Zhang H, Lomba P, Altpeter F (2007) Improved turf quality of transgenic bahiagrass (Paspalum notatum Flugge) constitutively expressing the ATHB16 gene, a repressor of cell expansion. Mol Breed 20:415–423

Genomics of Forage Crops

Sarmad Ali Qureshi, Amjad Ali, Parnaz Mortazavi, Salma Naimatullha Soomro, Sabeen Rehman Soomro, Mehmet Bedir, Muhammed Tatar, Meliha F. Sarıkaya, Muhammad Tanveer Altaf, Rada Sucur, and Kağan Kökten

Abstract

Forage crops, including ryegrass, alfalfa, and clover, are crucial for sustainable agriculture and livestock production. However, climate change threatens yield and nutritional quality, necessitating the integration of traditional and advanced breeding techniques. Genomics has revolutionized forage crop improvement by enabling precise identification and manipulation of genetic traits. Advances such as GWAS, QTL mapping, HTS, and CRISPR/Cas9 gene editing improve yield, quality, and stress resilience. Functional genomics, proteomics, and metabolomics are essential for understanding gene functions and metabolic pathways in forage crops. These strategies help develop forage varieties with enhanced biomass, digestibility, and stress adaptation, while also identifying resistance genes and stress–response pathways for breeding resilient cultivars. Genomic discoveries in forage improvement face challenges due to genetic complexity and limited resources, necessitating international collaboration and public–private partnerships to accelerate progress. The book chapter discusses the importance of integrating artificial intelligence (AI) and big data analytics in breeding programs, focusing on developing multi-trait forages for climate change and global livestock needs, and bridging traditional breeding with genomic technologies. Future research should integrate AI and big data analytics into breeding programs for

S. A. Qureshi · A. Ali (✉) · P. Mortazavi · S. N. Soomro · S. R. Soomro · M. Bedir · M. Tatar · M. F. Sarıkaya · K. Kökten
Faculty of Agricultural Sciences and Technologies, Sivas University of Science and Technology, Sivas, Turkey

M. T. Altaf
Department of Field Crops, Faculty of Agriculture and Natural Sciences, Recep Tayyip Erdoğan University, Pazar, Rize, Turkey

R. Sucur
University of Novi Sad, Faculty of Agriculture, Novi Sad, Serbia

F. S. Baloch et al. (eds.), *Forage Crops in the Omics Era*,
https://doi.org/10.1007/978-981-95-9249-4_5

improved efficiency and precision, focusing on developing multi-trait forages for climate adaptation and sustainable livestock production.

Keywords

Fodder crops · OMICS · GWAS · QTL mapping · Breeding

1 Introduction

Forages include plant parts that animals can consume directly or use as fodder for grazing (Allen et al. 2011). These crops, which include grasses, legumes, and other plants, play an important role in sustainable agriculture and livestock production. Forages encompass diverse species across several families, including grasses such as ryegrass and timothy from Poaceae; legumes such as alfalfa and clover from Fabaceae, and root crops such as fodder beet from Chenopodiaceae. Certain tree legumes, such as mulga (*Acacia aneura*) and lead tree (Leucaena leucocephala), are often planted in tropical grasslands and deserts (Muir et al. 2011). Globally, forage crops sustain around 1.52 billion cattle, 1.21 billion sheep, 1.02 billion goats, and 0.21 billion buffalo, facilitating the production of meat, wool, and dairy products, supporting farmer livelihoods in developing regions and offering essential ecosystem functions (FAOSTAT 2020). Beyond feeding, they are also essential in soil cover, reducing erosion, boosting nitrogen levels, and enhancing land quality (Haq et al. 2022).

Climate change significantly impacts the nutritional quality and production of crops (Sarıkaya and Tatar 2023) including forage, which also affects the well-being of livestock (Baumgard et al. 2012). Extreme weather conditions can stress forage crops, making them less productive and nutritious. Several studies have been conducted on the effects of climate change on forage crops (Godde et al. 2021). As one of agriculture's earliest sustainable practices, plant breeding has helped to enhance crop yield quality and stress tolerance. Despite its success, traditional breeding methods are limited by their slow pace, high labor demands, and potential biosafety risks. Genetic diversity is the cornerstone of plant breeding. Molecular markers are integral to genetic diversity research, aiding in DNA profiling, identifying genotypes, developing genetic linkage maps, QTL mapping, and exploring population diversity (Baloch et al. 2024). Molecular breeding emerged as a game-changer, utilizing tools such as marker-assisted selection and genomic selection. Various research efforts have explored marker-assisted selection and genomic selection in forage crops (Singh et al. 2022a, Singh et al. 2022b).

Building on this foundation, genomics offers an even more transformative approach by comprehensively understanding crop genomes. Genomics focuses on studying the complete set of DNA in a plant, including all its genes and their functions (Brummer and Wang 2020). Genomic techniques such as quantitative trait loci (QTL) analysis (Jiang et al. 2022), genome-wide association studies (GWAS) (Altaf et al. 2024a), genotyping-by-sequencing (GBS) (Gemeinholzer et al. 2023), RNA sequencing (Wang et al. 2024), genome sequencing (Franco et al. 2024), high-throughput sequencing (HTS) (Fraiture et al. 2023), and CRISPR/Cas9 gene editing (Zhao et al. 2024) have revolutionized the understanding and improvement of forage crops by enabling precise identification and manipulation of genes with key traits. This chapter aims to explore genomics advancements in forage crops, focusing on their genetic structures, enhancing traits such as yield, quality, and stress resilience, and applying genomics tools for breeding improvement.

2 Reference Genome Sequencing of Forage Crops

The reference genome is intended to serve as a standard, but its practical definition is not always clear. For instance, the allelic diversity in a reference genome does not represent an average of the global or any specific population. Instead, it typically reflects the genetic makeup of a single individual, containing long regions that are highly specific to that individual (Ballouz et al. 2019). A lot of effort is invested into assembling reference genomes. Reference assemblies provide a unified representation for comparing results and serve as the foundation for various downstream tools used in sequence alignment and annotation. Many tools that utilize reference assemblies generate independent resources that complement the original assembly (Stolarczyk et al. 2020). Draft genome sequences are available among some forage crops. Draft sequences are available for red clover (*T. pratense* L.) (De Vega et al. 2015) and the model legume species such as *Lotus japonicus* L. (Sato et al. 2008) and *Medicago truncatula* Gaertn (Hirakawa et al. 2016). In the absence of a good quality reference genome, most new genome assemblies are assessed based on the number of scaffolds and contigs needed to explain the genome, the percentage of assembled reads, and the length of contigs and scaffolds in absolute terms and relative to the size of the genome. This is because a genome assembly can only be accurate or complete only when it reflects not only the characteristics due to sequencing technology but also due to the inherent features of the genome being assembled (Baker 2012).

3 Genomics Technologies in Forage Crops

3.1 High-Throughput Sequencing (HTS)

The occurrence of high-throughput sequencing (HTS) technologies has revolutionized plant genomics, offering deep insights into the genetic makeup of forage crops. HTS allows for the rapid sequencing of large genomes, enabling the identification of genetic variations and understanding complex traits, such as disease resistance and drought tolerance. The process involves sequencing millions of small DNA fragments in parallel and computationally assembling them to reconstruct the original sequence (Jiang and Zhou 2024). High-throughput sequencing (HTS) is increasingly used in plant pathology to identify potential viral sequences, which are later confirmed using traditional methods. Most HTS applications focus on identifying unknown disease causes and detecting novel viruses in biosecurity efforts. HTS is also enhancing our understanding of virus population structure, ecology, and evolution. However, its widespread adoption in regulatory plant health has been slow due to challenges in generating validation data and assessing the risks of previously uncharacterized pathogens (Ibaba and Gubba 2020).

3.2 Advances in Sequencing Methods for Forage Crops

In recent years, new viral pathogens, including previously unreported ones, have been discovered in crops. This was made possible by advances in nucleic acid sequencing, which have replaced traditional detection methods with reliable high-throughput sequencing (HTS) technologies (Bejerman et al. 2020). HTS technologies have a number of advantages over traditional serological, molecular, and biological approaches for the diagnosis of diseases and virus detection in plants. One key benefit is that HTS does not require prior knowledge of the viral pathogens in a sample. Unlike PCR, which needs specific viral sequences to design primers, or ELISA, which relies on antibodies for detection, HTS can recognize unknown viruses that might be involved in plant diseases. This method speeds up the identification process and helps reveal the complete virome of a plant sample, often uncovering novel viral pathogens (Maliogka et al. 2018). Deep sequencing approaches in virus detection help overcome false negatives caused by high genetic variability or low viral titer, which are common in fruit tree viruses. The large number of reads generated by HTS allows for the complete assembly of viral genome sequences, enabling further studies on genetic variability (Katsiani et al. 2015). High-throughput sequencing has greatly improved the efficiency of genomic selection (GS) by enabling the fast and precise identification of genetic markers linked to desirable traits. Combining HTS with GS allows breeders to predict the performance of breeding lines based on their genomic profiles, speeding up the breeding cycle and enhancing genetic gains. For example, the development of genotyping by target sequencing (GBTS) platforms offers affordable and high-quality genotyping, which is essential for marker-assisted and genomic selection. Furthermore, HTS provides extensive genomic data, improving the accuracy of genomic predictions and optimizing breeding strategies (Wang and Zhang 2024).

High-throughput sequencing technologies such as Illumina (which is a short-read technology), Pacific Biosciences, and Oxford Nanopore for long reads have really revolutionized DNA and RNA sequencing. Short-read RNA sequencing (ShR RNA-seq) is now used in wide applications in the identification of GWAS events in plants and other organisms by combining gene expression quantification together with discovery. However, its application in polyploid plant species is limited when there is no reference genome or comprehensive transcript-level annotations. To address this, long-read RNA sequencing (LoR RNA-seq) technologies such as Pacific Biosciences (SMRT sequencing) and Oxford Nanopore (MinION) are being used to study complex AS in polyploids (Ardui et al. 2018). LoR RNA-seq can generate long reads (>10 kbp) that cover full-length transcripts, eliminating the need for transcript assembly. While LoR RNA-seq excels at resolving transcript structures, it has limitations such as lower sequencing depth, higher error rates (~15%), and difficulties in transcript quantification. These issues can be mitigated by combining LoR and ShR RNA-seq data (Bedre et al. 2019). Illumina offers nine sequencers capable of generating 4 million to 20 billion short reads (50–300 nucleotides), with applications in plant virus analysis using strategies such as RNASeq, shotgun sequencing, and siRNA sequencing. Similarly, Ion Torrent sequencers can produce 550,000 to 100 million short reads (200–400 nucleotides) and can be used for RNASeq and siRNA sequencing (Maliogka et al. 2018).

Despite advances in HTS technology, testing a large number of samples for plant virus epidemiology remains costly. There are no defined thresholds for read numbers to confirm virus existence, and low-read counts may indicate either low virus titer or sample contamination. Additionally, HTS detection depends on the chosen template, which can affect the viral pathogens identified. Unlike traditional methods, HTS requires complex bioinformatics tools and expertise, and bioinformatic challenges may arise depending on the template analyzed (Maliogka et al. 2018).

3.3 Genome Assembly and Annotation

Large eukaryotic genomes are hard to assemble due to their repetitive nature and also due to the computational load of handling millions of reads. Because of this, genome assembly is often divided into two distinct phases. In the first step, reads that show a clear overlap are assembled into "unitigs." Subsequent steps use mate-pair information to merge these unitigs into larger contigs and then connect the contigs into scaffolds. Scaffolds represent contigs connected by mate-pairs but separated by repetitive sequences or sequence gaps (Kane et al. 2011).

Genome annotation can be described as the identification of regions of interest within a genome, both structurally and functionally. Despite the availability of a number of online annotation servers, it is relevant to define the intended use of the curated data and to choose between two approaches: maximizing in-house efforts or outsourcing. This process demands advanced bioinformatics skills, specialized pipelines, and significant computational power. It involves three main steps: (1) detecting noncoding regions, (2) predicting coding regions or genes, and (3) associating these elements with their biological functions (Jung et al. 2020).

Assembling large eukaryotic genomes is challenging due to their repetitive structure and the computational demands of processing vast numbers of reads. To address these challenges, approaches have been innovated that divide the assembly process into phases. The early stage creates reads with clear overlaps into "unitigs," while subsequent phases use mate-pair information to combine these unitigs into larger contigs, and finally, scaffolds (which are contigs linked by mate pairs but separated by repeats or gaps) (Kane et al. 2011). Genome annotation involves identifying and characterizing biologically significant regions, both structurally and functionally.

Although there are several online annotation tools, it is important to define the goal of the data, considering whether to maximize in-house efforts or outsource.

3.4 Tools and Strategies for Building Forage Crop Genomes

RNA-seq is crucial for genome annotation, helping identify protein-coding genes using transcriptome data and predictive methods. However, its role in genome assembly is limited to scaffolding. Recent advances in single-molecule mRNA sequencing, such as Pacific Biosciences (PacBio) and Oxford Nanopore, have improved gene and genome annotation, offering powerful substitutes or enhancing methods (Tso et al. 2020).

The widespread use of whole-genome sequencing (WGS) has been driven by advancements in second-generation (SGS) and third-generation sequencing (TGS) technologies, which have lowered costs and simplified genome assembly. Researchers can choose from short reads (Illumina, 454, SOLiD, Ion Torrent), long reads (ONT, PacBio), or hybrid approaches. SGS technologies offer fast, accurate, and cost-effective reads (75–700 bp) but struggle with complex, repetitive, or heterozygous regions, leading to fragmented genome assemblies. Illumina's TruSeq PCR-free Library Prep Kit is suitable for large genomes with high DNA input, while the Nextera DNA Prep Kit works well for large, complex genomes with smaller inputs. Other Illumina kits cater to various genome sizes and sample types, and their methods have been thoroughly reviewed (Jung et al. 2020). TGS technologies generate long single-molecule reads (averaging >30 kb), enabling complete genome assembly. However, they face challenges with high costs per base and error rates. The PacBio RS II or SEQUEL system offers improved accuracy and up to 10–15 times more data than earlier systems, with HiFi reads reaching ABI Sanger quality for lengths up to 40 kb. PacBio recommends the SMRTbell Template Prep Kit for WGS with 20–40 kb templates using BluePippin Size Selection. For ONT, optimized methods such as the Rapid Sequencing Kit (SQK-RAD004) allow for longer reads, with some exceeding 2 Mb (Senol Cali et al. 2019).

Ab initio and evidence-based prediction methods are widely used. Both have strengths and weaknesses. Among the popular ab initio predictors are Augustus and SNAP but require information from close relatives of the gene model. In contrast, evidence-based prediction uses data from ESTs, protein sequences, and RNA-seq (especially Iso-Seq data from PacBio or ONT) aligned to a genome assembly. Gene predictors trained with tools such as Augustus and SNAP can be used in programs such as MAKER, BRAKER, and StringTie. For novel genome annotation, using RNA-seq data along with a genome sequence, MAKER and BRAKER are ideal for first-time users due to their inclusion of ab initio prediction, while StringTie focuses solely on evidence-based prediction. MAKER is especially useful for updating existing annotations. To refine gene structure, combining results from different prediction methods using tools such as Evidence Modeler, GLEAN, or GAAP is crucial. BRAKER, StringTie, PASA, and GAAP can also update annotations by correcting exon boundaries and adding untranslated regions or alternative splicing models based on transcriptomic data (Jung et al. 2020).

4 Genetic Diversity and Population Studies

4.1 Germplasm Resources

Germplasm resources represent the genetic foundation for crop improvement and biodiversity conservation. The majority of forage crops are sourced from wild genotypes and breeding efforts are minimal, protecting germplasm resources is vital for maintaining genetic diversity (Sandhu et al. 2015). Most crops benefit from plant genetic resource (PGR) repositories that are safeguarded and made accessible through various national and international gene banks. Only 182,000 accessions of forage species, including grasses, legumes, and fodder trees are preserved in 80 gene banks globally (www.genesys-pgr.org), compared to 7.4 million plant accessions in 1750 gene banks (Yu et al. 2016). These accessions act as reservoirs of forage genetic diversity, which are essential for breeding programs to challenge climate change.

4.2 Role of Genetic Diversity in Breeding Programs

Genetic diversity represents the variety of genetic traits and differences that exist among individuals within a population or a specific group of the same species (Altaf et al. 2024b). Breeders rely on genetic diversity in plant resources to create cultivars combining farmers' desired traits like yield, and breeder-focused traits such as stress and disease resistance (Uçer et al. 2024). Without genetic diversity, breeding programs face limitations. To conserve genetic diversity, efforts must focus on maintaining germplasm collections in genebanks, promoting in situ conservation in natural habitats, and encouraging the use of diverse genetic resources in breeding programs.

4.3 Molecular Markers

Molecular markers are DNA nucleotide sequences that are used to identify and track genetic variations within or between species. An effective DNA marker should exhibit codominance, be uniformly present across the genome, provide consistent and reliable outcomes, and have the capacity to reveal significant genetic diversity (Nadeem et al. 2018; Yalinkiliç et al. 2024; Altaf et al. 2024c). In plant genomics, molecular markers are valuable for mapping quantitative trait loci (QTL), marker-assisted selection (MAS), studying population genetics, and conserving genetic resources. DNA markers are valuable tools for exploring genetic variability, offering methods such as restriction fragment length polymorphism

(RFLP), amplified fragment length polymorphism (AFLP), simple sequence repeats (SSRs), single-nucleotide polymorphisms (SNPs), and diversity arrays technology (DArT) (Andersen 2013). Research on forage crops has established that morphological markers are influenced by environmental conditions. Molecular markers are stable substitutes for morphological traits, unbiased by environmental effects, and allow early and precise identification of various crop varieties (Singh et al. 2022a, Singh et al. 2022b). Extensive research has been carried out on forage crops using different kinds of molecular markers in order to detect genetic diversity and population structure.

In the genetic diversity study by Petolescu et al. (2024), 30 varieties of alfalfa were used, represented through the ISSR and RAPD markers. They proved that among ISSR primers, A17 was the most efficient, detecting 31 polymorphic bands, while L-07 had given 21 bands with the high resolving power of RAPD primer. The combined use of both markers improved the accuracy of clustering; "Pastoral" and "F1413–02" were identified as the best for breeding programs due to their low genetic similarity. Osterman et al. (2021) investigated the genetic diversity and structure of Nordic red clover (*Trifolium pratense*) with the use of SeqSNP-based markers. They identified 623 SNPs, including new variants, showing low but significant genetic variation (5.4%) among accessions.

Genetic diversity and population structure of 54 barley cultivars in Turkey and Bulgaria were investigated by Güngör et al. (2022) using iPBS-retrotransposon and SCoT markers. A total of 530 polymorphic bands with an average polymorphism rate of 94.6% was obtained. In this study, the obtained cultivars were divided into five subpopulations that exhibited striking genetic diversity, particularly between the most diverse cultivars Bayrak and Avcı-2002.

5 Functional Genomics in Forage Crops Present Status and Future Prospects

In the last 10 years, significant improvements in plant genomics have occurred, focusing on structural, comparative, and functional studies. Sequencing tools are unveiling plant genome organization, evolution, and gene functions, with ongoing efforts directed at agriculturally important species (Goff et al. 2002). The Poaceae family comprises major cereal crops and forage grasses, and consists of 10,000 species of which only 40 are used for forage and turf (Moser and Hoveland 1996). Grasses occupy more land than grain crops due to their adaptability, offering benefits such as livestock feed and environmental protection, making them economically valuable. Legumes, essential for forage, form symbiotic nitrogen-fixing relationships with *Rhizobium* bacteria, providing organic fertilizer and vegetable protein (Young et al. 2003). Traditionally, plant genetics focused on a "one gene at a time" approach, which advanced biotechnology but struggled with the complexity of grass genomes (Draper et al. 2001). Functional genomics now seeks to understand gene interactions and their roles in traits such as yield, stress tolerance, and disease resistance. Although functional genomics tools have become more routine, many remain underdeveloped for forage and turf grasses, limiting progress (Wang et al. 2001).

5.1 Expressed Sequence Tags (ESTs)

A general approach of function prediction relies on a bioinformatics-based method: sequence–structure comparison against the genes from known functions; otherwise, using a chromosome position can allow researchers to draw a prediction from similar functional gene groups (Holtorf et al. 2002). Many large-scale EST sequencing projects are underway, where grass complementary DNA (cDNA) clones of expressed genes create a sequence collection. EST data from grass genomes are often compared and refined because of their high conservation, and rice, with its relatively small genome size of 400 Mb, is a favorite in such comparisons (Arumuganathan and Earle 1991). Rice has fewer than 260,800 ESTs, while wheat (Triticum aestivum) has about 500,000 ESTs, making it the species with the highest number of ESTs in GenBank's dbEST database, following human, mouse, and rat.

In contrast, forage and turf species have fewer ESTs. Only nine species used in forage and turf are represented in the GenBank dbEST database, most with limited EST numbers, except for *Medicago truncatula. M. truncatula*, a nitrogen-fixing legume with a small genome (450 Mb), is a model for legume genomics (Bell et al. 2001). Despite its agronomic importance, until recently no ESTs were available for tall fescue. A large-scale EST project was initiated in the year 2000 at the Samuel Roberts Noble Foundation to catalog the majority of the expressed genes in tall fescue, including tissue, developmental stage, and stress-specific. To date, ~40,000 ESTs have been produced and are forthcoming in dbEST of GenBank.

5.2 Molecular Mapping and Reverse Genetics

Genetic maps are generated by the use of molecular markers that will give an idea about genome organization and complex traits, an essential phenomenon in plant breeding. Detailed maps have been developed for several forage grasses: alfalfa (Kaló et al. 2000), ryegrass (Jones et al. 2002), and meadow fescue (Alm et al. 2003). These maps facilitate gene transfer across species through comparative genetics, especially using the rice genome sequence as a reference. Comparative genetics helps identify quantitative trait loci (QTL) and genes of interest in forage and turf species, which have limited genomic data. A significant study between perennial ryegrass and *Triticeae* cereals (Jones et al. 2002) revealed high conservation in genetic maps, while the molecular genetic map of *Medicago truncatula* aids in legume genomics and gene isolation. Reverse genetics, such as gene knockout techniques, is used to study gene functions by generating mutant plants. T-DNA and transposon insertion lines in rice (Hirochika 2001), co-suppression (Wesley et al.

2001), virus-induced gene silencing (Lacomme et al. 2003), and RNA interference (Waterhouse and Helliwell 2003) are some of the techniques employed to discover the gene functions with high-throughput approaches.

5.3 Gene Expression Analyses on Microarrays

Array-based technologies such as microarrays allow systematic and high-throughput analyses of gene expression and interactions in metabolic processes (Young 2000). cDNA and oligonucleotide microarrays are both applied in functional genomics for monitoring gene expression in both forage and turf. For example, a microarray study on *Medicago truncatula* identified nodule-specific transcripts (Fedorova et al. 2002), and another study analyzed gene expression in the saponin biosynthesis pathway (Suzuki et al. 2002). cDNA microarrays are flexible and applicable to non-model organisms, with a 16 K optimized oligonucleotide array set for *M. truncatula* commercially available. However, for the majority of forage and turf species, comprehensive arrays are not yet available due to large genome sizes and limited amount of sequence data. Dedicated arrays can be generated from enriched cDNA libraries targeting specific biological processes using techniques such as suppression subtractive hybridization (SSH) (Diatchenko et al. 1996) or representational difference analysis (RDA) (Welford et al. 1998). These methods focus on tissue-specific gene expression and can identify genes not present in public databases. For instance, SSH, expression profiling, and EST sequencing identified genes involved in the arbuscular mycorrhizal symbiosis between *M. truncatula* and *Glomus mosseae* (Brechenmacher et al. 2003) and regulated genes in *M. truncatula* (Wulf et al. 2003). Recently, SSH was applied to study heat tolerance in tall fescue, generating 2495 ESTs for gene expression profiling of heat-related genes (unpublished data). Although gene expression arrays are valuable, the impact of genes also depends on the protein levels and modifications in specific cellular contexts. Therefore, combining transcriptome data with protein and metabolite analyses is essential for a complete understanding of gene function.

5.4 Proteomics and Metabolomics

Proteomics involves studying proteins expressed under specific conditions (Wilkins 1997) and is categorized into three areas: large-scale protein identification, protein responses to biological changes, and protein–protein interactions (Holtorf et al. 2002). For example, a proteome reference map for *Medicago truncatula* root proteins was created using two-dimensional gel electrophoresis and peptide mass fingerprinting to study nodulation and root development (Mathesius et al. 2001). Proteomics also helped identify seed development-related proteins in *M. truncatula* (Gallardo et al. 2003) and symbiosis-related proteins using mass spectrometry (Bestel-Corre et al. 2002). Metabolomics, the study of metabolites, identifies how gene functions are linked to metabolites (Fiehn 2002). Unlike other functional genomics, metabolomics has faced challenges due to technical difficulties in quantifying and identifying metabolites. Recent advancements, such as gas chromatography-mass spectrometry (GC-MS) and high-performance liquid chromatography (HPLC), have improved metabolite profiling (Bailey et al. 2000; Glassbrook and Ryals 2001). For example, metabolites such as lolicines were identified in *Lolium perenne* seeds infected with the endophytic fungus *Neotyphodium lolii* (Munday-Finch et al. 1998). Additionally, HPLC/PDA/ESI/MS was used to identify triterpene saponins in *M. truncatula* and alfalfa (Huhman and Sumner 2002).

6 *Brachypodium distachyon*: A New Model System for Grass Functional Genomics

Model organisms have greatly enhanced the study of plant genomics in species of commercial or biological importance. The work done on *Arabidopsis thaliana* and rice has contributed significantly to research on many agricultural species. *Medicago truncatula*, a relative of alfalfa, is a model for legume genomics because of its small genome, short generation time, self-fertility, and high transformation efficiency (Cook 1999). Rice (*Oryza sativa*), with its compact genome, has been a model for cereal genomics (Goff 1999), but its phylogenetic distance from forage grasses limits its application to traits such as pathogen resistance and cold tolerance (Catalán and Olmstead 2000). *Brachypodium distachyon* has emerged as a new model for grass genomics. It is closely related to the Pooideae subfamily, which includes many forage and turf grasses (Draper et al. 2001). Its small genome and characteristics, such as self-fertility, fast life cycle, and resistance to various pathogens, make it ideal for high-throughput genetics (Shi et al. 1993). Unlike rice, *B. distachyon* exhibits traits relevant to forage grasses, such as disease resistance, freezing tolerance, and postharvest biochemistry, making it a complementary model for studying traits not well represented in rice (Draper et al. 2001; Khan and Stace 1999).

7 Genetic and Genomic Tools for Improvement

The integration of molecular breeding techniques into traditional breeding programs has revolutionized the field of forage crop improvement, enabling breeders to efficiently identify and incorporate beneficial traits into novel varieties. Techniques such as quantitative trait Loci mapping and genome-wide association studies, among others, have provided insights into the genetic basis of complex traits, leading to the development of more precise, targeted breeding strategies.

7.1 Quantitative Trait Loci (QTL) Mapping

Quantitative trait loci (QTL) mapping is one of the statistical techniques widely used in genetics for locating those regions of the genome that control quantitative traits. Quantitative traits are those influenced by multiple genes, which show continuous variation and exist in a continuum of phenotypic expressions among different populations. QTL mapping involves the establishment of genetic markers associated with the phenotypic trait in order to draw inferences on the genetic architecture of the complex trait (Kassie et al. 2023). For the past two decades, linkage mapping has proven to be a highly effective genetic mapping tool in diverse plant species, including forage crops. The establishment of the first QTL map for abiotic stress resistance in forage grasses by Alm et al. (2003) was a breakthrough that laid the groundwork for the widespread adoption of QTL mapping as a crucial tool in the genetic analysis of forage crops. Subsequently, QTL mapping has become a vital tool for understanding the genetic basis of important agronomic characteristics in forage crops, including yield potential, forage quality, and stress resistance (Chen 2024). QTLs associated with abiotic stress (e.g., drought, salinity) and biotic stress (e.g., pests, pathogens) have been identified in various forage species. The studies on switchgrass biomass and forage quality used a combination of near-infrared spectroscopy and linkage mapping to identify several QTLs that are associated with switchgrass biomass and forage quality traits, providing valuable insights into the genetic underpinnings of these important crop characteristics. By understanding the genetic basis of these traits, breeders can utilize this knowledge to develop improved switchgrass varieties with superior performance as a forage crop (Razar et al. 2021). In a recent study, Cai et al. (2024) conducted a quantitative trait locus analysis of tiller-related traits in wheat, identifying 38 additive QTLs and 55 epistatic QTLs for spike number, maximum tiller number, and ear-bearing tiller rate in eight environments. Among these, three major and stable QTLs were discovered, characterized, and found to impact yield-related traits. By identifying and characterizing major and stable QTLs for tiller-related traits in wheat that impact yield-related traits, this study offers valuable insights that can be applied to breeding programs for forage crops, potentially improving their resilience and productivity.

7.2 Genome-Wide Association Studies (GWAS)

Genome-wide association studies (GWAS) have proven invaluable in revealing genetic loci associated with key traits in crop varieties. This, in turn, has led to the successful introduction of novel traits and strengthened the statistical relationship between genetic markers and phenotypic traits in natural populations (Khan et al. 2021). On the other hand, GWAS allows breeders and scientists to delve into the complex genetic network of traits, cataloging variants associated with specific phenotypes. By unlocking these intricate relationships, GWAS provides a unique tool for understanding functional genomics and developing targeted breeding strategies. This information serves as a valuable resource for breeders and scientists seeking to understand crop functional genomics and can be harnessed to develop targeted breeding strategies aimed at introgressing beneficial alleles into breeding lines and varieties. By revealing the number and distribution of causative genes, alleles, and their interactions, GWAS provides a holistic view of the genetic architecture of traits and illuminates potential avenues for genetic improvement (Huguenin-Elie et al. 2019).

GWAS have emerged as a powerful tool for unraveling the genetic architecture of complex traits in forage species, surpassing the limitations of biparental QTL mapping. Unlike QTL mapping, which is constrained by the choice of parental genotypes and their phenotypic range, GWAS enables the assessment of broad-based panels that better capture the genetic variability present in typical forage breeding programs (Huguenin-Elie et al. 2019). The advent of GWAS has led to groundbreaking advancements in forage crop breeding, offering a comprehensive understanding of the genetic factors regulating complex traits such as quality, yield, and stress resistance. This technique has enabled breeders to pinpoint genetic markers linked to desired phenotypes, thereby accelerating the development of high-performing, stress-tolerant forage crops (Chen 2024). Recently, a number of GWAS studies on various forage crops such as alfalfa, buffel grass ramie, sorghum, triticale, and others have revealed valuable information about the genetic factors that influence these crops' adaptive traits (Niu et al. 2020; Somegowda et al. 2021; Bai et al. 2022; Behera et al. 2024; Negawo et al. 2024). Recent studies have demonstrated that sorghum has been the subject of more GWAS investigations than other forage crops due to its favorable attributes, such as a compact genome size and intermediate linkage disequilibrium (Hamblin et al. 2005), which make it an ideal species for efficiently identifying natural genetic variations and establishing associations between markers and agronomic traits.

8 Abiotic and Biotic Stress Tolerance in Forage Crops from a Genomic Perspective

Forage crops play a pivotal role in global agriculture by serving as essential feed for livestock, contributing to food security, and maintaining ecosystem balance. However, their productivity and quality are often compromised due to various abiotic and biotic stresses. The integration of genomic tools and approaches offers promising avenues for understanding and enhancing stress tolerance in forage crops, enabling their sustainable production in the face of environmental challenges (Sustek-Sánchez et al. 2023).

Abiotic stresses such as drought, salinity, temperature extremes, and nutrient deficiencies significantly affect forage crop growth and yield. Among these, drought stress stands out as a major limiting factor, especially in arid and semiarid regions (Mehta et al. 2023). Forage crops such as alfalfa (*Medicago sativa*), sorghum (*Sorghum bicolor*), and perennial ryegrass (*Lolium perenne*) are frequently subjected to water scarcity. Genomic studies have identified key quantitative trait loci (QTLs) and genes associated with drought tolerance. For instance, in alfalfa, transcription factors such as DREB (dehydration responsive element binding) and NAC (NAM, ATAF, and CUC) have been shown to regulate stress–response pathways (Erpen et al. 2018). Marker-assisted selection (MAS) and genome-wide association studies (GWAS) have further facilitated the identification of superior drought-tolerant genotypes (Tang et al. 2024).

Salinity stress, caused by excessive salt accumulation in soils, poses another challenge for forage crops. High salinity disrupts ion homeostasis and reduces water uptake, leading to impaired plant growth (Ali et al. 2022). Genomic research has highlighted the role of ion transporters such as HKT (high-affinity potassium transporters) and SOS (salt overly sensitive) pathways in salinity tolerance (Shah et al. 2021). In perennial ryegrass, transcriptomic studies have revealed stress-responsive genes that enhance ion compartmentalization and osmotic adjustment. Advances in genomic selection (GS) and CRISPR-Cas9 genome editing have opened new possibilities for developing salinity-tolerant forage varieties (Jattan et al. 2023). Temperature extremes, including both heat and cold stress, also affect forage crop productivity. Heat stress during critical growth stages can lead to poor flowering, reduced biomass, and compromised seed set (Norton et al. 2016). Genomic analyses have identified heat shock proteins (HSPs) and heat-responsive transcription factors in sorghum and pearl millet. Similarly, cold stress tolerance in temperate forage crops such as ryegrass has been linked to antifreeze proteins and CBF (C-repeat binding factor) transcription factors. Genomic approaches, including transcriptomics and metabolomics, are helping unravel complex stress–response networks (Sahito et al. 2024). In addition to abiotic stresses, biotic stresses such as pests and diseases severely impact forage crop yields and quality. Fungal pathogens such as rusts and blights, bacterial diseases, and viral infections are common threats (Sun et al. 2020). For example, anthracnose caused by *Colletotrichum* species affects alfalfa, while rust diseases are widespread in perennial ryegrass. Genomic studies have identified resistance (R) genes and defense pathways that confer biotic stress tolerance (Balconi et al. 2012). Efforts to breed resistant forage crops have been accelerated by integrating molecular markers and genomic editing techniques.

Insect pests, including aphids and armyworms, also reduce forage crop productivity (McNeill et al. 2021). Genomic research has revealed the role of secondary metabolites, such as phenolic compounds and alkaloids, in pest resistance. Transcriptomic studies have provided insights into pest-induced gene expression changes, enabling the development of insect-resistant forage varieties. Furthermore, beneficial endophytes and plant-associated microbes have emerged as natural allies in enhancing biotic stress tolerance, and genomic tools are being employed to understand plant–microbe interactions (Njaci et al. 2020).

Recent advancements in genomics, such as next-generation sequencing (NGS) and high-throughput phenotyping, have revolutionized stress tolerance research in forage crops (Ashraf et al. 2022; Harris-Shultz et al. 2019). GWAS and QTL mapping have become essential tools for dissecting complex traits associated with abiotic and biotic stresses. For instance, GWAS in alfalfa has identified genomic regions linked to drought and pest resistance, while QTL mapping in sorghum has revealed loci associated with salinity and heat tolerance (Harris-Shultz et al. 2019). Functional genomics approaches, including transcriptomics, proteomics, and metabolomics, have provided deeper insights into stress–response mechanisms. RNA sequencing (RNA-seq) has been particularly instrumental in identifying stress-responsive genes and pathways. For example, transcriptomic studies in alfalfa have highlighted the role of ABA (abscisic acid)-mediated pathways in drought tolerance, while metabolomic analyses in ryegrass have identified key osmoprotectants involved in salinity stress adaptation. Genome editing technologies, especially CRISPR-Cas9, have opened new avenues for precise manipulation of stress-responsive genes in forage crops. For instance, CRISPR-mediated knockouts of negative regulators of drought and salinity tolerance have shown promising results in model plants and are being extended to forage crops (Razzaq et al. 2022). Similarly, transgenic approaches have enabled the overexpression of stress-related genes, such as DREB and HSPs, to enhance tolerance to abiotic and biotic stresses.

The integration of genomic selection (GS) in forage crop breeding programs has also gained momentum. GS combines genomic data with phenotypic traits to predict the breeding value of genotypes, enabling the rapid development of stress-tolerant varieties. In sorghum and alfalfa, GS has been effectively used to improve traits such as drought tolerance, disease resistance, and forage quality. Systems biology approaches are emerging as powerful tools to understand the intricate networks governing stress tolerance in forage crops. By integrating multi-omics data, researchers can construct predictive models to identify key regulatory hubs and candidate genes (Roychowdhury et al. 2023). These insights can guide the design of breeding strategies for enhanced stress resilience. Despite significant progress, challenges remain in translating genomic discoveries into field-ready forage crop varieties. The genetic complexity of stress tolerance traits, coupled with environmental variability, often complicates breeding efforts. Additionally, the limited availability of genomic resources for many forage species hinders research progress. Collaborative efforts, including international consortia and public–private partnerships, are essential to overcome these challenges and accelerate genomic innovations in forage crops.

9 Forage Quality and Yield Traits

9.1 Importance of Forage Crops in Agriculture and Livestock Systems

Forage crops play a pivotal role in agriculture and livestock systems, serving as the foundation of ruminant diets and contributing to sustainable agriculture. High-quality forage ensures optimal animal growth, milk production, and reproductive performance, which directly impacts economic outcomes for farmers (Rao et al. 2015). Yield and quality traits of forages determine their efficiency as feed resources. Forage quality encompasses parameters such as digestibility, crude protein content, fiber quality, and mineral availability, while yield traits include biomass production, regrowth potential, and stress tolerance (Wright et al. 2012). Increasing global demand for livestock products has intensified the focus on enhancing forage quality and yield through traditional breeding and advanced genomic techniques. Environmental sustainability also plays a crucial role, as improved forages reduce methane emissions per unit of milk or meat produced, contributing to climate change mitigation (Rauf et al. 2016).

9.2 Key Determinants of Forage Quality

The quality of forage is influenced by several intrinsic and extrinsic factors, including plant species, growth stage, and environmental conditions (Leng 1990). Digestibility, a cornerstone of forage quality, is largely dictated by the lignin content in plant cell walls; lower lignin levels correlate with higher digestibility. Nutritional composition, particularly crude protein and fiber, is another determinant, with legumes such as alfalfa often outperforming grasses in protein content (Amigot et al. 2006). Palatability, defined by the animal's willingness to consume the forage, is influenced by taste, texture, and chemical composition (Forbes 2010). Anti-nutritional factors, such as tannins and alkaloids, can negatively affect nutrient absorption and livestock health, necessitating mitigation strategies. Breeding efforts have focused on enhancing these quality traits, utilizing genetic diversity within forage species to create superior cultivars with improved digestibility and nutritional profiles.

9.3 Factors Influencing Forage Yield

Forage yield traits are shaped by genetic, environmental, and management factors. Biomass production, the primary indicator of yield, is influenced by photosynthetic efficiency, growth habit, and nutrient uptake. Stress tolerance, encompassing drought, salinity, and disease resistance, is crucial for sustaining yields under adverse conditions (Putnam 2021). Regrowth potential, particularly for perennial species, ensures long-term productivity and is a vital trait for rotational grazing systems. Management practices, including fertilization, irrigation, and pest control, further modulate yield outcomes (Behera et al. 2024; Rauf et al. 2016). Advances in genomics and molecular biology have enabled the identification of quantitative trait loci (QTLs) associated with yield traits, providing opportunities for targeted breeding programs to enhance biomass production and stress resilience.

9.4 Integrating Genomic Tools for Forage Improvement

The integration of genomic technologies has revolutionized forage breeding, offering precision tools for the enhancement of quality and yield traits. Marker-assisted selection (MAS) and genomic selection (GS) facilitate the identification and incorporation of desirable traits into new cultivars. For instance, the application of single nucleotide polymorphism (SNP) markers has streamlined the breeding of alfalfa for improved protein content and digestibility (Lin et al. 2021; Sinha et al. 2023). Transcriptomic and proteomic analyses have revealed key pathways regulating lignin biosynthesis, providing targets for genetic modification to enhance forage digestibility. Furthermore, genome-wide association studies (GWAS) have identified critical loci for stress tolerance, enabling the development of resilient forage varieties suited to diverse agroclimatic conditions. The integration of these tools has significantly accelerated the breeding pipeline, ensuring the rapid deployment of improved forages. Despite significant progress, challenges remain in optimizing forage quality and yield traits. Climate change poses a major threat, altering growth patterns and increasing abiotic stress factors such as drought and salinity (Chaudhry and Sidhu 2022). The genetic complexity of forage species, characterized by polyploidy and high heterozygosity, complicates breeding efforts. Additionally, ensuring farmer adoption of improved cultivars requires collaboration among breeders, agronomists, and extension services to demonstrate economic and environmental benefits. Future research should prioritize the development of multi-trait forages that combine high quality, yield, and stress tolerance. Integrating artificial intelligence (AI) and big data analytics into breeding programs could further enhance the precision and efficiency of forage improvement efforts. A holistic approach, combining traditional knowledge with cutting-edge technologies, will be essential to meet the growing demand for sustainable livestock production systems.

10 Conclusion and Future Direction

The genomic revolution has significantly improved forage crops' quality, yield, and stress resilience, using advanced molecular tools such as high-throughput sequencing, GWAS, and QTL mapping to understand complex traits such as abiotic and biotic stress tolerance. Climate change threatens forage crops, affecting productivity and nutritional value. Genomic studies reveal stress adaptation pathways, enabling resilient species such as alfalfa, sorghum, and perennial ryegrass. Functional genomics offers targeted genetic interventions. The chapter emphasizes the significance of genetic diversity and germplasm conservation in breeding programs, preserving wild and underutilized forage species. The use of molecular markers has improved diversity characterization. Challenges include polyploid nature, genetic complexity, environmental variability, genomic technology costs, and limited resources. CRISPR-Cas9 genome editing and genomic selection are promising technologies for forage crop improvement. They enable precise gene editing and accelerate breeding, meeting modern agriculture demands. Integrating AI and big data can enhance efficiency. Focusing on multi-trait forage varieties, expanding genomic resources, and incorporating sustainable practices can contribute to global food security.

Author Contributions

Conceptualization of the chapter: S.A.Q., A.A., M.T.A.
Defining the scope and identifying key themes: M.T.A., R.S., K.K.
Literature review: P.M., S.N.S., S.R.S., M.B., M.T., M.F.S.
First draft preparation: S.A.Q., A.A., P.M., M.F.S., R.S.
Editing and proofreading: All authors contributed equally.

References

Ali A, Altaf MT, Nadeem MA, Karaköy T, Shah AN, Azeem H et al (2022) Recent advancement in OMICS approaches to enhance abiotic stress tolerance in legumes. Front Plant Sci 13:952759

Allen VG, Batello C, Berretta EJ, Hodgson J, Kothmann M, Li X et al (2011) An international terminology for grazing lands and grazing animals. Grass Forage Sci 66(1):2

Alm V, Fang C, Busso CS, Devos KM, Vollan K, Grieg ZROA, Rognli OA (2003) A linkage map of meadow fescue (Festuca pratensis Huds.) and comparative mapping with other Poaceae species. Theor Appl Genet 108:25–40

Altaf MT, Tatar M, Ali A, Liaqat W, Mortazvi P, Kayihan C, Baloch FS (2024a) Advancements in QTL mapping and GWAS application in plant improvement. Turk J Bot 48:376–426

Altaf MT, Nadeem MA, Ali A, Liaqat W, Bedir M, Baran N et al (2024b) Applicability of start codon targeted (SCoT) markers for the assessment of genetic diversity in bread wheat germplasm. Genet Resour Crop Evol 72:1–14

Altaf MT, Liaqat W, Ali A, Jamil A, Fahad M, Rahman MAU et al (2024c) Advancing chickpea breeding: omics insights for targeted abiotic stress mitigation and genetic enhancement. Biochem Genet 63:1–53

Amigot SL, Fulgueira CL, Bottai H, Basflico JC (2006) New parameters to evaluate forage quality. Postharvest Biol Technol 41(2):215–224

Andersen SB (ed) (2013) Plant breeding from laboratories to fields. IntechOpen

Ardui S, Ameur A, Vermeesch JR, Hestand MS (2018) Single molecule real-time (SMRT) sequencing comes of age: applications and utilities for medical diagnostics. Nucleic Acids Res 46(5):2159–2168

Arumuganathan K, Earle ED (1991) Nuclear DNA content of some important plant species. Plant Mol Biol Report 9:208–218

Ashraf MF, Hou D, Hussain Q, Imran M, Pei J, Ali M et al (2022) Entailing the next-generation sequencing and metabolome for sustainable agriculture by improving plant tolerance. Int J Mol Sci 23(2):651

Bai X, Wang X, Wang Y, Wei Y, Fu Y, Rao J et al (2022) Genome-wide association study of six forage traits in ramie (Boehmeria nivea L. Gaud). Plants 11(11):1443

Bailey NJ, Stanley PD, Hadfield ST, Lindon JC, Nicholson JK (2000) Mass spectrometrically detected directly coupled high performance liquid chromatography/nuclear magnetic resonance spectroscopy/mass spectrometry for the identification of xenobiotic metabolites in maize plants. Rapid Commun Mass Spectrom 14(8):679–684

Baker M (2012) De novo genome assembly: what every biologist should know. Nat Methods 9(4):333–337

Balconi C, Stevanato P, Motto M, Biancardi E (2012) Breeding for biotic stress resistance/tolerance in plants. In: Crop production for agricultural improvement. Springer, pp 57–114

Ballouz S, Dobin A, Gillis JA (2019) Is it time to change the reference genome? Genome Biol 20(1):159

Baloch FS, Ali A, Tajibayev D, Nadeem MA, Ölmez F, Aktaş H et al (2024) Stripe rust resistance gene Yr15 in Turkish and Kazakhstan wheat germplasms and the potential of Turkish wild emmer for stripe rust breeding. Genet Resour Crop Evol 71(6):2699–2719

Baumgard LH, Rhoads RP, Rhoads ML, Gabler NK, Ross JW, Keating AF et al (2012) Impact of climate change on livestock production. In: Environmental stress and amelioration in livestock production. Springer, pp 413–468

Bedre R, Irigoyen S, Petrillo E, Mandadi KK (2019) New era in plant alternative splicing analysis enabled by advances in high-throughput sequencing (HTS) technologies. Front Plant Sci 10:740

Behera PP, Singode A, Bhat BV, Borah N, Verma H, Supriya P, Sarma RN (2024) Identifying genetic determinants of forage sorghum [Sorghum bicolor (Moench)] adaptation through GWAS. BMC Plant Biol 24(1):1–18

Bejerman N, Roumagnac P, Nemchinov LG (2020) High-throughput sequencing for deciphering the virome of alfalfa (Medicago sativa L.). Front Microbiol 11:553109

Bell CJ, Dixon RA, Farmer AD, Flores R, Inman J, Gonzales RA et al (2001) The Medicago Genome Initiative: a model legume database. Nucleic Acids Res 29(1):114–117

Bestel-Corre G, Dumas-Gaudot E, Poinsot V, Dieu M, Dierick JF, Tuinen DV et al (2002) Proteome analysis and identification of symbiosis-related proteins from Medicago truncatula Gaertn. By two-dimensional electrophoresis and mass spectrometry. Electrophoresis 23(1):122–137

Brechenmacher L, Weidmann S, Van Tuinen D, Chatagnier O, Gianinazzi S, Franken P, Gianinazzi-Pearson V (2003) Expression profiling of up-regulated plant and fungal genes in early and late stages of Medicago truncatula-Glomus mosseae interactions. Mycorrhiza 14:253–262

Brummer EC, Wang ZY (2020) Biotechnology and molecular approaches to forage improvement. Forages Sci Grassland Agric 2:567–579

Cai Y, Zhou X, Wang C, Liu A, Sun Z, Li S et al (2024) Quantitative trait loci detection for three tiller-related traits and the effects on wheat (Triticum aestivum L.) yields. Theor Appl Genet 137(4):87

Catalán P, Olmstead RG (2000) Phylogenetic reconstruction of the genus Brachypodium P. Beauv. (Poaceae) from combined sequences of chloroplast ndhF gene and nuclear ITS. Plant Syst Evol 220:1–19

Chaudhry S, Sidhu GPS (2022) Climate change regulated abiotic stress mechanisms in plants: a comprehensive review. Plant Cell Rep 41(1):1–31

Chen S (2024) Advances in molecular breeding of forage crops: technologies, applications and prospects. *Agriculture* 14(2):279

Cook DR (1999) Medicago truncatula-a model in the making! Curr Opin Plant Biol 2(4):301–304

De Vega JJ, Ayling S, Hegarty M, Kudrna D, Goicoechea JL, Ergon Å et al (2015) Red clover (Trifolium pratense L.) draft genome provides a platform for trait improvement. Sci Rep 5(1):17394

Diatchenko L, Lau YF, Campbell AP, Chenchik A, Moqadam F, Huang B et al (1996) Suppression subtractive hybridization: a method for generating differentially regulated or tissue-specific cDNA probes and libraries. Proc Natl Acad Sci 93(12):6025–6030

Draper J, Mur LA, Jenkins G, Ghosh-Biswas GC, Bablak P, Hasterok R, Routledge AP (2001) Brachypodium distachyon. A new model system for functional genomics in grasses. Plant Physiol 127(4):1539–1555

Erpen L, Devi HS, Grosser JW, Dutt M (2018) Potential use of the DREB/ERF, MYB, NAC and WRKY transcription factors to improve abiotic and biotic stress in transgenic plants. Plant Cell Tissue Organ Cult (PCTOC) 132:1–25

FAOSTAT (2020) Statistics division. Retrieved from http://www.fao.org/statistics/en/

Fedorova M, Van de Mortel J, Matsumoto PA, Cho J, Town CD, VandenBosch KA et al (2002) Genome-wide identification of nodule-specific transcripts in the model legume Medicago truncatula. Plant Physiol 130(2):519–537

Fiehn O (2002) Metabolomics—the link between genotypes and phenotypes. Plant Mol Biol 48:155–171

Forbes JM (2010) Palatability: principles, methodology and practice for farm animals. CABI Rev 5:1–15

Fraiture MA, D'aes J, Guiderdoni E, Meunier AC, Delcourt T, Hoffman S et al (2023) Targeted high-throughput sequencing enables the detection of single nucleotide variations in CRISPR/Cas9 gene-edited organisms. Foods 12(3):455

Franco AL, Azevedo ALS, Reis AC, Matos EM, Fonseca MA, Pereira AV et al (2024) Uncovering the genomic diversity of the wild forage crop Setaria sphacelata. Crop Sci 64(6):3381–3398

Gallardo K, Le Signor C, Vandekerckhove J, Thompson RD, Burstin J (2003) Proteomics of Medicago truncatula seed development establishes the time frame of diverse metabolic processes related to reserve accumulation. Plant Physiol 133(2):664–682

Gemeinholzer B, Rupp O, Becker A, Strickert M, Müller CM (2023) Genotyping by sequencing and a newly developed mRNA-GBS approach to link population genetic and transcriptome analyses reveal pattern differences between sites and treatments in red clover (Trifolium pratense L.). Front Ecol Evol 10:1003057

Glassbrook N, Ryals J (2001) A systematic approach to biochemical profiling. Curr Opin Plant Biol 4(3):186–190

Godde CM, Mason-D'Croz D, Mayberry DE, Thornton PK, Herrero M (2021) Impacts of climate change on the livestock food supply chain; a review of the evidence. Glob Food Sec 28:100488

Goff SA (1999) Rice as a model for cereal genomics. Curr Opin Plant Biol 2(2):86–89

Goff SA, Ricke D, Lan TH, Presting G, Wang R, Dunn M et al (2002) A draft sequence of the rice genome (Oryza sativa L. ssp. japonica). Science 296(5565):92–100

Güngör H, Ilhan E, Kasapoğlu AG, Filiz E, Pour AH, Valchev D et al (2022) Genetic diversity and population structure of barley cultivars released in Turkey and Bulgaria using iPBS-retrotransposon and SCoT markers. J Agric Sci 28:239–250

Hamblin MT, Salas Fernandez MG, Casa AM, Mitchell SE, Paterson AH, Kresovich S (2005) Equilibrium processes cannot explain high levels of short-and medium-range linkage disequilibrium in the domesticated grass Sorghum bicolor. Genetics 171(3):1247–1256

Haq SIU, Zheng D, Feng N, Jiang X, Qiao F, He JS, Qiu QS (2022) Progresses of CRISPR/Cas9 genome editing in forage crops. J Plant Physiol 279:153860

Harris-Shultz KR, Hayes CM, Knoll JE (2019) Mapping QTLs and identification of genes associated with drought resistance in sorghum. In: Sorghum: methods and protocols. Springer, pp 11–40

Hirakawa H, Kaur P, Shirasawa K, Nichols P, Nagano S, Appels R et al (2016) Draft genome sequence of subterranean clover, a reference for genus trifolium. Sci Rep 6(1):30358

Hirochika H (2001) Contribution of the Tos17 retrotransposon to rice functional genomics. Curr Opin Plant Biol 4(2):118–122

Holtorf H, Guitton MC, Reski R (2002) Plant functional genomics. Naturwissenschaften 89:235–249

Huguenin-Elie O, Studer B, Kölliker R, Reheul D, Probo M, Barre P et al (2019) Improving sown grasslands through breeding and management. In: Proceedings of the joint 20th symposium of the European Grassland Federation and the 33rd meeting of the EUCARPIA section 'Fodder Crop Amen Grasses'

Huhman DV, Sumner LW (2002) Metabolic profiling of saponins in Medicago sativa and Medicago truncatula using HPLC coupled to an electrospray ion-trap mass spectrometer. Phytochemistry 59(3):347–360

Ibaba JD, Gubba A (2020) High-throughput sequencing application in the diagnosis and discovery of plant-infecting viruses in Africa, a decade later. Plants 9(10):1376

Jattan M, Kumari N, Rani B, Phogat DS, Kumari S, Ravika et al (2023) Conventional breeding and advance approaches to mitigate drought and salt stress in crop plants. In: Salinity and drought tolerance in plants: physiological perspectives. Singapore, Springer Nature Singapore, pp 121–135

Jiang L, Zhou L (2024) High-throughput sequencing in maize: a gateway to precision breeding. Maize Genomics Genet 15:247–256

Jiang X, Yu A, Zhang F, Yang T, Wang C, Gao T et al (2022) Identification of QTL and candidate genes associated with biomass yield and feed quality in response to water deficit in alfalfa (Medicago sativa L.) using linkage mapping and RNA-Seq. Front Plant Sci 13:996672

Jones ES, Mahoney NL, Hayward MD, Armstead IP, Jones JG, Humphreys MO et al (2002) An enhanced molecular marker based genetic map of perennial ryegrass (Lolium perenne) reveals comparative relationships with other Poaceae genomes. Genome 45(2):282–295

Jung H, Ventura T, Chung JS, Kim WJ, Nam BH, Kong HJ et al (2020) Twelve quick steps for genome assembly and annotation in the classroom. PLoS Comput Biol 16(11):e1008325

Kaló P, Endre G, Zimanyi L, Csanádi G, Kiss GB (2000) Construction of an improved linkage map of diploid alfalfa (Medicago sativa). Theor Appl Genet 100:641–657

Kane NC, Gill N, King MG, Bowers JE, Berges H, Gouzy J et al (2011) Progress towards a reference genome for sunflower. Botany 89(7):429–437

Kassie FC, Nguepjop JR, Ngalle HB, Assaha DV, Gessese MK, Abtew WG et al (2023) An overview of mapping quantitative trait loci in peanut (Arachis hypogaea L.). Genes 14(6):1176

Katsiani AT, Maliogka VI, Amoutzias GD, Efthimiou KE, Katis NI (2015) Insights into the genetic diversity and evolution of little cherry virus 1. Plant Pathol 64(4):817–824

Khan MA, Stace CA (1999) Breeding relationships in the genus Brachypodium (Poaceae: Pooideae). Nord J Bot 19(3):257–269

Khan SU, Saeed S, Khan MHU, Fan C, Ahmar S, Arriagada O et al (2021) Advances and challenges for QTL analysis and GWAS in the plant-breeding of high-yielding: a focus on rapeseed. Biomolecules 11(10):1516

Lacomme C, Hrubikova K, Hein I (2003) Enhancement of virus-induced gene silencing through viral-based production of inverted-repeats. Plant J 34(4):543–553

Leng R (1990) Factors affecting the utilization of 'poor-quality' forages by ruminants particularly under tropical conditions. Nutr Res Rev 3(1):277–303

Lin S, Norberg S, Combs D (2021) Genomics of forage quality in alfalfa. In: The alfalfa genome. Springer, Cham, pp 129–143

Maliogka VI, Minafra A, Saldarelli P, Ruiz-García AB, Glasa M, Katis N, Olmos A (2018) Recent advances on detection and characterization of fruit tree viruses using high-throughput sequencing technologies. Viruses 10(8):436

Mathesius U, Keijzers G, Natera SH, Weinman JJ, Djordjevic MA, Rolfe BG (2001) Establishment of a root proteome reference map for the model legume Medicago truncatula using the expressed sequence tag database for peptide mass fingerprinting. Proteom Int Ed 1(11):1424–1440

McNeill M, Tu X, Ferguson C, Ban L, Hardwick S, Rong Z et al (2021) Diversity and impacts of key grassland and forage arthropod pests in China and New Zealand: an overview of IPM and biosecurity opportunities. NeoBiota 65:137

Mehta BK, Meena SK, Dikshit N, Shashikumara P, Kumar A, Kumar P et al (2023) Forage genetic resources and scope for allele Mining of Abiotic Stress Tolerance. In: Molecular interventions for developing climate-smart crops: a forage perspective. Springer Nature Singapore, Singapore, pp 35–56

Moser LE, Hoveland CS (1996) Cool-season grass overview. In: Moser et al (eds) Cool-season forage grasses. American Society of Agronomy, Crop Science Society of America, Soil Science Society of America, Madison, pp 1–14

Muir JP, Pitman WD, Foster JL (2011) Sustainable, low-input, warm-season, grass–legume grassland mixtures: Mission (nearly) impossible? Grass Forage Sci 66(3):301–315

Munday-Finch SC, Wilkins AL, Miles CO (1998) Isolation of lolicine A, lolicine B, lolitriol, and lolitrem N from Lolium perenne infected with Neotyphodium lolii and evidence for the natural occurrence of 31-epilolitrem N and 31-epilolitrem F. J Agric Food Chem 46(2):590–598

Nadeem MA, Nawaz MA, Shahid MQ, Doğan Y, Comertpay G, Yıldız M et al (2018) DNA molecular markers in plant breeding: current status and recent advancements in genomic selection and genome editing. Biotechnol Biotechnol Equip 32(2):261–285

Negawo AT, Muktar MS, Gutiérrez RAS, Habte E, Muchugi A, Jones CS (2024) A genome-wide association study of biomass yield and feed quality in Buffel grass (Cenchrus ciliaris L.). Agriculture 14(2):257

Niu H, Ping J, Wang Y, Lv X, Li H, Zhang F et al (2020) Population genomic and genome-wide association analysis of lignin content in a global collection of 206 forage sorghum accessions. Mol Breed 40:1–13

Njaci I, Ngugi-Dawit A, Oduor RO, Kago L, Williams B, Hoang LTM et al (2020) Comparative analysis delineates the transcriptional resistance mechanisms for pod borer resistance in the pigeonpea wild relative Cajanus scarabaeoides (L.) thouars. Int J Mol Sci 22(1):309

Norton MR, Malinowski DP, Volaire F (2016) Plant drought survival under climate change and strategies to improve perennial grasses. A review. Agron Sustain Dev 36:1–15

Osterman J, Hammenhag C, Ortiz R, Geleta M (2021) Insights into the genetic diversity of Nordic red clover (Trifolium pratense) revealed by SeqSNP-based genic markers. Front Plant Sci 12:748750

Petolescu C, Sarac I, Popescu S, Tenche-Constantinescu AM, Petrescu I, Camen D et al (2024) Assessment of genetic diversity in alfalfa using DNA polymorphism analysis and statistical tools. Plants 13(20):2853

Putnam DH (2021) Factors influencing yield and quality in alfalfa. In: The alfalfa genome. Springer, Cham, pp 13–27
Rao IM, Peters M, Castro A, Schultze-Kraft R, White D, Fisher M et al (2015) LivestockPlus: the sustainable intensification of forage-based agricultural systems to improve livelihoods and ecosystem services in the tropics. Trop Grasslands 3(2):59–82
Rauf S, Sienkiewicz-Paderewska D, Malinowski DP, Hussain MM, Niazi IAK, Kausar M (2016) Forages: ecology, breeding objectives and procedures. In: Advances in plant breeding strategies: agronomic, abiotic biotic stress traits. Springer, pp 149–201
Razar RM, Makaju S, Missaoui AM (2021) QTL mapping of biomass and forage quality traits measured using near-infrared reflectance spectroscopy (NIRS) in switchgrass. Euphytica 217(3):51
Razzaq MK, Akhter M, Ahmad RM, Cheema KL, Hina A, Karikari B et al (2022) CRISPR-Cas9 based stress tolerance: new hope for abiotic stress tolerance in chickpea (Cicer arietinum). Mol Biol Rep 49(9):8977–8985
Roychowdhury R, Das SP, Gupta A, Parihar P, Chandrasekhar K, Sarker U et al (2023) Multi-omics pipeline and omics-integration approach to decipher plant's abiotic stress tolerance responses. Genes 14(6):1281
Sahito JH, Zhang H, Gishkori ZGN, Ma C, Wang Z, Ding D et al (2024) Advancements and prospects of genome-wide association studies (GWAS) in maize. Int J Mol Sci 25(3):1918
Sandhu JS, Kumar D, Yadav VK, Singh, Sah RP, Radhakrishna A (2015) Recent trends in breeding of tropical grass and forage species. In: Proceedings of 23rd International Grassland Congress 2015-Keynote Lectures, pp 337–348
Sarıkaya MF, Tatar M (2023) Global climate change and crop production. Uluslararası Sivas Bilim ve Teknoloji Üniversitesi Dergisi 2(2):65–78
Sato S, Nakamura Y, Kaneko T, Asamizu E, Kato T, Nakao M et al (2008) Genome structure of the legume, Lotus japonicus. DNA Res 15(4):227–239
Senol Cali D, Kim JS, Ghose S, Alkan C, Mutlu O (2019) Nanopore sequencing technology and tools for genome assembly: computational analysis of the current state, bottlenecks and future directions. Brief Bioinform 20(4):1542–1559
Shah WH, Rasool A, Saleem S, Mushtaq NU, Tahir I, Hakeem KR, Rehman RU (2021) Understanding the integrated pathways and mechanisms of transporters, protein kinases, and transcription factors in plants under salt stress. Int J Genomics 2021(1):5578727
Shi Y, Draper J, Stace C (1993) Ribosomal DNA variation and its phylogenetic implication in the genus Brachypodium (Poaceae). Plant Syst Evol 188:125–138
Singh L, Pierce C, Santantonio N, Steiner R, Miller D, Reich J, Ray I (2022a) Validation of DNA marker-assisted selection for forage biomass productivity under deficit irrigation in alfalfa. Plant Genome 15(1):e20195
Singh S, Singh B, Sharma VR, Kumar M, Sirohi U (2022b) Assessment of genetic diversity and population structure in pea (Pisum sativum L.) germplasm based on morphological traits and SSR markers. Legume Res Int J 45(6):683–688
Sinha D, Maurya AK, Abdi G, Majeed M, Agarwal R, Mukherjee R et al (2023) Integrated genomic selection for accelerating breeding programs of climate-smart cereals. Genes 14(7):1484
Somegowda VK, Rayaprolu L, Rathore A, Deshpande SP, Gupta R (2021) Genome-wide association studies (GWAS) for traits related to fodder quality and biofuel in sorghum: Progress and prospects. Protein Pept Lett 28(8):843–854
Stolarczyk M, Reuter VP, Smith JP, Magee NE, Sheffield NC (2020) Refgenie: a reference genome resource manager. Gigascience 9(2):giz149
Sun S, Hoy MJ, Heitman J (2020) Fungal pathogens. Curr Biol 30(19):R1163–R1169
Sustek-Sánchez F, Rognli OA, Rostoks N, Sõmera M, Jaškūnė K, Kovi MR et al (2023) Improving abiotic stress tolerance of forage grasses–prospects of using genome editing. Front Plant Sci 14:1127532
Suzuki H, Achnine L, Xu R, Matsuda SP, Dixon RA (2002) A genomics approach to the early stages of triterpene saponin biosynthesis in Medicago truncatula. Plant J 32(6):1033–1048

Tang X, Xue Y, Cao D, Luan X, Liu Q, Zhu Z, Liu X (2024) Marker-assisted selection strategies for drought tolerance in soybean and future perspectives. Legume Genomics Genet 15:296–302
Tso CH, Wu JL, Lu MW (2020) Blast2Fish: a reference-based annotation web tool for transcriptome analysis of non-model teleost fish. BMC Bioinformatics 21:1–11
Uçer VA, Aglar E, Mortazavi P, Qureshi SA, Ali A, Tatar M et al (2024) Exploring genetic diversity of Turkish fig (Ficus carica L.) germplasm using inter-primer binding site (iPBS) retrotransposon markers. Genet Resour Crop Evol 72:1–12
Wang Y, Zhang L (2024) Gene-driven future: breakthroughs and applications of marker-assisted selection in tree breeding. Mole Plant Breed 15:132–143
Wang Z, Hopkins A, Mian R (2001) Forage and turf grass biotechnology. Crit Rev Plant Sci 20(6):573–619
Wang Y, Liu Y, Yu C, Chen S, Li Y, Wei L et al (2024) Identification of a Rye spring mutant derived from a winter Rye variety by high-altitude environment screening using RNA sequencing technology. Genes 15(5):572
Waterhouse PM, Helliwell CA (2003) Exploring plant genomes by RNA-induced gene silencing. Nat Rev Genet 4(1):29–38
Welford SM, Gregg J, Chen E, Garrison D, Sorensen PH, Denny CT, Nelson SF (1998) Detection of differentially expressed genes in primary tumor tissues using representational differences analysis coupled to microarray hybridization. Nucleic Acids Res 26(12):3059–3065
Wesley SV, Helliwell CA, Smith NA, Wang M, Rouse DT, Liu Q et al (2001) Construct design for efficient, effective and high-throughput gene silencing in plants. Plant J 27(6):581–590
Wilkins MR (ed) (1997) Proteome research: new frontiers in functional genomics. Springer Science & Business Media
Wright IA, Tarawali S, Blümmel M, Gerard B, Teufel N, Herrero M (2012) Integrating crops and livestock in subtropical agricultural systems. J Sci Food Agric 92(5):1010–1015
Wulf A, Manthey K, Doll J, Perlick AM, Linke B, Bekel T et al (2003) Transcriptional changes in response to arbuscular mycorrhiza development in the model plant Medicago truncatula. Mol Plant-Microbe Interact 16(4):306–314
Yalinkiliç NA, Başbağ S, Altaf MT, Ali A, Nadeem MA, Baloch FS (2024) Applicability of SCoT markers in unraveling genetic variation and population structure among sugar beet (Beta vulgaris L.) germplasm. Mol Biol Rep 51(1):584
Young R (2000) Biomedical discovery with DNA arrays. Cell 102:9–16
Young ND, Mudge J, Ellis TN (2003) Legume genomes: more than peas in a pod. Curr Opin Plant Biol 6(2):199–204
Yu X, Li X, Guo T, Zhu C, Wu Y, Mitchell SE et al (2016) Genomic prediction contributing to a promising global strategy to turbocharge gene banks. Nat Plants 2(10):1–7
Zhao S, Han X, Zhu Y, Han Y, Liu H, Chen Z et al (2024) CRISPR/CasΦ2-mediated gene editing in wheat and rye. J Integr Plant Biol 66(4):638–641

[illegible]

Insights of Transgenic Technology in Improving Forage Crops

M. Vennela, D. Dinesh Varma, Maya Chandra, and Rashmi Bonoriya

Abstract

The importance of forage crops/fodder crops in the present scenario has drastically increased as they provide high-quality produce to livestock and maintain sustainability in nature. Indeed, they also improve soil health, which in the long run helps to increase the productivity of agricultural systems. Transferring the beneficial traits that enhance the genetic potential of the forage crops for sustainable food production could be only possible through genetic engineering technology. The traits that can be transferred through transgenesis majorly include improving the forage quality besides their adaption to different biotic and abiotic stresses. By lowering the amount of lignin in the majority of forage crops, transgenic technology has increased biomass yield and improved animal digestibility. The technique has also proven effective in creating short-lived, healthy plants. The addition of genes from the germplasm collection will effectively boost the possibility of creating transgenic variants that meet the growing demand for animal products. Forage crops that have undergone genetic engineering (GE) hold promise for mitigating the effects of climate change, improving animal health and productivity, and improving human health when consumed in moderation. Furthermore, many studies were conducted concurrently to find out if there is any adverse effect of using genetically modified crops to feed animals, which turned out to be negative.

Keywords

Forage · Genetic engineering · Livestock · Quality · Transgenes

M. Vennela (✉) · D. Dinesh Varma · M. Chandra · R. Bonoriya
Dr. Rajendra Prasad Central Agricultural University, Pusa, Samastipur, Bihar, India

F. S. Baloch et al. (eds.), *Forage Crops in the Omics Era*,
https://doi.org/10.1007/978-981-95-9249-4_6

1 Introduction

Forage crops are plant species cultivated specifically for feeding livestock. These crops are harvested in various forms such as fresh forage, silage (preserved under anaerobic conditions), or hay (dried fodder). They play a crucial role in providing essential nutrients to animals, supporting sustainable livestock production and agricultural practices. Forage grasslands are utilized for feeding livestock and are estimated to cover 26% of the world's land area, accounting for 70% of the agricultural land area globally (FAO 2010). Forage crops sustain around 1.52 billion cattle, 1.21 billion sheep, 1.02 billion goats, and 0.21 billion buffalo worldwide. In most of the developing countries, the income of farmers can be enhanced by these crops as they help in overall production of various dairy products, meat, and wool (FAOSTAT 2020).

Most of the forage crops belong to Poaceae and Fabaceae families as they include grasses and legumes. Poaceae includes Napier grass (*Pennisetum purpureum*), Brachiaria (*Brachiaria* spp.), Panicum (*Panicum* spp.), Bentgrass (*Agrostis* spp.), Fescue (*Festuca* spp.), Ryegrass (*Lolium* spp.), Orchard grass (*Dactylis* spp.), Maize (*Zea mays*), Festulolium (*Festuca* × *Lolium hybrid*), whereas Fabaceae consists of Mulga (*Acacia aneura*), Leadtree (*Leucaena leucocephala*), Trefoil (*Lotus corniculatus*), Medics (*Medicago* spp.), Alfalfa (*Medicago sativa*), Clover (*Trifolium* spp.), Vetches (*Vicia* spp.) (Kendir et al. 2009). Other family includes Brassicaceae which has Oilseed rape (*Brassica napus*), Kale (*Brassica oleracea*), Amaranthaceae (Amaranth Family), and Fodder beet (Beta vulgaris). The essential nutrients like proteins, carbohydrates, vitamins, and minerals are provided to animals through these forage crops (Capstaff and Miller 2018). Alfalfa, a leguminous fodder crop, has rich sources of crude protein, macronutrients, and micronutrients, all of which are very essential for dairy cattle (Radovic et al. 2009).

The healthy digestive systems of the animals can be maintained by balancing the sources of energy and fiber, which can be provided through grasses. Thus, the nutritional value of forage crops is important and it greatly depends on the growth stage, environmental conditions, and also on stage of harvesting as younger leaves have higher concentration of nutritional factors than mature fibrous stems. The importance of forage crops is not only restricted to animal health but also have crucial role in improving soil health, structure, and nutrient cycling. It also contributes in enhancing the biodiversity and reducing the soil erosion (Pulina et al. 2016). With all these beneficial aspects, the demand for forage crops has been steadily increasing. However, the productivity of the livestock sources has not kept up with the increasing population. This gap between the supply and demand leads to nutritional insecurity of livestock and overall sustainability of animal agriculture sector. So, a better alternative to mitigate this issue is adopting advanced breeding and biotechniques that enhance the accessibility of livestock products and ease of demand on limited land resources (Singer et al. 2017; Capstaff and Miller 2018; Fuglie et al. 2021).

The central focus of any researcher is to put efforts to improve forage crops by enhancing the productivity, nutritional value, and long-term availability of livestock production systems. The first and foremost approach to mitigate the imbalance between demand and supply is using selective breeding strategy that helps in increasing the desirable traits, whereas the other approach could be utilizing genetic engineering techniques (Said et al. 2020). New crop varieties with increased biomass yield, improved nutrient content, and higher tolerance to environmental stresses have been developed through selective breeding by the researchers. In addition to this, the adoption of precision farming techniques and rotational grazing helped in stabilizing the cultivation and utilization of forage crops (Kulkarni et al. 2018).

Though there was certain improvement in the production of these crops, the traditional breeding completely relied on natural genetic variability and sexual hybridization for developing new lines (Wang and Brummer 2012). Thus, the use of chemical or radiation mutagenesis and artificial hybridization progressed for introducing beneficial traits, but the hybridization technique requires sexual compatibility and chromosome pairing. To overcome this limitation, techniques such as embryo rescue and protoplast culture are used as they allow creation of variability through asexual means. However, through advancement in technology, the improvement of forage crops has made possible also through sexually incompatible species by means of genetic engineering where the desired genes can be transferred from one organism to another (Wang and Brummer 2012). Many scientists are working to create transgenic forage crops by introducing foreign genes for nutritional qualities, resistance or tolerance to biotic and abiotic factors, and greater adoptability to climatic changes. The high-quality livestock feed can be made available through successful development and adoption of these transgenic forage crops (Kulkarni et al. 2018; Bao et al. 2022). The various genetic engineering techniques and recent advancements to improve forage yield and other beneficial traits were discussed in detail in this chapter.

2 Status of Transgenic Forage Crops Worldwide

The total area under transgenic crops during the year 2023 has increased to 206.3 mha, which is 1.9% more than previous year. The top three countries that contribute to this increase are the United States, Brazil, and Argentina, with 74.4, 66.9, and 23.1 mha, respectively. One of the important forage crops, i.e., alfalfa, is grown in an area of 1.2 mha having greater than 90% adoption in 2023. Genetically modified (GM) crops are deliberately used all over the world in terms of food and feed. The livestock alone contributes 70–80% of GM crop biomass. Net economic returns to the grower are known to differ by crop product, location, and year. These variations are attributed to various factors, including the density of weeds, the degree of pest infestation, and the spread of disease. The paucity of datasets detailing the agronomic or nutritional equivalency of genetically modified perennial forages and the non-GM cultivars that the transgenic event has been crossed into is not surprising, considering the small number of GM perennial forages that have advanced to feed trials. Fig. 1 represents the DNA delivery methods.

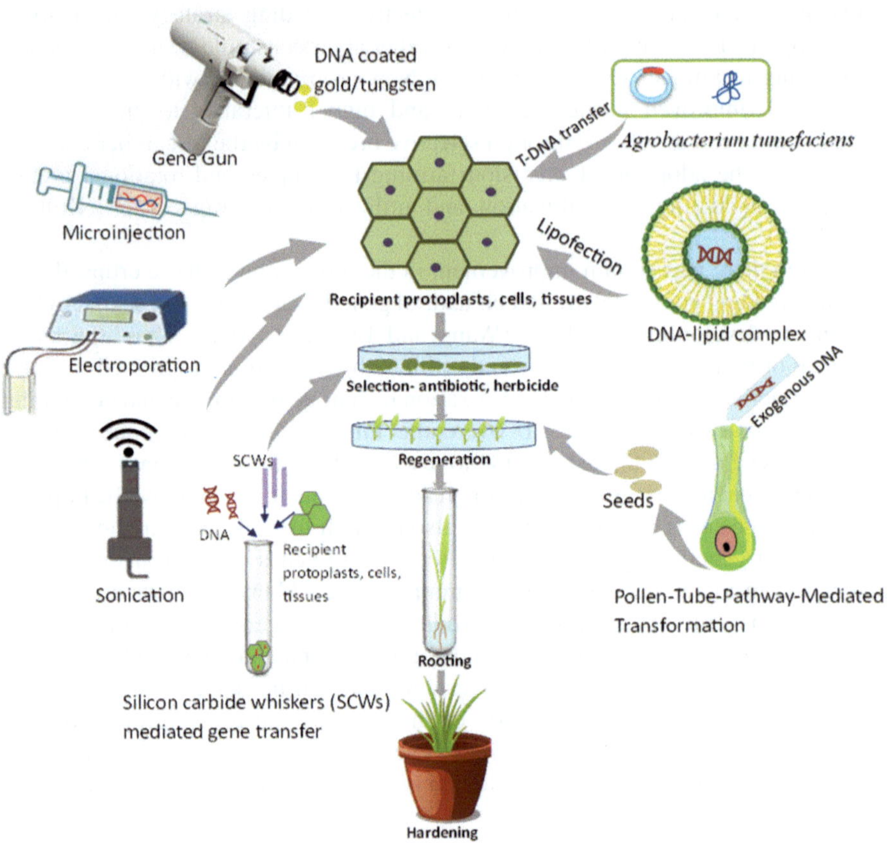

Fig. 1 DNA delivery methods

2.1 Indirect DNA Delivery Method

Agrobacterium-mediated gene transfer is a widely used method for introducing foreign genes into plants. This process exploits the natural ability of the pathogenic gram-negative soil bacterium *Agrobacterium tumefaciens* to transfer part of its DNA (i.e., T-DNA) to plant cells, which then integrate this DNA into their genome, and hence it is known as natural genetic engineer (Nester 2015). For the purpose, the gene of interest is inserted into the T-DNA region of a modified Ti plasmid (tumor-inducing plasmid) of *Agrobacterium tumefaciens* of size 200 kbp. The Ti plasmid is engineered so that the genes responsible for producing plant hormones (i.e., auxin and cytokinin) causing plant tumors (crown gall disease) are removed and replaced with the desired gene and a selectable marker gene (e.g., antibiotic resistance). The recombinant plasmid is introduced into *Agrobacterium tumefaciens* cells via transformation. The bacterium will then carry the gene of interest within its T-DNA region (Rogowsky et al. 1987). The transformed *Agrobacterium* cells are used to infect plant cells. This can be done by several methods, including the following: (1) *Leaf disk method*: Small disks of plant leaves are cut and incubated with the *Agrobacterium* containing the recombinant plasmid. (2) *Cocultivation method*: Plant tissues or explants (such as leaf pieces, stem segments, or roots) are cocultivated with the *Agrobacterium* on a nutrient medium. *Agrobacterium* transfers the T-DNA, including the gene of interest, into the plant cell by action of virulence (*vir*) genes present in Ti plasmid. Once inside, the T-DNA integrates randomly into the plant's genome. The stable integration into plant's host genome requires 10–14 days after transformation. This integration is stable and heritable, meaning it will be passed on to future plant generations. The bacterium has pathogenic host range from dicotyledonous to monocotyledons (Hooykaas-Van Slogteren et al. 1984; Kado 2014). For instance, the method has been successfully employed to improve many forage crops such as overexpression of hv1 gene of barley showed increased drought tolerance in creeping bentgrass (*Agrotis stolonifera* var. *palustris*) (Fu et al. 2007). *Arabidopsis thaliana* have been transformed with transcription factor genes *WXP1* and *WXP2* from *Medicago truncatula* showed improved drought tolerance (Zhang et al. 2007). Dollar-spot resistance in creeping bentgrass is achieved by overexpression of *TLPD34* gene of rice. Overexpression of miR156 gene in a bioenergy crop switchgrass (*Panicum virgatum* L.) improved biomass production (Fu et al. 2012). Saccharification in switchgrass is improved by downregulation of cinnamyl alcohol dehydrogenase (CAD) gene (Fu et al. 2011). Expression of bacteriophage T4 lysozyme gene in tall fescue showed resistance to gray leaf spot and brown patch disease (Dong et al. 2008). *Medicago sativa* showing delayed senescence due to introduction of *ipt* gene (Calderini et al. 2007). Drought tolerance in white clover (*Trifolium repens*) is improved by expression of transcription factor gene *WXP1* (Jiang et al. 2010). Phosphate uptake in *Medicago sativa* is improved by expression of phytase and acid phosphatase gene (Ma et al. 2012).

2.2 Direct DNA Delivery Methods

1. *Biolistic gene gun method*: It is also known as particle bombardment or the microprojectile method of gene transfer. It is a physical method to directly deliver the foreign DNA into the cell. The method involves cloning the gene of interest in the cloning vector, followed by coating it around the high-velocity gold or tungsten particles of a diameter of around 1–2 μm. After loading the coated metal particles in the gene gun apparatus, the gene gun is triggered to fire, which causes a burst of compressed gas (usually helium) to accelerate these particles to a high velocity toward the plant tissue. The small size of particles coated with foreign DNA allows them to penetrate the cell walls and membranes without causing significant damage. Once inside the cell, the DNA gets released and integrates with the plant's genome (Sandal et al. 2015). The approach has been utilized to increase the digestibility of tall fescue (*Festuca arundinacea* Schreb.) by transferring the cinnamyl alcohol dehydrogenase gene (Chen et al. 2003). Similarly, wheat fructosyltransferase genes wtf1 and wtf2 are transferred

into perennial ryegrass (*Lolium prenne*), which improve tolerance for cell level freezing (Hisano et al. 2004), transfer the AtGA2ox1 gene to improve turf quality of bahiagrass (*Paspalum notatum* Flugge) (Agharkar et al. 2007), and for the downregulation of the main pollen allergens in ryegrass.

2. *Lipofection*: The method utilizes lipid-based carriers known as liposomes that deliver the DNA or RNA into the cells. Liposomes and DNA/lipid complex are two main components of this method. Liposomes are spherical vesicles of 30 nm to several micrometers in size, made up of one or more phospholipid bilayer membranes that can encapsulate the nucleic acids (Akbarzadeh et al. 2013), whereas the DNA/lipid complex is formed due to the electrostatic interactions between cationic lipids in the liposome and negatively charged nucleic acids that are bound to the cell membrane. This complex protects the nucleic acids from degradation and facilitates the entry of nucleic acid inside the cell via endocytosis (i.e., a process in which the cell engulfs the complex and forms an endosome) (Van Wordragen et al. 1997). After entering the cell, the lipid–DNA complex escapes from the endosome to release the DNA into the cytoplasm. Further, the delivered DNA enters the nucleus, where it integrates into the host genome for further expression through transcription and translation.
3. *Other chemical-mediated gene transfer methods*: The chemical methods of genetic transformation other than lipofection include calcium phosphate-mediated transfer, DEAE-Dextran-mediated transfer, and polyethylene glycol (PEG)-mediated transfer. In the case of calcium phosphate-mediated transformation, the DNA containing the gene of interest is mixed with calcium chloride and potassium phosphate solutions to form a calcium phosphate precipitate. When the plant tissue is incubated with the precipitate, it is taken up by cells through endocytosis (Kingston et al. 1999). DEAE-Dextran-mediated transfer DEAE-Dextran is utilized as a medium for transfection, which is suitable for transfection in cells as well as cost-effective (Uchiumi et al. 2014). The PEG-mediated delivery protoplasts are treated with solution containing PEG, which promotes the uptake of DNA with gene of interest into the protoplast. For instance, the tall fescue (*Festuca arundinacea* Schreb.), which was the first forage transgenic, has been developed using PEG-mediated gene delivery methods to introduce hygromycin phosphotransferase and phosphinothricin acetyl transferase genes.
4. *Protoplast fusion*: It is a process that involves the fusion of protoplasts (cells without cell walls) from various plant species to create the somatic hybrids with desirable traits (Shepard et al. 1983). The process begins with enzymatic digestion of the cell wall, which releases the protoplast. Furthermore, protoplasts containing the complete genetic complement of the plant's genetic material are purified and prepared for fusion triggered by PEG or by generating an electric field in the case of electrofusion (Tempelaar and Jones 1985). PEG promotes fusion by drying the protoplasts and facilitating merger of membranes, whereas electrofusion involves aligning protoplasts in an electric field and then providing a pulse to cause the membranes to fuse. When the protoplasts get fused, the resulting hybrid cells possess nuclei from both parent lines. The primary benefit

of this approach is that it allows direct transfer of cell organelles (e.g., mitochondria and chloroplast) between different species, facilitating the creation of hybrids with desirable traits governed by organellar genome. Additionally, it allows the creation of somatic hybrids (i.e., hybrids having the nuclear genome of one species and the cytoplasm of another species), which can be utilized to introduce cytoplasmic male sterility or other cytoplasmic traits into crop plants.

5. *Electroporation*: The technique utilizes the electric field to temporarily increase the permeability of the cell membrane to facilitate the uptake of DNA into the cell (Ozyigit 2020). The process starts with the enzymatic digestion of the cell wall, which increases the susceptibility of the cell to uptake the DNA during the application of an electric field. The DNA will come into close contact with the protoplast, increasing the likelihood of successful uptake. The mixture of desired DNA with the recipient protoplast is placed in the electroporation cuvette, which is a tiny chamber with electrodes. A short but alternating electrical pulse of high frequency is applied across the cell membrane, which creates transient pores on it and results in increased permeability (Ozyigit 2020). The water pores induced by an electric pulse form in two stages. First, water molecules penetrate the bilayer, creating an aqueous channel. Then, the polar head groups of adjacent lipids reorient toward these water molecules, forming hydrophilic channels. During this brief period, the DNA can enter the cells through these pores (Kotnik et al. 2019). The main disadvantage of this method is that it is difficult to transform cells with thick walls. The method has been used to transform the alfalfa (*Medicago sativa* L.) to increase the digestibility by targeted downregulation of cytochrome P450 enzymes (Reddy et al. 2005). Bacterial citrate synthase gene confers improved soil aluminum tolerance in alfalfa (Barone et al. 2008).
6. *Silicon carbide whiskers (SCWs)-mediated gene transfer*: Silicon carbide whiskers are needle-like microstructures, approximately 0.5 μm in diameter and 10–80 μm in length. These whiskers are strong yet easily cleaved, forming sharp edges capable of penetrating the cell wall and eventually reaching the nucleus (Akram et al. 2016). In SCW-mediated plant genetic transformation, suspended cells or embryogenic calli, along with DNA, are placed in a centrifuge tube containing SCWs, which, due to their negatively charged surface, do not bind to DNA (Mutsuddy 1990). During vortexing, SCWs create needle-like pores in the cell membrane, allowing the entry of exogenous DNA into the target cells (Mizuno et al. 2004).
7. *Microinjection*: Microinjection is a precise gene delivery method where DNA or other genetic material is directly injected into the single plant nucleus or cytoplasm of a target cell using microcapillary injection pipette (Zhang and Yu 2008). In this technique, the target cell is positioned under a microscope, where it is held in place by a holding pipette. A second micromanipulator with a microcapillary tube containing a small amount of DNA solution is then used to penetrate the cell membrane or nuclear membrane. The DNA is injected directly into the cytoplasm or nucleus of plant cells or protoplasts through the microcapillary pipette, which has a tip size ranging from 0.5 to 10 μm. After the gene transfer is

completed, the transformed cells are cultured and developed into transgenic plants.

8. *The pollen–tube–pathway (PTP)-mediated transformation method* is a gene transfer technique where foreign DNA is introduced into plants through the pollen tube after fertilization. In this process, DNA is applied to the stigma, and as the pollen tube grows toward the ovary, it carries the DNA into the ovule, where it integrates with the undivided but fertilized egg, resulting in incorporation of foreign DNA in plant's genome, leading to development of transgenic seeds (Ali et al. 2015). The method has been used in Kentucky bluegrass (*Poa pratensis* L.), a temperate turfgrass.
9. *Sonication*: It is a technique developed to enhance the permeability of biological barriers, such as cell membranes and endothelial layers, using ultrasound (US) and microbubbles (MBs). Protoplasts are exposed to ultrasound (sound waves of frequency more than 20 kHz) while in the presence of a plasmid containing the desired gene, MBs oscillate near cells, temporarily increasing membrane permeability facilitating the noninvasive and efficient delivery of genes into cells, which results in the development of transgenic plants (Beranova et al. 2008).

3 Methods to Produce Marker-Free Transgenic Plants

To produce transgenic plants, selection systems are employed to promote the growth of transformed cells. Genes conferring resistance to specific antibiotics or herbicides are commonly used, enabling the rapid identification of transformed tissues. While these selectable marker genes (SMGs) are effective, their integration into the plant genome raises concerns about the potential spread of transgenes, such as antibiotic resistance, in new ecosystems. Although horizontal gene transfer to bacteria and human beings is deemed unlikely, the hazards are not fully addressed, resulting in continued regulatory and public concerns globally. Aside from the potential danger of horizontal gene transfer, there is chance of "vertical cross-species" transmission, which may result in emergence of even more invasive or weedy species of plants (Dale et al. 2002). In this context, the production of marker-free transgenic crops can eliminate the risk of horizontal gene transfer and decrease the chances of vertical gene transfer. There are two primary techniques to accomplish this: the first includes utilizing selectable markers that are independent of antibiotic or herbicide resistance genes (e.g., morphological markers), and the second involves elimination of segregating marker genes from the host genome once the transgenic plants have been regenerated (Zuo et al. 2002). This includes methods such as co-transformation (separate introgression of marker and transgene), site-specific recombinase systems (e.g., Cre/loxP, R/RS), transposon-based removal systems (e.g., Ac transposon), intrachromosomal recombination for gene excision, and using marker genes that do not depend on selection for herbicide or antibiotic resistance.

4 Advances in Molecular Breeding Transgenic Forage Crops

Molecular breeding of transgenic forage crops has transformed the production of plants with higher nutritional value, biomass yield, and resistance to abiotic stresses. Genome-wide association studies (GWASs), genomic selection (GS), quantitative trait loci (QTL) mapping, and genome editing techniques such as CRISPER/Cas9 are the some of the significant advancements in this area (Chen 2024; Altaf et al. 2024).

4.1 Quantitative Trait Locus (QTL) Mapping

This is one of the most important methods in the molecular breeding of transgenic crops. The methods enable identification of precise genomic loci linked to important traits such as nutritional quality, yield, and tolerance to biotic and abiotic stresses. Using this technique, the researchers can precisely develop the plants with specific traits (Talukder and Saha 2017).

1. *Key trait identification*: Using QTL mapping the researchers can identify genetic areas linked to the specific trait of interest for forage crops such as higher biomass, better digestibility or higher nutritional content (Talukder and Saha 2017).
2. *Marker-assisted selection*: Genetic markers linked to QTLs associated with important traits can be utilized in marker-assisted selection after establishment of QTLs. This method allows the early and precise selection of plants with desired traits, even before they full development of plants, which speeds up the breeding of transgenic forage crops.
3. *Stress tolerance enhancement*: Abiotic stresses such as drought, heat, salinity, cold are common for forage crops. Using QTL mapping, breeders can introduce genes that give tolerance to these stresses (Larson et al. 2019).
4. *Nutritional quality improvement*: Nutritional traits such as lignin, protein, and carbohydrate content are important for forage crops. Genes responsible for these traits can be introduced in plants, which will lead to improved nutritional profiles that can promote the growth of livestock sector.
5. *Improvement in yield*: High biomass yield is always one of the most important objectives of forage breeding. The QTL mapping facilitates the identification of genetic loci linked with phenotypes that directly affects the yield such as leaf size and plant height. Transgenic crops using QTLs can then be developed into more productive types.
6. *Resistance to diseases*: Disease has the potential to severely affect crop yields and quality. QTLs associated with traits that confer resistance against specific microbes enables breeders to develop transgenic varieties with improved resistance, which will ultimately reduce the reliance on chemical inputs and further its effect on the environment.

Example

- *Alfalfa with low lignin content*: Lignin content is one of the major factors for the digestibility of fodder crops. Zhang et al. (2019) identified QTLs associated with lignin content and developed a transgenic lucerne variety with less lignin content and increased digestibility, as well as effectiveness of nutrient absorption.
- *Grasses resistant to drought*: QTL mapping has identified the genomic regions conferring drought tolerance in switchgrass and ryegrass (Chen et al. 2003). These loci have been utilized to develop transgenic verities with increased tolerance to water stress conditions (Cropano et al. 2021).
- *Forage crops with enhanced protein content*: QTL mapping has also been found beneficial in the identification of genes linked to protein content in legume fodders such as lucerne and clover. Using these genes, breeders have developed transgenic forage crops with higher protein content, which is essential for enhancing the nutritional intake of cattle (Zhang et al. 2019).

5 Genetic Engineering of Transgenic Forage Crops

The genetic engineering techniques is an essential tool to improve genetic variability in forage crops as it makes possible to create new variations by transferring essential traits from different species or genera. The first industry to make significant investments in the use of genetic modifications was agriculture (Raman 2017). The integration of GM crops into sustainable food production systems has become possible through biotechnology aspects that helps in introduction of beneficial traits into crops. The creation of the first transgenic fodder crop was carried out in tall fescue plants as reported by Wang et al. (1992) and it paved a significant advancement in developing genetically modified forage crops. The following text presents the state of the research and developments regarding genetic engineering in fodder crops. Table 1 represents the usage of transgenic technology for improving different traits in various fodder crops.

5.1 Alfalfa

The perennial crop lucerne (*Medicago sativa* L.) has a tetraploid genome. Alfalfa is widely used as livestock feed because of its high animal-friendly protein content (Kumar et al. 2013). Phenotypic selection, the foundation of conventional breeding for lucerne quality improvement, has been shown to be time-consuming (Resende et al. 2014). Transgenic technology has been used to add new traits or improve existing ones in lucerne, and a number of studies have been published on various attributes (Wang et al. 2016; Kumar et al. 2018). There are less ethical or public concerns when using transgenic lucerne plants to produce new proteins because they are a nonfood plant for humans. Numerous genes for the production and expression of novel proteins have been included into lucerne in recent years.

Table 1 Trait improvement in different crops through transgenic technology

S. no.	Crop	Trait improved	Gene transferred	Reference
1.	Alfalfa	Biomass yield and quality	LjmiR156a (three SPL genes: *MsSPL6*, *MsSPL12*, and *MsSPL13*)	Aung et al. (2015)
		Root regenerative capacity and nitrogen fixation activity	MsmiR156	Aung et al. (2017)
		Increase in digestibility and lignocellulose processing ability	Hydroxycinnamoyl coenzyme A: shikimate hydroxycinnamoyl transferase (HCT)	Gallego-Giraldo et al. (2014)
		Salt tolerance	AtNXH1-expressing transgenic alfalfa	Stritzler et al. (2018).
2.	Tall fescue	Heat tolerance	A2-type heat stress transcription factor	Wang et al. (2017)
		Disease resistance (*R. solani*)	RNA polymerase + importin beta-1 subunit and cohesin complex subunit Psm1 + a ubiquitin E3 ligase	Zhou et al. (2016)
		Forage quality	*faeA* (*Aspergillus niger*)	Buanafina et al. (2010)
		Herbicide resistance	*OsALS* (*sm*)	Sato et al. (2013)
3.	Bermuda grass	Glufosinate herbicide resistance	*bar* gene	Hu et al. (2005)
		Insect resistance (black cutworm)	*cry1Ac* gene	Salehi et al. (2005)
		Photosynthetic biosynthesis	*CdHEMA1* gene	Xu et al. (2022a, b)
		Salt tolerance	CdWRKY50	Huang et al. (2021)
4.	Switchgrass	Reduced lignin content and improved biomass saccharification	3-Dehydroshikimate dehydratase (QsuB)	Liu et al. (2024)
		Increase in dry weight biomass	*PvWOX3a*	Yang et al. (2021)
5.	White clover	Increased phosphorous acquisition	*MtPHY1*	Ma et al. (2009)
		Drought tolerance	*TrMYB33-TrSAMS1*	Zhang et al. (2024)

5.2 Tall Fescue

In the United States, tall fescue (*Festuca arundinacea* Schreb.) is the most common perennial grass for the cool season. It is extensively utilized for turf as well as fodder (Ge and Wang 2015). Being an allohexaploid outcrossing species with self-incompatibility, tall fescue's reproductive traits and genetic complexity frequently make it challenging to enhance its agronomic traits through traditional breeding. The quality trait in tall fescue was improved using one indirect transformation method, i.e., *Agrobacterium*-mediated transformation (Dong and Qu 2005) and direct method of transformation, i.e., particle bombardment (Cho et al. 2000).

5.3 Bermuda Grass

Bermuda grass (*Cynodon dactylon* L. Pers) is grown mostly in warmer climatic conditions and can be widely used as turf and forage grass that paved the way to improve research in genetic analysis of this crop. The first transgenic plant was developed in the year 2003 by transferring phosphotransferase marker gene through biolistic or gene gun method of transformation (Zhang et al. 2003). However, the most used transformation method is *Agrobacterium* mediated as it produces better transgenic expression stability in later generations. The use of transgenic technology helped in creating varieties that have tolerance to stress conditions and increased quality and productivity (Wang and Brummer 2012).

5.4 Switchgrass

For the last 60 years, switchgrass is extensively grown as a forage grass in pastures and rangelands on the great plains of the United States. Its spread has also continued in the central and eastern United States in the last 20 years (Vogel 2004). Within each ecotype, i.e., lowland and upland, there are two common ploidy levels: octoploid and tetraploid. While lowland switchgrass cultivars frequently undergo genetic transformation, upland cultivars resist this process (Xu et al. 2022a, b). Upland switchgrass transformation efforts have yielded very little positive results. A notable progress in shoot regeneration in upland switchgrass was documented by Liu et al. (2015). By maximizing the cocultivation and preculture conditions of type I callus, Ogawa et al. (2016) reported the successful transformation of an upland cultivar called "Trailblazer," but the cultivar Blackwell was not successfully transformed.

5.5 White Clover

In mixed-sward environments, white clover is cultivated alongside pasture grass to become a significant pasture legume in temperate regions. In addition to offering grazing animals highly nutritious herbage, white clover benefits the environment by biologically fixing nitrogen (Vala 2012). Transgenic technology can be used to modify a number of important agronomic traits in white clover (White 2000). Effective techniques for transforming white clover through the action of *Agrobacterium* have been established. The tetraploid white clover species has a diploid breeding system and $2n = 4x = 32$ chromosomes. The organism possesses a gametophytic incompatibility system and is primarily self-incompatible. Because of this self-incompatibility, individuals in natural populations are heterozygous at numerous genetic loci.

6 Composition and Performance of Genetically Modified Feeds and Forages

The composition and performance of genetically modified (GM) feeds and forages have been extensively studied to assess their nutritional benefits, safety, and overall impact on livestock production. The increase in nutritional quality, resistance to various biotic and environmental stresses has become the primary goal of GM feeds and forages. The overview of their performance along with compositions are represented as follows.

6.1 Nutritional Composition

Protein Content and Quality:

- *Enhanced amino acid profiles*: The amino acids that are crucial for the growth and health of animals include lysine and methionine and this essentiality was improved through GM forages by creating higher levels of amino acid profiles.
- *Increased crude protein*: Some GM crops are engineered to have higher protein content, enhancing their nutritional value as animal feed.

Fatty Acid Content:

- *Improved omega-3 fatty acids*: Omega-3 fatty acids improve the quality of meat and milk in livestock and this trait was massively improved in GM crops such as alfalfa and clover.
- *Modified oil content*: The altered oil profiles in crops such as soybean has made possible the levels of unsaturated fatty acids through genetic engineering, which is found to be highly beneficial for animal health and quality of product.

Carbohydrate Composition:

- *Reduced lignin*: The presence of lignin in the cell wall of plants makes the feed to be less digestible, and hence there is a need to develop GM crops with reduced lignin content. In this aspect, a low-lignin alfalfa was developed, which has high digestibility and energy content for feeding animals.
- *Enhanced soluble sugars*: To enhance the palatability of animals, some GM forage crops with increased soluble sugar content were introduced.

Vitamins and Minerals:

- *Biofortification*: Through transgenesis, increasing the levels of vitamins and minerals in forages could be possible, which contribute to better health and productivity of animals. The focus could be mainly on increasing the content of iron or zinc.

6.2 Performance in Livestock Production

Growth and Feed Efficiency:

- *Improved weight gain*: Those animals fed with improved nutritional content has observed to have increased weight as compared to animals fed with conventional forages. They also show better nutrient absorption and utilization.
- *Better feed conversion ratios (FCR)*: The production and cost-efficiency can be increased when animals are fed with better FCR transgenic plants as they require less feed to gain per unit body weight.

Milk Production and Quality:

- *Increased milk yield*: GM crops such as corn silage or protein-rich alfalfa when fed to cows resulted in production of quality milk, which in turn supports increase levels of lactation.
- *Improved milk composition*: Through genetically modified technology, we can improve fatty acid and protein content in forage crops, which in turn increases the essential protein and fatty acid content in animals. This leads to the production of quality milk.

Animal Health and Welfare:

- *Disease resistance*: Feed-related sickness can be reduced by developing GM crops that have added disease resistance genes to fungal toxicity, which helps in improving the health conditions of animals.
- *Stress tolerance*: GM forages help to withstand environmental stresses that help to reduce the fluctuations in feed quality, confirming stable health and performance of livestock.

Reproductive Performance:

- *Enhanced fertility*: The enhanced nutrient profiles through GM forages support better reproductive health in livestock, thus helping in higher fertility rates and improved offspring viability.

6.3 Environmental and Economic Impact

Sustainability:

- *Reduced need for chemical inputs*: The requirement of both fungicides and pesticides can be reduced by using transgenic crops, which also lowers the impact of polluting the environment.
- *Lower nitrogen requirements*: Transgenic forage crops help to reduce the artificial fertilizers usage as the technology helps in developing increased nitrogen-efficient varieties, which also contributes to more sustainability.

Economic Viability:

- *Cost-effectiveness*: Although the initial cost of transgenic seeds is more, the yield and quality increase along with reduced input costs often help in overall profitability of farmers.
- *Market acceptance*: The market acceptance and regulatory approvals are utmost important as the economic impact of GM crops is dependent on these factors. The approvals may vary according to the region.

7 Future Prospects and Challenges

New Traits Development:

- *Climate resilience*: Climate change is becoming a huge threat to our agriculture systems by posing various biotic and biotic stresses (Sarıkaya and Tatar 2023). Therefore, the development of transgenic forage crops should also include climate resilient changes in addition to imparting tolerance to adverse weather conditions, as they further improvise the performance of livestock with changing environmental conditions.
- *Precision breeding*: Advances in gene editing tools such as CRISPR could lead to even more precise modifications, allowing for the development of GM forages tailored to specific regional or species needs.

Public Perception and Adoption:

- *Consumer acceptance*: The public perception and regulatory issues are still the major challenges that have to be faced in adopting GM feeds and forages, which also lead to greater impact on the market dynamics and overall acceptance of the technology in livestock production.

8 Conclusion

The challenges imposed by conventional forage crop breeding can be overcome by utilizing genetic engineering and biotechnology. This technology can be applied to develop better forage cultivation techniques. Due to a lack of arable land for forage crops and the need to meet rising demand, new approaches must be taken to increase forage productivity. These GM crops have numerous advantages as they help in increasing the nutrient composition for the feeds and also bring environmental sustainability by reducing the adverse impact of using fertilizers and pesticides. The advances in molecular breeding and biotechnology helps to overcome the factors such as regulatory approvals and public acceptance by improving the efficiency in the livestock production systems. Genetically modified feeds and forages offer significant advantages in terms of enhanced nutritional composition, improved livestock performance, and environmental sustainability. While their adoption depends on various factors including economic viability, regulatory approvals, and public acceptance, the ongoing advancements in molecular breeding and biotechnology hold promise for further improving the efficiency and sustainability of animal production systems.

References

Agharkar M, Lomba P, Altpeter F, Zhang H, Kenworthy K, Lange T (2007) Stable expression of AtGA2ox1 in a low-input turfgrass (Paspalum notatum Flugge) reduces bioactive gibberellin levels and improves turf quality under field conditions. Plant Biotechnol J 5(6):791–801. https://doi.org/10.1111/j.1467-7652.2007.00284.x

Akbarzadeh A, Rezaei-Sadabady R, Davaran S, Joo SW, Zarghami N, Hanifehpour Y, Samiei M, Kouhi M, Nejati-Koshki K (2013) Liposome: classification, preparation, and applications. Nanoscale Res 8:102

Akram Z, Ali S, Ali GM, Zafar Y, Shah ZH, Alghabari F (2016) Whisker-mediated transformation of peanut with chitinase gene enhances resistance to leaf spot disease. Crop Breed Appl Biotechnol 16:108–114. https://doi.org/10.1590/1984-70332016v16n2a17

Ali A, Bang SW, Chung SM, Staub JE (2015) Plant transformation via pollen tube-mediated gene transfer. Plant Mol Biol Report 33:742–747. https://doi.org/10.1007/s11105-014-0839-5

Altaf MT, Tatar M, Ali A, Liaqat W, Mortazvi P, Kayihan C et al (2024) Advancements in QTL mapping and GWAS application in plant improvement. Turk J Bot 48(7):376–426

Aung B, Gruber M, Amyot L, Omari K, Bertrand A, Hannoufa A (2015) Ectopic expression of LjmiR156 delays flowering, enhances shoot branching, and improves forage quality in alfalfa. Plant Biotechnol Rep 9:1–15

Aung B, Gao R, Gruber MY, Yuan ZC, Sumarah M, Hannoufa A (2017) MsmiR156 affects global gene expression and promotes root regenerative capacity and nitrogen fixation activity in alfalfa. Transgenic Res 26:541–557

Bao J, Wang L, Yu Z (2022) Effects of different moisture levels and additives on the ensiling characteristics and in vitro digestibility of Stylosanthes Silage. Animals 12(12):1555–1555. https://doi.org/10.3390/ani12121555

Barone P, Rosellini D, LaFayette P, Bouton J, Veronesi F, Parrott W (2008) Bacterial citrate synthase expression and soil aluminum tolerance in transgenic alfalfa. Plant Cell Rep 27:893–901. https://doi.org/10.1007/s00299-008-0517-x

Beranova M, Rakouský S, Vávrová Z et al (2008) Sonication assisted *Agrobacterium*-mediated transformation enhances the transformation efficiency in flax (*Linum usitatissimum* L.). Plant Cell Tissue Organ Cult 94:253–259

Buanafina MM, Langdon T, Hauck B, Dalton S, Timms-Taravella E, Morris P (2010) Targeting expression of a fungal ferulic acid esterase to the apoplast, endoplasmic reticulum or Golgi can disrupt feruloylation of the growing cell wall and increase the biodegradability of tall fescue (Festuca arundinacea). Plant Biotechnol J 8(3):316–331. https://doi.org/10.1111/j.1467-7652.2009.00485.x

Calderini O, Bovone T, Scotti C, Pupilli F, Piano E, Arcioni S (2007) Delay of leaf senescence in Medicago sativa transformed with the ipt gene controlled by the senescence-specific promoter SAG12. Plant Cell Rep 26(5):611–615. https://doi.org/10.1007/s00299-006-0262-y

Capstaff NM, Miller AJ (2018) Improving the yield and nutritional quality of forage crops. Front Plant Sci 9:535. https://doi.org/10.3389/fpls.2018.00535

Chen S (2024) Advances in molecular breeding of forage crops: technologies, applications and prospects. Agriculture 14:279. https://doi.org/10.3390/agriculture14020279

Chen L, Auh CK, Dowling P, Bell J, Chen F, Hopkins A, Dixon RA, Wang ZY (2003) Improved forage digestibility of tall fescue (Festuca arundinacea) by transgenic down-regulation of cinnamyl alcohol dehydrogenase. Plant Biotechnol J 1(6):437–449. https://doi.org/10.1046/j.1467-7652.2003.00040.x

Cho MJ, Ha CD, Lemaux PG (2000) Production of transgenic tall fescue and red fescue plants by particle bombardment of mature seed-derived highly regenerative tissues. Plant Cell Rep 19(11):1084–1089. https://doi.org/10.1007/s002990000238

Cropano C, Manzanares C, Yates S, Copetti D, Canto JD, Lübberstedt T, Koch M, Studer B (2021) Identification of candidate genes for self-compatibility in perennial ryegrass (*Lolium perenne* L.). Front Plant Sci 12:707901

Dale PJ, Clarke B, Fontes EM (2002) Potential for the environmental impact of transgenic crops. Nat Biotechnol 20(6):567–574. https://doi.org/10.1038/nbt0602-567

Dong S, Qu R (2005) High efficiency transformation of tall fescue with Agrobacterium tumefaciens. Plant Sci 168:1453–1458

Dong S, Shew HD, Tredway LP, Lu J, Sivamani E, Miller ES, Qu R (2008) Expression of the bacteriophage T4 lysozyme gene in tall fescue confers resistance to gray leaf spot and brown patch diseases. Transgenic Res 17:47–57. https://doi.org/10.1007/s11248-007-9073-3

FAO (2010) Challenges and opportunities for carbon sequestration in grassland systems: a technical report on grassland management and climate mitigation. Food and Agriculture Organization of the United Nations, Rome

FAOSTAT (2020) Statistics division. Retrieved from http://www.fao.org/statistics/en/

Fu D, Huang B, Xiao Y, Muthukrishnan S, Liang GH (2007) Overexpression of barley hva1 gene in creeping bentgrass for improving drought tolerance. Plant Cell Rep 26:467–477. https://doi.org/10.1007/s00299-006-0258-7

Fu C, Xiao X, Xi Y, Ge Y, Chen F, Bouton J, Dixon RA, Wang ZY (2011) Downregulation of cinnamyl alcohol dehydrogenase (CAD) leads to improved saccharification efficiency in switchgrass. Bioenergy Res 4:153–164. https://doi.org/10.1007/s12155-010-9109-z

Fu C, Sunkar R, Zhou C, Shen H, Zhang JY, Matts J, Wolf J, Mann DG, Stewart CN Jr, Tang Y, Wang ZY (2012) Overexpression of miR156 in switchgrass (Panicum virgatum L.) results in various morphological alterations and leads to improved biomass production. Plant Biotechnol J 10(4):443–452. https://doi.org/10.1111/j.1467-7652.2011.00677.x

Fuglie K, Peters M, Burkart S (2021) The extent and economic significance of cultivated forage crops in developing countries. Front Sustain Food Syst 5:712136. https://doi.org/10.3389/fsufs.2021.712136

Gallego-Giraldo L, Bhattarai K, Pislariu CI, Nakashima J, Jikumaru Y, Kamiya Y, Udvardi MK, Monteros MJ, Dixon RA (2014) Lignin modification leads to increased nodule numbers in alfalfa. Plant Physiol 164:1139–1150. https://doi.org/10.1104/pp.113.232421

Ge Y, Wang ZY (2015) Tall fescue (Festuca arundinacea Schreb.). Methods Mol Biol 1224:365–372. https://doi.org/10.1007/978-1-4939-1658-0_29

Hisano H, Kanazawa A, Kawakami A, Yoshida M, Shimamoto Y, Yamada T (2004) Transgenic perennial ryegrass plants expressing wheat fructosyltransferase genes accumulate increased amounts of fructan and acquire increased tolerance on a cellular level to freezing. Plant Sci 167(4):861–868. https://doi.org/10.1016/j.plantsci.2004.05.037

Hooykaas-Van Slogteren GMS, Hooykaas PJJ, Schilperoort RA (1984) Expression of Ti plasmid genes in monocotyledonous plants infected with Agrobacterium tumefaciens. Nature 311(5988):763–764. https://doi.org/10.1038/311763a0

Hu F, Zhang L, Wang X (2005) Agrobacterium-mediated transformed transgenic triploid bermudagrass (Cynodon dactylon × C. transvaalensis) plants are highly resistant to the glufosinate herbicide liberty. Plant Cell Tissue Organ Cult 83(1):13–19

Huang X, Amee M, Chen L (2021) Bermuda grass *CdWRKY50* gene negatively regulates plants' response to salt stress. Environ Exp Bot 188:104513

Jiang Q, Zhang JY, Guo X, Bedair M, Sumner L, Bouton J, Wang ZY (2010) Improvement of drought tolerance in white clover (Trifolium repens) by transgenic expression of a transcription factor gene WXP1. Funct Plant Biol 37(2):157–165. https://doi.org/10.1071/FP09177

Kado CI (2014) Historical account on gaining insights on the mechanism of crown gall tumorigenesis induced by agrobacterium tumefaciens. Front Microbiol 5:340. https://doi.org/10.3389/fmicb.2014.00340

Kendir H, Sahin-Demirbag N, Aasim M, Khawar KM (2009) In vitro plant regeneration from Turkish Narbon Vetch (Vicia narbonensis L. var. narbonensis L.). Afr J Biotechnol 8(4):614–618
Kingston RE, Chen CA, Okayama H (1999) Calcium phosphate transfection. Curr Protoc Immunol 31(1):10–13. https://doi.org/10.1002/0471142735.im1013s31
Kotnik T, Rems L, Tarek M, Miklavčič D (2019) Membrane electroporation and electropermeabilization: mechanisms and models. Annu Rev Biophys 48(1):63–91. https://doi.org/10.1146/annurev-biophys-052118-115451
Kulkarni KP, Tayade R, Asekova S, Song JT, Shannon JG, Lee J (2018) Harnessing the potential of forage legumes, alfalfa, soybean, and cowpea for sustainable agriculture and global food security. Front Media 9:67–69. https://doi.org/10.3389/fpls.2018.01314
Kumar D, Pannu R, Kumar S, Arya RK (2013) A review: recent advances in forage crop improvement through biotechnology. Biomirror 4:97–101. https://doi.org/10.1080/08935699908685571
Kumar T, Ai-Ke B, Zhulatai B, Fei W, Li G, Suo-Min W (2018) The progress of genetic improvement in alfalfa (*Medicago sativa* L.). Czech J Genet Plant Breed 54(2):41–51
Larson S, DeHaan L, Poland J, Zhang X, Dorn K, Kantarski T, Anderson J, Schmutz J, Grimwood J, Jenkins J (2019) Genome mapping of quantitative trait loci (QTL) controlling domestication traits of intermediate wheatgrass (*Thinopyrum intermedium*). Theor Appl Genet 132:2325–2351
Liu S, Chou M-Y, Benucci GMN, Eudes A, Bonito G (2024) Genetic modification of the shikimate pathway to reduce lignin content in switchgrass (*Panicum virgatum* L.) significantly impacts plant microbiomes. bioRxiv:592240. https://doi.org/10.1101/2024.05.02.592240
Ma X-F, Wright E, Ge Y, Bell J, Xi Y, Bouton JH, Wang Z-Y (2009) Improving phosphorus acquisition of white clover (*Trifolium repens* L.) by transgenic expression of plant-derived phytase and acid phosphatase genes. Plant Sci 176:479–488
Ma XF, Tudor S, Butler T, Ge Y, Xi Y, Bouton J, Harrison M, Wang ZY (2012) Transgenic expression of phytase and acid phosphatase genes in alfalfa (Medicago sativa) leads to improved phosphate uptake in natural soils. Mol Breed 30:377–391. https://doi.org/10.1007/s11032-011-9628-0
Mizuno K, Takahashi W, Ohyama T, Shimada T, Tanaka O (2004) Improvement of the aluminum borate whisker-mediated method of DNA delivery into rice callus. Plant Prod Sci 7(1):45–49. https://doi.org/10.1626/pps.7.45
Mutsuddy BC (1990) Electrokinetic behavior of aqueous silicon carbide whisker suspensions. J Am Ceram Soc 73;2747–2749. https://doi.org/10.1111/j.1151-2916.1990.tb06757.x
Nester EW (2015) Agrobacterium: nature's genetic engineer. Front Plant Sci 5:730. https://doi.org/10.3389/fpls.2014.00730
Ogawa Y, Honda M, Kondo Y, Hara-Nishimura I (2016) An efficient *Agrobacterium*-mediated transformation method for switchgrass genotypes using type I callus. Plant Biotechnol 33:19–26. https://doi.org/10.5511/plantbiotechnology.15.1127b
Ozyigit II (2020) Gene transfer to plants by electroporation: methods and applications. Mol Biol Rep 47(4):3195–3210. https://doi.org/10.1007/s11033-020-05343-4
Pulina G, Francesconi AHD, Stefanon B, Sevi A, Calamari L, Lacetera N, Dell'Orto V, Pilla F, Ajmone-Marsan P, Mele M, Rossi F, Bertoni G, Crovetto GM, Ronchi B (2016) Sustainable ruminant production to help feed the planet. Ital J Anim Sci 16(1):140–171. https://doi.org/10.1080/1828051x.2016.1260500
Radovic J, Sokolovic D, Markovic JJBAH (2009) Alfalfa-most important perennial forage legume in animal husbandry. Biotechnol Anim Husb 25(5-6-1):465–475. https://doi.org/10.2298/BAH0906465R
Raman R (2017) The impact of genetically modified (GM) crops in modern agriculture: a review. GM Crops Food 8(4):195–208. https://doi.org/10.1080/21645698.2017.1413522
Reddy MS, Chen F, Shadle G, Jackson L, Aljoe H, Dixon RA (2005) Targeted down-regulation of cytochrome P450 enzymes for forage quality improvement in alfalfa (*Medicago sativa* L.). Proc Natl Acad Sci 102(46):16573–16578. https://doi.org/10.1073/pnas.0505749102
Resende S, Maria R, Casler MD, Vilela de Resende MD (2014) Genomic selection in forage breeding: accuracy and methods. Crop Sci 54:143–156. https://doi.org/10.2135/cropsci2013.05.0353

Rogowsky PM, Close TJ, Chimera JA, Shaw JJ, Kado CI (1987) Regulation of the vir genes of Agrobacterium tumefaciens plasmid pTiC58. J Bacteriol 169(11):5101–5112. https://doi.org/10.1128/jb.169.11.5101-5112.1987

Said S, Agung PP, Putra WPB, Kaiin EM (2020) The role of biotechnology in animal production. IOP Conf Ser Earth Environ Sci 492(1):012035. https://doi.org/10.1088/1755-1315/492/1/012035

Salehi H, Seddighi Z, Kravchenko AN (2005) Expression of the cry1Ac in 'Arizona common' common bermudagrass *via* Agrobacterium-mediated transformation and control of black cutworm. J Am Soc Hortic Sci 130(4):619–623

Sandal I, Koul R, Saini U, Mehta M, Dhiman N, Kumar N, Ahuja PS, Bhattacharya A (2015) Development of transgenic tea plants from leaf explants by the biolistic gun method and their evaluation. Plant Cell Tissue Organ Cult (PCTOC) 123:245–255. https://doi.org/10.1007/s11240-015-0828-x

Sarıkaya MF, Tatar M (2023) Global climate change and crop production. Uluslararası Sivas Bilim ve Teknoloji Üniversitesi Dergisi 2(2):65–78

Sato H, Shimizu T, Kawai K, Kaku K, Arakawa A, Tachibana T, Takamizo T (2013) Herbicide-resistant tall fescue with cytoplasmic male sterility selected by a mutated rice acetolactate synthase gene. Crop Sci 53(1):201–207

Shepard JF, Bidney D, Barsby T, Kemble R (1983) Genetic transfer in plants through interspecific protoplast fusion. Science 219(4585):683–688. https://doi.org/10.1126/science.219.4585.683

Singer SD, Hannoufa A, Acharya SN (2017) Molecular improvement of alfalfa for enhanced productivity and adaptability in a changing environment. Plant Cell Environ 41(9):1955–1971. https://doi.org/10.1111/pce.13090

Stritzler M, Elba P, Berini C, Gomez C, Ayub N, Soto G (2018) High-quality forage production under salinity by using a salt-tolerant AtNXH1-expressing transgenic alfalfa combined with a natural stress-resistant nitrogen-fixing bacterium. J Biotechnol 276:42–45. https://doi.org/10.1016/j.jbiotec.2018.04.013

Talukder SK, Saha MC (2017) Toward genomics-based breeding in C3 cool-season perennial grasses. Front Plant Sci 8:1317

Tempelaar MJ, Jones MGK (1985) Fusion characteristics of plant protoplasts in electric fields. Planta 165:205–216. https://doi.org/10.1007/BF00395043

Uchiumi F, Larsen S, Tanuma SI, Figgs GP (2014) Application of DEAE-dextran to an efficient gene transfer system. Seikagaku 86:532–537

Vala BH (2012) Transgenic white clover for enhanced yield and performance: trait dissection, trait stacking and phenomics. Thesis, La Trobe. https://doi.org/10.26181/21857706.v1

Van Wordragen M, Shakya R, Verkerk R, Peytavis R, Van Kammen A, Zabel P (1997) Liposome-mediated transfer of YAC DNA to tobacco cells. Plant Mol Biol Report 15:170–178. https://doi.org/10.1007/BF02812268

Vogel KP (2004) Switchgrass. In: Moser LE, Burson BL, Sollenberger LE (eds) Warm-season (C4) grasses. Agronomy monograph 45. American Society of Agronomy, Crop Science Society of America, Soil Science Society of America, Madison, pp 561–588

Wang ZY, Brummer EC (2012) Is genetic engineering ever going to take off in forage, turf and bioenergy crop breeding? Ann Bot 110(6):1317–1325. https://doi.org/10.1093/aob/mcs027

Wang Z, Takamizo T, Iglesias V (1992) Transgenic plants of tall fescue (*Festuca arundinacea* Schreb.) obtained by direct gene transfer to protoplasts. Nat Biotechnol 10:691–696. https://doi.org/10.1038/nbt0692-691

Wang D, Khurshid M, Sun ZM, Tang XY, Zhou ML, Wu YM (2016) Genetic engineering of alfalfa (*Medicago sativa* L.). Protein Pept Lett 23:495–502. https://doi.org/10.2174/0929866523666160314152618

Wang X, Huang W, Liu J, Yang Z, Huang B (2017) Molecular regulation and physiological functions of a novel FaHsfA2c cloned from tall fescue conferring plant tolerance to heat stress. Plant Biotechnol J 15(2):237–248. https://doi.org/10.1111/pbi.12609

White DWR (2000) Transgenic white clover (*Trifolium repens*). In: Bajaj YPS (ed) Transgenic crops I. Biotechnology in agriculture and forestry, vol 46. Springer, Heidelberg, pp 567–570. https://doi.org/10.1007/978-3-642-59612-4_24

Xu N, Kang M, Zobrist JD, Wang K, Fei S (2022a) Genetic transformation of recalcitrant upland switchgrass using morphogenic genes. Front Plant Sci 12:781565. https://doi.org/10.3389/fpls.2021.781565

Xu X, Wenwen L, Xiaoyan L, Yingping C, Xiaoning L, Guangyang W, Chunxiang F, Jinmin F (2022b) Genetic manipulation of bermudagrass photosynthetic biosynthesis using *Agrobacterium-mediated* transformation. Physiol Plant 174(3):e13710

Yang R, Wu Z, Bai C (2021) Overexpression of *PvWOX3a* in switchgrass promotes stem development and increases plant height. Hortic Res 8:252. https://doi.org/10.1038/s41438-021-00678-w

Zhang Y, Yu LC (2008) Microinjection as a tool of mechanical delivery. Curr Opin Biotechnol 19(5):506–510. https://doi.org/10.1016/j.copbio.2008.07.005

Zhang G, Lu S, Chen TA, Funk CR, Meyer WA (2003) Transformation of triploid bermudagrass (Cynodon dactylo n × C. transvaalensis cv. TifEagle) by means of biolistic bombardment. Plant Cell Rep 21(9):860–864

Zhang JY, Broeckling CD, Sumner LW, Wang ZY (2007) Heterologous expression of two Medicago truncatula putative ERF transcription factor genes, WXP1 and WXP2, in Arabidopsis led to increased leaf wax accumulation and improved drought tolerance, but differential response in freezing tolerance. Plant Mol Biol 64:265–278. https://doi.org/10.1007/s11103-007-9150-2

Zhang F, Kang J, Long R, Yu LX, Wang Z, Zhao Z, Zhang T, Yang Q (2019) High-density linkage map construction and mapping QTL for yield and yield components in autotetraploid alfalfa using RAD-seq. BMC Plant Biol 19:165

Zhang Y, Qin X, He Z, Zhang Y, Li Z, Nie G, Zhao J, Feng G, Peng Y (2024) The White clover TrMYB33-TrSAMS1 module contributes to drought tolerance by modulation of spermidine biosynthesis via an ABA-dependent pathway. Int J Mol Sci 25(13):6974. https://doi.org/10.3390/ijms25136974

Zhou B, Bailey A, Niblett CL, Qu R (2016) Control of brown patch (rhizoctonia solani) in tall fescue (Festuca arundinacea Schreb.) by host induced gene silencing. Plant Cell Rep 35(4):791–802. https://doi.org/10.1007/s00299-015-1921-7

Zuo J, Niu QW, Ikeda Y, Chua NH (2002) Marker-free transformation: increasing transformation frequency by the use of regeneration-promoting genes. Curr Opin Biotechnol 13(2):173–180. https://doi.org/10.1016/S0958-1669(02)00301-4

High-Throughput Phenotyping and Digital Agriculture of Forages

Shanvi Nautiyal, Monika Kalia, Sonika Kalia, Indra Rautela, Ruchi, Anamika Thakur, and Vinay Sharma

Abstract

Forage crops form the most cultivated species for pasture and animal rearing. With the evolution of climatic changes, degrading soil health, and reducing relative water content in the soils, the growth of forage crops has been limited and compromised. To cultivate crops with greater persistence, productivity, and yield, genotype analysis of these species is an essential activity. Traditionally, the viability and quality of the plant was estimated based on subjective and labor-intensive methods such as measuring plant growth and counting surviving plants manually. With the evolution of high-resolution image-based new technologies such as high-throughput phenotyping (HTP), screening and monitoring of plants has become a much simpler process. It involves installation of various ground and air-based sensors and devices to keep track of the growing forage. Apart from this, HTP can be used in studying dominant traits and characteristics, botanical traits, ageing, and resistance against pathogens and pests. This forms the basis of the digital agriculture landscape that is currently under development stages. It involves monitoring, control, tracking of plant health using biosensors

S. Nautiyal · S. Kalia (✉) · I. Rautela · Ruchi
Department of Biotechnology, School of Applied and Life Sciences, Uttaranchal University, Dehradun, Uttarakhand, India

M. Kalia
Department of Entomology, Dr. YS Parmar University of Horticulture & Forestry, COH & F Neri, Hamirpur, Himachal Pradesh, India

A. Thakur
ICAR-Vivekananda Parvatiya Krishi Anusandhan Sansthan, Hawalbhag, Almora, Uttarakhand, India

V. Sharma
Center of Excellence in Genomics and Systems Biology, International Crops Research Institute for the Semi-Arid Tropics (ICRISAT), Patancheru, Hyderabad, India

F. S. Baloch et al. (eds.), *Forage Crops in the Omics Era*,
https://doi.org/10.1007/978-981-95-9249-4_7

and artificial intelligence. These systems have evolved for efficient water irrigation management and reducing dependence on fertilizers. Renewal technologies such as agrivoltaics and solar-powered meters can be deployed for enhancing plant growth and productivity. Efficient use of these technologies will help us redefine the future of agricultural of forage crops in an economically and socially sustainable way.

Keywords

High-throughput phenotyping · Genetic variability · Forage crops · Productivity

1 Introduction

Forage crops are principal group of plants that play a crucial role in the development of sustainable agriculture. They are used for feeding livestock and form an essential part of ruminant diet. Apart from this, their role has been investigated in soil conservation, enhancing nutrient efficiency of the soil, conservation of biodiversity, and sequestration of carbon. These crops are primarily cultivated for their dry matter yield. Therefore, over the past few years, greater influence has been made in studying techniques and evolving mechanisms to increase the crop yield of these forage crops. After conducting several studies and analysis, it was found that these crops have a diverse set of genotypic- and phenotypic-related genes and variation (Yang et al. 2020). The one common shortcoming in the assessment of these genes is that multiple long procedures and complex machines are involved based on traditional knowledge and methods prone to human errors. Apart from this data managing, analysis and storage remains to be another disadvantage associated with this approach in the breeding programs. Traditional methods involve sample clipping, collection, and processing that requires cost input and trained personnels. This work mechanism can be made more efficient and climate resilient by introduction of sensors, computerized information systems, and other digital interfaces to enhance the proficiency of plant breeding and phenotyping (Araus 2014). Phenotyping broadly involves the estimation and phenotype prediction that are characteristics or traits common to one specific plant or species and their utilization for better decision-making with respect to breeding programs to get a better crop yield. It helps in the measurement of dry matter yield (DMY), which is in turn estimated using leaf mass, shoot and root dry mass, and so on (Yang et al. 2020). With technological advancements and growing demand for phenotyping, using machine algorithms, artificial intelligence, aerial vehicles, sensors, and drones in the form of high-throughput phenotyping can result in greater yield and produce. Unmanned aerial vehicles are also deployed in this purpose as they are economically viable, accurate, and dependable (Gehan 2017). The use of professional high-quality cameras and lenses can further enhance the phenotyping results. These digital agricultural technologies (DAT) are an emerging concept in the recent times. They provide a much simpler and integrated approach for precision agriculture. Communication technologies, use of internet, crop monitoring, specific analysis, information sharing, analysis, and review are some important components of digital agricultural practices. Although it is a transformative process, it has major challenges as well. To harness its full potential, policymakers and researchers need to perform detailed analysis of their shortcomings, including economic and environmental concerns (Gehan 2017).

2 Developing Need for Crop Phenotyping in Modern Agriculture

With the ever-growing world population, it is expected that the crop yield needs to potentially double by the year 2050 or earlier to meet the rising food demands of the population worldwide. Moreover, climate change is becoming huge threat for agriculture by lowering crop yield through various biotic and abiotic stresses (Sarıkaya and Tatar 2023). Increasing instances of urbanization, deforestation, soil degradation, and limited availability of agricultural land have raised concerns about the need to develop new plant breeding techniques and tools to enhance crop yield and production of perennial forage crops to meet demands and profitability of large number of grazing industries and farmers over time. Traditionally followed plant breeding techniques are no longer suitable to achieve this target due to multiple associated disadvantages such as high cost, loss of variation and genetic diversity, time consuming, and so on (Andrade-Sanchez et al. 2013).

The exploration of genetic variability and genome of forage crops has been in focus recently by researchers to identify and investigate new crop varieties that can have desirable traits if subjected to that possess desirable traits for genetic improvement using modern techniques. Dry matter yield (DMY) of the forages is linked to phenotyping, which involves assessing genetic traits (Gehan 2017). One aspect of this kind of genetic selection has been practiced by the farmers for decades where healthier, improved varieties of crop were identified based on physical phenotyping, i.e., identifying desired traits (Andrade-Sanchez et al. 2013). The success of this approach truly depends on the accuracy and ability to identify the most suitable plant variety exhibiting more pronounced genetic variation compared to others (Gebremedhin et al. 2019). Conventionally breeding techniques were based on the principle that increasing the number of crosses in varying environmental conditions would increase the chance of developing plant varieties with superior genetic traits. Manually selecting these phenotypes not only reduces the efficiency of identifying new varieties but also is a tedious and time-consuming task for the farmers (Walter 2012).

Breeding techniques such as marker-assisted selection and genomic selection that are primarily based on genotyping of crop varieties always rely on phenotypic analysis for greater accuracy in results. It has been reported that dry matter yield in ryegrass potentially increased by enhancing phenotyping techniques to assess the genetic expression. Thus, development of sensor-based data collection methods and technologies such as image-based high-throughput phenotyping has emerged as a revolutionary approach in agriculture (Gebremedhin et al. 2019). It includes a wide range of equipment ranging from ground-based tools to airborne platforms for collection of crucial research data for identification, analysis, and development of new botanical traits among forage crops (Walter 2012).

3 Concept of Phenotyping

Phenotyping of forage crops refers to the process of identifying quality and quantity, developmental architecture, biomass, and overall growth of an individual plant in a particular field area. Traditionally, this approach has been utilized worldwide for investigation of superior genotypes and physiological traits or genes in a plant. This forms a crucial step in the plant breeding programs (Mir 2019). Visual external characteristics, biochemical examination, observations, and measurements of growing plant parts are some of the classical techniques employed for this purpose.

With the advancement in technology and research in phenotyping have enabled the use of more targeted and automated procedures such as field monitoring, imaging, and refining tools for faster and more accurate phenotype selection (Araus 2014). However, the use of these technologies currently is limited to greenhouses, and laboratory setups as quantitative imaging in the fields is much more hindered due to the presence of arid environmental conditions such as overlapping canopies, varying illumination, sunlight division, and so on. This can be better understood by considering the example of estimation of dry matter yield of forage crops. Several challenges faced during phenotyping of it ranges from lack of appropriate number of plants available for sampling, which may not include the best superior traits, to sample harvesting limitations such as weighing of dry matter that is time-consuming (De Swaef et al. 2021). Other factors such as transportation and operation cost may add to existing challenges. Much of these challenges can be potentially dealt by the development of precise phenotyping methods that are cost-effective as well as less destructive. In case of forage crops, the estimation of yield is relate to estimation of the biomass that involves morphological and functional analysis and record making. It has been observed that development or expression of superior trait is not only dependent on genetic factors but also on the genetic potential, which can be enhanced if subjected to better maintenance and farming methods (Cabrera-Bosquet et al. 2012).

The aim of phenotyping is arranging or ranking the breeding plants based on quantifiable and quality specific markers, which in turn reflect the quality or traits related to a particular desirable genotype. It promotes identification of most superior phenotype and comparing it with others for better result interpretation (Cabrera-Bosquet et al. 2012). Phenotyping results are based on detailed transcriptome, proteomic, and metabolic analysis, which in turn enhances its accuracy.

4 Current Trends and Approaches in the Field of High-Throughput Phenotyping

Phenotyping forage crops involves assessment of key agricultural indicators such as dry matter yield (DMR), nutritional quality, and tolerance to various biotic and abiotic stresses, all of which impact over crop yield. Based on traditional methods and techniques, it can make this process complicated and time-consuming. To ensure high quality and reliable research, multiple trials must be conducted across the various locations and environmental conditions in different seasons. However, this approach is not only time-consuming but also requires the deployment of labor, which significantly increases the cost (Salas Fernandez 2017).

All these challenges and process hurdles can be overcome using automated high-throughput phenotyping (HTP) that has greatly revolutionized the process in the past decade. Automated analysis of physical traits of the plants forms the basis of this technique that results in generation of phenotype-based data. This data after collection is revised, reviewed, and acquired further. Apart from effective screening, another advantage associated with this method is that the physical attributes can be detected at the early growing stage of the crop life cycle (Cabrera-Bosquet et al. 2012). As of now, HTP has been deployed in laboratory trials for detection of disease-specific traits, senescence, and other mortality-related destructive traits. Phenomics is emerged as the new discipline under research recently and some ongoing HTP methods used for this purpose are discussed in the following section (Hund et al. 2019).

5 Imaging and Sensor-Based Technologies for Plant Phenotyping

Imaging technologies involve around the use electromagnetic radiations and electromatic interactions such as absorption, reflection, transmission, emission, and florescence for detecting and differentiating a healthy plant from a diseased or unhealthy plant (Table 1). Deploying these techniques facilitates detection of those characters or traits that cannot be detected with naked eyes.

It involves the recording of spectral data collected from recurrent interactions between the radio waves and the plant parts/tissues and calculating the reflected light intensity ratio to the light that has been illuminated. It mostly uses lights such as infrared or light near to wavelength of infrared for this purpose (Hund et al. 2019). Leaf reflectance is defined as the part of light that has been reflected back. Various constituents and pigments present in the leaf of plant such as chlorophyll, carotenoids, and anthocyanin are photoactive in nature and leads to stronger absorbance.

Table 1 Various types of imaging techniques and specifications

S. no.	Technique involved	Sensors deployed	Specific parameters for phenotyping	Merits	Demerits
1	Thermal image phenotyping	Infrared cameras	Severity of disease, area of leaf index, canopy and shoot temperatures, conductance of stomata, status of water of the plant leaves	Processing of images allowed, fast and accurate, less destructive	Soil backgrounds need corrections, weather conditions such as wind speed and directions need to be taken care of
2	Visible light imaging phenotyping	Suitable standard cameras using visible light	Dry matter yield, biomass yield, growth dynamics, morphological attributes of growth such as height, color, shape, seed germination indicators	Less complex procedures and cost-effective	Sensitive can be costly and sensitive; wavelength specificity may hinder better results
3	Hyperspectral phenotyping	Thermal cameras, specifically hyperspectral camera and imaging equipment	Water content of leaves and adjacent canopies, growth of leaves based on length of panicle, quality of the gain, soil health and presence of phytopigments	Better efficient prediction and accuracy, measurement of diverse parameters and specific disease	Data interpretation may be difficult due to sophistication of involved processes
4	Fluorescence imaging mapping	Specially designed florescent cameras	Leaf diseases and their severity, pigment levels and phytohormones levels, quantum yield, pigment composition	Stress can be monitored and detected even before it is visible externally	Data analysis and productivity constraints and strict adherence to protocols is needed
5	Three-dimensional techniques	Flight and stereo-based cameras and imaging devices	Root and shoot architecture and development along with canopy height and health detection	Real-time imaging with upmost accuracy is possible	Expensive equipment and data interpretation can be difficult

(continued)

Table 1 (continued)

S. no.	Technique involved	Sensors deployed	Specific parameters for phenotyping	Merits	Demerits
6	Magnetic resonance imaging	Magnetic resonance-based imaging devices and cameras	Physical attributes and morphology of plant structures, metabolic rates and health	Sample preparation is quite easy and applicable on single or whole plant cultures	If the cells are bigger, resolution of images can be compromised
7	Positron topography emission	Specific devices	Water transport and speed of flow	Qualitative and quantitative analysis is possible	Limited resolution

6 Use of Sensors in Phenotyping

Sensors are typically involved in high-throughput phenotyping based on ground (GB-HTB). Usually, passive sensors such as light-based (red, green, blue sensors), hyperspectral, or visible light imaging-based sensors are deployed for this purpose. With technological advancements aiding the developments, active sensors such as Green Seeker sensors, ultrasonic, and other laser-based sensors have been developed for better data collection and analysis. Depending on the kind of light source used, the sensors can be classified as active sensors and passive sensors (Table 2). Active sensors usually use light or sound energy while the passive sensors depend on natural source such as sunlight (Neilson 2015).

7 Light Detection and Ranging (Popularly Known as LiDAR)

LiDAR is an HTP technology that is exclusively based on the application of laser light to measure the emission time and the distance between the light source, which is usually the LiDAR scanner, and the target object. Various aspects of plant architecture such as the volume, height, width, and structure of canopy can be efficiently measured using this technique, as it creates a 3D cloud of collected data specifically filtering out desired results and outputs (Hund et al. 2019).

Initially, this technique was used by the researchers worldwide for collecting high-resolution and accurate phenotyping of the forest crops, fruit trees, and cereals. Forage crops such as ryegrass have also been analyzed for accurate measurement of biomass and dry matter yield along with other parameters such as plant height, shape, and texture of the plant parts. Researchers have reported that LiDAR has much less error rate as compared to manually applied procedures in both static and dynamic modes. The addition of other modifications in the LiDAR system such as NDVI can potentially enhance the biomass estimation in the forage crops (Kipp et al. 2014).

Table 2 Differences between active and passive sensors

S. no.	Parameters	Active sensors	Passive sensors
1	Source of illumination	Possess own sound or light emitting sources or unit	Depend on external sources such as sunlight
2	Operation feasibility	They can operate in cloudy or gloomy weather conditions without depending on radiation differences for visibility	Usually depend on the radiation difference of the incoming light for visibility in cloudy weather conditions
3	Field of view	The field of view (FOV) is narrow usually	Depending on the source of the light and the distance from the object, the FOV may be limited
4	Process complexity	The data processing is less complex	More number of steps required, making the process time-consuming and complex
5	Radiometric calibration	Images derived usually do not require radiometric calibration	Passive sensors greatly require post-radiometric calibration for better quality and resolution
6	Advantages	Better accuracy in measurement and operational during day and night	Very cheaply available and easy to handle
7	Disadvantages	Very costly and prone to breakage due to mishandling. Require extra operational care as sensitive	Operational during only the daytime. Lacks detailed information-gathering tools

Although the existing challenges with respect to traditionally used phenotyping technologies can be greatly overcome using LiDAR, it too has certain disadvantages associated with its functioning such as high cost, complexity in working procedures, high installation and maintenance costs, and result interpretation limited to the physiological characters (Singh et al. 2016).

8 Types of High-Throughput Phenotyping Platforms

Depending on the physical location of the sensors and other remote sensing devices deployed for screening and monitoring purposes, the phenotyping platforms can be broadly classified into three types.

Controlled environment phenotyping platforms

Controlled environment phenotyping platforms represent the third major type of high-throughput phenotyping systems and are designed to study plant traits under precisely regulated environmental conditions. These platforms are typically established in greenhouses or growth chambers where factors such as temperature, light intensity, humidity, and CO_2 concentration can be carefully controlled to minimize external variability. They often utilize automated conveyor systems that transport plants to imaging and sensing stations, enabling repeated, non-destructive measurements of growth, morphology, and physiological responses over time. Advanced imaging techniques, including RGB, thermal, and fluorescence imaging, are integrated to capture detailed phenotypic data with high accuracy and reproducibility.

Table 3 Types of ground-based platforms for high-throughput phenotyping

S. no.	Type of ground-based system	Description
1	Tractor-based systems	Tractors with greater clearance mounted with multiple payloads and sensors that need to be driver manually. Developed first in Arizona consisting of infrared thermometers and ultrasonic sensors to record the temperature changes and canopy morphology. High payload capacity compared to others along with additional features such as imaging chambers. Disadvantages include precision depends on speed of tractor and sensors such as hyperspectral cameras are very sensitive.
2	Motorized cart and pushcart carts	Can be used as an alternative to tractors as involved low cost and the speed can be controlled manually. Motor carts use motors to drive the cart through the fields and involves a self-propelling system. Potential risk involves damaging the crops if not controlled properly.
3	Cable-driven systems and gantry	A gantry refers to the bridge-like structure to carry the sensors and other remote devices that is developed to support a crane moving linearly. Examples include the field Scanalyzer of University of Arizona that has a payload capacity of more than 500 and equipped with better fluorescence devices. Cable-driven systems are similarly developed instead of gantry, and cables are used as support systems. These systems are more accurate as they avoid physical contact and allow multiple scanning and monitoring of the crops. Disadvantages include less coverage capacity and fixed site of detection usually focus only on upper canopies.
4	Robots for phenotyping	Recently evolved in the past decades to overcome the labor-associated disadvantages. They are automated devices capable of functioning in all environments without getting tired and provide better outputs as compared to humans. Attempts are being made to replace tractors and carts with robots in the coming years.

Such platforms are particularly valuable for studying plant responses to abiotic stresses like drought, salinity, and heat, as well as for genetic analysis and functional genomics. By ensuring uniform experimental conditions, controlled environment platforms enhance data reliability and are widely used in Plant Phenomics and crop improvement programs (Fiorani and Schurr, 2013; Araus et al., 2018; Cobb et al., 2013).

8.1 Platforms Based on Ground

A platform in simple term refers to a physical facility or carrier where the remote sensing devices are usually placed for the purpose of data collection and monitoring. More the proximity between the remote devices and the target crop or plant, more are the chances of accuracy and efficiency in data collection. Platforms thus form a very crucial part of the research under study. The ground-based sensors used for ground platforms are usually automated or semiautomated in nature (Table 3). Various crops such as lucerne, ryegrass, and red clover have been monitored using these platforms (Kipp et al. 2014).

The sensors here are usually placed in top of bicycles, pushcart and payload cart, gantry and cable driven systems, and all-terrain vehicles such as tractors and robots. Multiple synchronous sensors are mounted on one bicycle and can capture the data from the entire farmland, specific area of the plot, and even on an individual plat or crop. The major advantage associated with these platforms is that they are capable of handling multiple payloads at one time and minimum postprocessing modifications and processing are required (Honsdorf et al. 2014). As compared to other platforms, the ground-based phenotyping platforms have better data quality as they control the environmental factors such as light source for better monitoring.

8.2 Aerial-Based Platforms

This method involves application of UAV or unmanned aerial vehicles, which are specialized aircrafts in the form of drones that can operate without a human pilot. They belong to a group of serial systems known as unmanned aerial systems (UAS). It has various components ranging from controller, sensors, communication systems, and recording and imaging devices. Conventionally deployed method and even the ground-based methods were unable to collect high temporal and special resolution of the images, which is now possible using the aerial systems (Wang et al. 2018). They are efficient in collecting data with respect to biomass, plant height, ground cover, and so on. Apart from this, various satellites specialized designed to large areas under cultivation can be employed as part of aerial systems.

However, using satellites for imaging purposes can have multiple challenges such as cloud interference, rains, and other environmental hurdles that can hinder the process. Also, the resolution of the satellites is much lower in terms of spatial resolution. UAS are increasingly used for aerial investigations and monitoring because they operate at much lower altitudes than satellites and fly in proximity to the crops, giving better results (Lane 2021). Also, the unmanned aerial vehicles are much more flexible in working and low in cost as compared to the satellite systems. Aerial methods have produced promising results in the fields of forage crops, but further research and development are needed for better results and data collection accuracy. Considering the changing environmental conditions and hazards, calibration methods need to be updated to ensure that phenotyping remains both timely and cost-effective (Wang et al. 2018). Another technical hurdle that persists in these platforms is that they lead to generation of tremendous amount of complex data that firstly needs to be stored, curated, revised, and processed. This requires large memory and storage spaces, which are currently not available. To utilize the true potential of these platforms, these limitations need to be overcome timely (Lane 2021).

Table 4 List of worldwide commercial associations and facilities relating to HTP

S. no.	Networking facility	Description
1	European Plant Phenotyping Network (EPPN)	Funded by the European Union, this research infrastructure project allows the European scientists access to 23 types of phenotyping platforms present at various institutes across the European continent
2	Australian Plant Phenomics Facility (APPF)	This diverse platform forms a group of various national infrastructure projects, allowing open access to all state tools and techniques relating to phenomics exclusively to its population
3	International Plant Phenomics Centre (IPPC)	This web-based platform aims at providing open unrestricted access to all the major stakeholders, such as researchers, government enterprises, and industries, and also to the general public
4	LEPSE	Working for more than 15 years now, this platform consists of a panel of expert scientists who have expertise in phenomics and crop breeding. They work together to develop and analyze various plant traits under altering conditions. This in turn leads to development of new phenotyping platforms and foster research
5	Lemma Tech	Equipped with latest technology and precision methods, this platform aims at precise calculation and monitoring of specific plant traits

9 Private and Commercial Enterprises Available Currently for Phenotyping

The major limitation associated with the use of automated phenotyping facilities revolves around the high operational and maintenance cost that these technologies demand. At present, few communities in the West such as Europe and America have come up with these facilities in the lab for commercial use (Table 4) (Wang et al. 2018).

10 High-Throughput Phenotyping Platforms and Current Facilities in India and South Asia

With the developing world population and India being the most populous country in the world, India faces the most risk of food shortage and limited forage crop production can be a great concern in the coming years. Developing countries such as India and China have considered the seriousness of these challenges and made efforts to develop phenotyping platforms to support agriculture and the small, marginalized farmers of the country. In India the main organization related to HTP was developed in 2015, recently in the form of a Plant Phenomics National Facility at ICAR (Araus et al. 2018). This facility is aimed at identification of resistant and tolerant phenotypes and suspending them in adverse conditions to select the most advanced plant (Fig. 1). A high-throughput phenotyping facility was developed in 2014 at ICRISAT, which had the potential to monitor and phenotype more than 10,000 plants with an efficiency rate of 500 plants per hectare (Chawade et al. 2019).

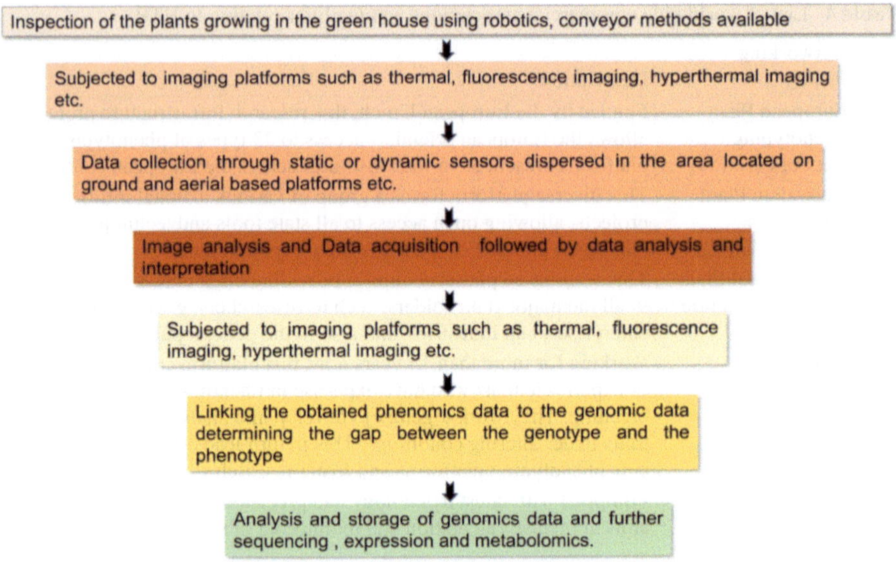

Fig. 1 Steps involved in high-throughput plant phenotyping

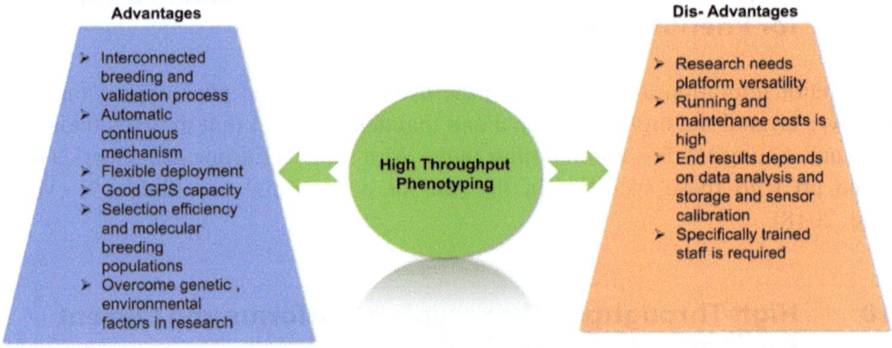

Fig. 2 Advantages and disadvantages of HTTP

Platform named Leasy Scan inaugurated at the Central Research Institute for Dry Land Agriculture involves the use of 3D cameras and imaging devices capable of monitoring the development of the leaves, their conductance, and measure their drought stress tolerance (Fig. 2) (Feng et al. 2021).

11 Digital Agriculture as the Future of Forage Agriculture

Digital agricultural practices, also referred as the e-agriculture or smart farming, have acquired worldwide attention in the recent years as a result of ever-evolving threats of low agricultural productivity and related socio-technical issues (Aharon et al. 2020). It refers to the tools and techniques that are developed for collection,

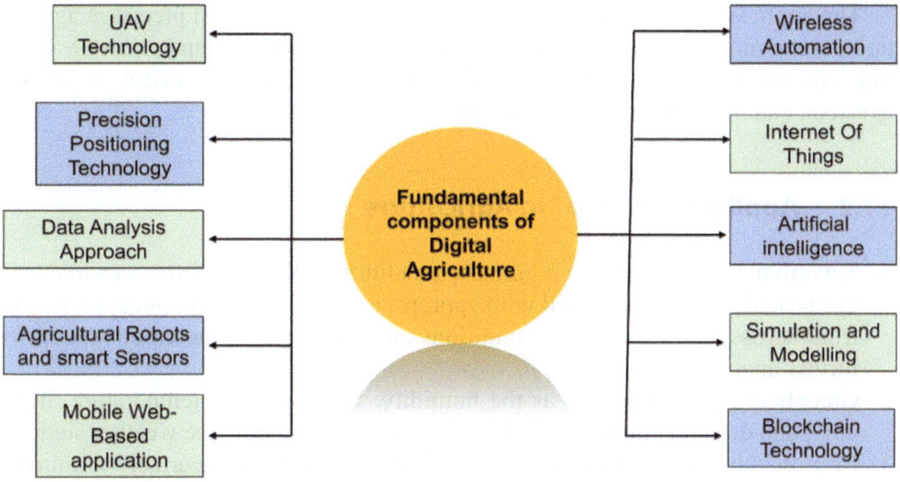

Fig. 3 Components of digital agriculture in forage crops

storage, and sharing of electronically acquired media. This digitalization is not only limited to agriculture and food forages but also extends to fiber, bioenergy, and other interconnected supply chains.

India being the key stakeholder to developing agricultural challenges and more than 45% of workforce employed in agricultural practices has come up with the "Digital Agriculture Mission" launched by the agriculture ministry to achieve use of latest digital technologies, artificial intelligence, drone technology in agriculture, and promote digital infrastructure for the same by 2025. This initiative is no less than a revolution and is regarded as Agricultural Revolution 4.0 (Balasundram 2023).

Digital techniques, when applied to farming and animal systems, allow for optimum utilization and management of resources without any wastage. Various components of digital technology include internet of things (IOT), artificial intelligence (AI), robotics, UAS, and so on (Aharon et al. 2020).

The modern-day digital agriculture methods comprise of the following components, which is shown in Fig. 3.

11.1 Internet of Things (IoT)

IoT consists of physical infrastructure such as machines and computers that are embedded with certain software, tools, and technologies that enable the exchange of information, communication, and connection of devices with other systems over the internet. This acts as a bridge connecting the physical world with the digital world. The IoT sensors devices collect the data and send it through gateway of internet for analysis and interpretation. The results again travel through the same pathway to reach the output fronts and record as results (Mehmet Ali et al. 2021).

The application of this system has been diverse, ranging from precision agriculture to crop monitoring, soil health management, warehousing, and smart pest control. Animal tracking, health monitoring, and early illness detection in animal livestock can be made possible using these tools (Reis 2020).

11.2 Applications of IoT in Agriculture

1. Surveillance and monitoring climatic conditions: Weather stations available at the agriculture sites equipped with appropriate sensors help in collection of data from across varying climatic conditions and sending it to the cloud station for further analysis.
2. Greenhouse activities such as the humidity, soil condition, temperature, lightning, and drought conditions can be accurately adjusted by the weather stations.
3. Crop health and the presence of pest or diseases can be accurately monitored before its development.
4. Use of robots, UAV, drones along with remote sensors connected to artificial intelligence and machine learning platforms helps in surveying, mapping the fields, and promote management of the farm in a rational logistical way (Rotz et al. 2019).

12 Artificial Intelligence and Machine Learning in Digital Agriculture

Artificial intelligence is an emerging field that deals with the use of computer science to create models having the potential to execute and perform functions associated with human cognizance and intelligence such as rationality and reasoning ability. It comprises analysis, decision-making, and data acquisition to produce human-like outputs. The use of AI promotes better precision and problem-solving, planning, organizing, learning, and speech recognition. They help in real-time monitoring of the fields, diseases, crop production, weather prediction, soil health, and water usage by using remote technology, robotics, sensors, deep learning software, and artificial networks (Fountas et al. 2020).

Presently, agricultural practices are deeply labor intensive, and automation of agriculture will assist farmers in better prediction and crop management to guarantee better quality and quantity yield.

The efficiency of AI tools in problem solving depends on quality of data made available to it, which will help in producing problem solving outcomes. It utilizes datasets and measurements to determine the probable implication of fertilizer dosage on the crop and soil health and how it will affect the productivity and yield in the future (Reis 2020).

Remote sensing involves the use of meteorological data and algorithms for enhancing the accuracy and reliability of the results. Data that lacks accuracy and precision can be easily deleted or removed. Machine learning platforms are widely used for determination and calculation of production and yield, detection of weed, crop quality, and identifying new crop traits (Rotz et al. 2019).

A machine learning algorithm consists of methods and special arrangement for optimization of a task by analysis of experience and use them in the upcoming best ways possible. The efficiency depends on the amount of data available: the more the data available, the better will be the functioning of the algorithm. Using ML techniques to determine the relationship between various factors such as environment, temperature, water availability, soil health management practices, and the yield production can help in predicting the production of the crop (Fountas et al. 2020).

Artificial intelligence uses a combination of various techniques such as robotics, deep learning, artificial neutral networks to collectively overcome the agricultural challenges faced today. Continuous evaluation and monitoring of the nutrient levels in the soils have been made possible using these modern techniques. With the growing scarcity of water resources, it has become a challenge to utilize each drop of water without any wastage (Balasundram 2023). Artificial intelligence tools have made it easier monitoring water uptake by the plants while reducing the amount of wastewater generated. Depending on the type of crop under cultivation, AI has the potential to decide which fertilizer shall be best suited for optimal growth and provide the accurate exact dosage amount that needs to be given to the crops (Rotz et al. 2019).

Predictive analysis tools and method of deep learning can be deployed for monitoring the transpiration, evaporation, and respiration rates of the plants. These techniques minimize the number of labors needed in agriculture as most of the processes involved are automated or semiautomated. This helps in meeting the urban demands such as lack of human resource and workforce in agriculture (Feng et al. 2021). The data is collected in the original raw form using these methods, which is further passed on to neutral wireless networks that are connected to weather analysis or soil health prediction equipment and give appropriate outcomes as desired. It can even predict the best weather and time for sowing the seeds in the soil for greater quality yield (Tedeschi 2021). Weed elimination is a major challenge in agriculture today; deep learning has enabled the use of algorithms and datasets to monitor the growth of weed and apply needed herbicide to suppress the growth. Such companies have even developed robots and aerial systems to monitor the weeds and spray herbicide accordingly (Aharon et al. 2020).

Intelligent irrigation systems have been developed that combine the use of machine learning algorithms dedicated to calculating the soil moisture levels using various deep learning and IoT platforms. Satellite imaging and precision technologies are later deployed for quick and accurate data collection and monitoring (Rotz et al. 2019).

Everyday application of these modern technologies lead to the development of large quantity of data, which is often referred to as the big data. Various sources of data include sensors, drones, satellites, and ground and aerial systems, which are employed for data monitoring and research purpose. Adequate management and storage of this data is needed for efficient future research and analysis (Smith et al. 2020).

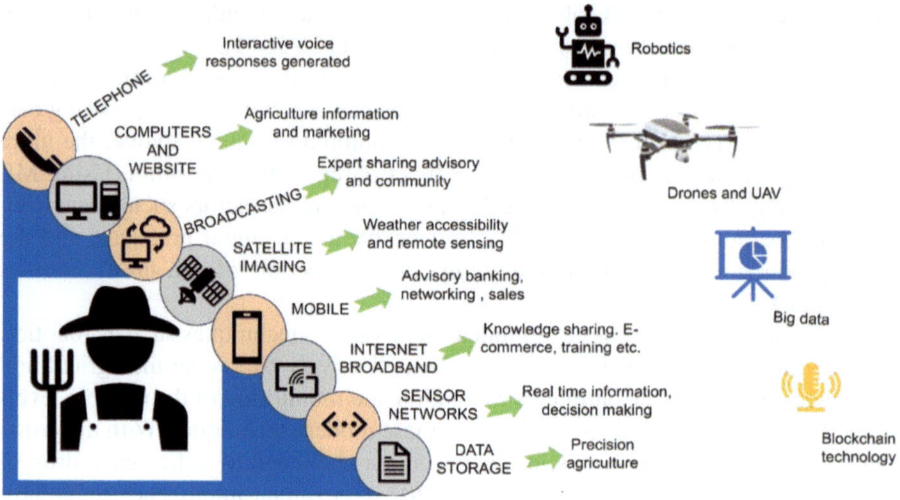

Fig. 4 Role of various digital technologies in agriculture

Applications of artificial intelligence in digital agriculture are as follows (Fig. 4):

1. Weather prediction, land quality, crop cycle, multispectral imaging, precision farming, output efficiency.
2. Creative farming, digital realm, remote sensing and testing, data analysis.
3. Reduced workload, job creation, operational cost reduction, maintenance cost reduction.
4. Detect disease and plant illness, better agronomic products, automated harvesting, reduction in postharvesting cost.
5. Self-driving machinery and tractors, automated trimming and maintenance of plants.
6. Crop and soil monitoring, better decision-making, proper information sharing, identification of wasteful resources, and suggestion of alternatives (Trendov 2019).

13 Conclusion

Digitalization and application of new farming techniques in agriculture have revolutionized the entire forage agriculture landscape in the world today. They are emerging fields with numerous applications extending from the use of HTP in plant physiology to sensing technology that are targeted to identify complex and new traits. However, these methods are still in infancy and require robust developments to be used in large-scale agricultural practices. The protocols currently available are limited to the use in laboratories and smaller setups. With the rising population and climatic changes, HTP will be in great focus in the upcoming years due to its ability to identify high-yielding crop varieties. Currently, the phenotyping methods have been limited to analyzing the shoot-related external traits only (Baath 2020). In the future, focus needs to be made on exploring the root-related plant traits and their phenotyping. With respect to the use of digital method of technology, the major drawback associated with them is the high cost of application, their sensitivity, and less availability (Beyaz et al. 2023). The big data generated by all these processes needs to be managed and stored at reliable destinations for the future use, otherwise it could lead to unwanted leakages and piracy issues (Nunes et al. 2021).

Author Contributions Writing—original draft preparation, writing, and conceptualization: S.K. and S.N.; formal analysis and writing—review and editing: M.K., R., and I.R.; review and editing: A.T. and V.S.

References

Aharon S et al (2020) Image based high-throughput phenotyping of cereals early vigor and weed-competitiveness traits. Remote Sens 12(23):3877
Andrade-Sanchez P et al (2013) Development and evaluation of a field-based high-throughput phenotyping platform. Funct Plant Biol 41(1):68–79
Araus JL (2014) Field high-throughput phenotyping: the new crop breeding frontier. Trends Plant Sci 19(1):52–61
Araus JL et al (2018) Translating high-throughput phenotyping into genetic gain. Trends in Plant Science 23(5):451–466
Baath GS (2020) Predicting forage quality of warm-season legumes by near infrared spectroscopy coupled with machine learning techniques. Sensors 20(3):867
Balasundram SK (2023) The role of digital agriculture in mitigating climate change and ensuring food security: an overview. Sustainability 15(6):5325
Beyaz A et al (2023) YOLOv4 and tiny YOLOv4 based forage crop detection with an artificial intelligence board. Braz Arch Biol Technol 66:e23220803
Cabrera-Bosquet L et al (2012) High-throughput phenomic selection; the frontiers of crop breeding converge F. J Integr Plant Biol 54(5):312–320
Chawade A et al (2019) High throughput field-phenotyping tools for plant breeding and precision agriculture. Agronomy 9(5):258
Cobb JN et al (2013) Next-generation phenotyping: Requirements and strategies for enhancing our understanding of genotype–phenotype relationships. Frontiers in Plant Science 4:1–9
De Swaef T et al (2021) Applying RGB-and thermal-based vegetation indices from UAVs for high-throughput field phenotyping of drought tolerance in forage grasses. Remote Sens 13(1):147
Feng L et al (2021) A comprehensive review on recent applications of unmanned aerial vehicle remote sensing with various sensors for high-throughput plant phenotyping. Comput Electron Agric 182:106033
Fiorani FF, Schurr U (2013) Future scenarios for plant phenotyping. Annu Rev Plant Biol 64:267–291
Fountas S et al (2020) The future of digital agriculture: technologies and opportunities. IT Prof 22(1):24–28
Gebremedhin A et al (2019) Prospects for measurement of dry matter yield in forage breeding programs using sensor technologies. Agronomy 9(2):65
Gehan MA (2017) High-throughput phenotyping. Am J Bot 104(4):505–508
Honsdorf N et al (2014) High-throughput phenotyping to detect drought tolerance QTL in wild barley introgression lines. PLoS One 9(5):e97047
Hund A et al (2019) Methodological advances, challenges and perspectives in field phenotyping and its application to forage crops. In: Improving sown grasslands through breeding and management, vol 24. Wageningen Academic Publishers, Zurich, Wageningen, pp 295–305

Kipp S et al (2014) High-throughput phenotyping early plant vigour of winter wheat. Eur J Agron 52:271–278
Lane HM (2021) High throughput can produce better decisions than high accuracy when phenotyping plant populations. Crop Sci 61(5):3301–3313
Mehmet Ali M, Al-Ani A, Eamus D, Tan DKY (2021) A review of remote sensing applications in crop phenotyping. Remote Sensing 13(2):1–25
Mir RR (2019) High-throughput phenotyping for crop improvement in the genomics era. Plant Sci 282:60–72
Neilson EH (2015) Utilization of a high-throughput shoot imaging system to examine the dynamic phenotypic responses of a C4 cereal crop plant to nitrogen and water deficiency over time. J Exp Bot 66(7):1817–1811
Nunes L et al (2021) Horse foraging behavior detection using sound recognition techniques and artificial intelligence. Comput Electron Agric 183:106080
Reis ÂV (2020) Technological trends in digital agriculture and their impact on agricultural machinery development practices. Rev Ciênc Agron 51(5):e20207740
Rotz S et al (2019) The politics of digital agricultural technologies: a preliminary review. Sociol Rural 59(2):203–229
Salas Fernandez MG (2017) A high-throughput phenotyping technology for tall biomass crops. Plant Physiol 174(4):2008–2022
Sarıkaya MF, Tatar M (2023) Global climate change and crop production. Uluslararası Sivas Bilim ve Teknoloji Üniversitesi Dergisi 2(2):65–78
Singh A et al (2016) Machine learning for high-throughput stress phenotyping in plants. Trends Plant Sci 21(2):110–124
Smith C et al (2020) Machine learning algorithms to predict forage nutritive value of in situ perennial ryegrass plants using hyperspectral canopy reflectance data. Remote Sens 12(6):928
Tedeschi LO (2021) Advancements in sensor technology and decision support intelligent tools to assist smart livestock farming. J Anim Sci 99(2):skab038
Trendov NM (2019) Digital technologies in agriculture and rural areas. FAO
Walter AB (2012) Advanced phenotyping offers opportunities for improved breeding of forage and turf species. Ann Bot 110(6):1271–1279
Wang X et al (2018) Field-based high-throughput phenotyping of plant height in sorghum using different sensing technologies. Plant Methods 14:1–16
Yang W et al (2020) Crop phenomics and high-throughput phenotyping: past decades, current challenges, and future perspectives. Mol Plant 13(2):187–214

Recent Advancements in Breeding and Genomics of Alfalfa

Aina Inam, Sadia Afzal, Lalarukh Azhar, Nadia Ghaffar, Sumera Javad, and Ajit Singh

Abstract

This chapter delves into the advancements in alfalfa (*Medicago sativa* L.) breeding and genomics, exploring the historical evolution of alfalfa cultivars and the shift from traditional breeding methods to modern genomic techniques. Alfalfa, known as the "Queen of Forages," is a critical crop for livestock due to its high protein content and environmental adaptability. The chapter outlines the integration of molecular breeding techniques such as marker-assisted selection (MAS) and genomic selection (GS), which have revolutionized the breeding process by enabling the precise identification and selection of desirable traits. It also highlights the role of advanced biotechnological approaches, including CRISPR/Cas9 genome editing, in enhancing stress tolerance, yield, and nutritional quality in alfalfa. The challenges of implementing these technologies, including high costs, complex trait architectures, and regulatory hurdles, are discussed. Additionally, the chapter emphasizes the ethical and environmental considerations related to using genetically modified organisms (GMOs) in agriculture. Overall, this comprehensive review underscores the importance of integrating modern genomic tools with traditional breeding strategies to improve alfalfa's resilience and productivity, ensuring its continued value in sustainable agriculture.

Keywords

Agriculture · Alfalfa · Breeding · *Medicago sativa* · Queen of Forages

A. Inam · S. Afzal · L. Azhar · N. Ghaffar · S. Javad (✉)
Department of Botany, Lahore College for Women University, Lahore, Pakistan

A. Singh
School of Biological and Environmental Sciences, University of Nottingham Malaysia, Semenyih, Selangor, Malaysia

F. S. Baloch et al. (eds.), *Forage Crops in the Omics Era*,
https://doi.org/10.1007/978-981-95-9249-4_8

1 Introduction to Alfalfa Breeding and Genomics

1.1 Introduction to Crop

Medicago sativa L., well known by its common name alfalfa, is familiar in the world as the "Queen of Forages." Some features make it an ideal crop for livestock as it contains sufficient amount of protein and nitrogen content. Shi et al. (2017) mentioned that due to the unique features of alfalfa such as stress tolerance against cold and salinity, acceptance to the environment, high yield capacity, amelioration of soil, and good economic benefit to the country made *Medicago sativa* L. more cultivated forage plant (Azad et al. 2023; Shi et al. 2017). Alfalfa is cultivated worldwide having autotetraploidy with a basic number of chromosome 8 and a genome size of 800–1000 Mb (Hawkins and Yu 2018).

1.2 Relationship Between Breeding and Genomics

Crop breeding has a unique relationship with the genetic data of the crop. In the traditional approach of plant breeding, breeders mostly used natural selection of traits such as selection of phenotypic traits for better crop productivity. However, many environmental factors influenced the developing varieties. To make reliability in the process of breeding, breeders made a relationship between breeding and genomics. In other words, we can conclude that phenomics and genomics are very important in the development and flourishment of crops (Krishna et al. 2023).

A lot of work has been done in the past in the search for alfalfa genomics. Multiple genetic markers have been used to identify the alfalfa genome and for its transcriptomic studies. In these genetic studies, several genetic markers are used from which single nucleotide polymorphism (SNP) is comprehensively used (Hawkins and Yu 2018).

2 Historical Perspective

2.1 Evolution of Alfalfa Cultivars

The historical perspective of alfalfa and its evolution is very sequential and quite informative. According to written literature and manuscripts, it was suggested that this forage plant first originated from Iran (which was known earlier as Persia). It was present in several provinces of Anatolia, South Caucasus, Kashmir, and Balochistan as the wild indigenous plant.

2.2 From Persia to Arabia

When the forage plant entered the Arabic region, then people named it "Alfalfa," meaning "Best forage." The exact sample from Persia was later on distributed into Greece (490 BC) and the Roman Empire (200 BC). The utilization of forage by Romans continued and they were involved in its spread to Italy and areas of the Mediterranean region. Romans also contributed to the introduction of alfalfa into Western Europe. In the period of the Dark Ages, alfalfa had started to disappear from Europe, even in the records. At this time, Arabs were the only agriculturists who cultivated the alfalfa among them all.

2.3 Introduction to the New World

Alfalfa was introduced to Mexico and Peru by Spanish victors at the start of the sixteenth century. The next destiny was Chile and Argentina (1600) and extended to Uruguay (1776).

2.4 Introduction of Alfalfa into the South America

The literature review suggested that Chile and Mexico were the major sources of its entry into southwestern America. Today, the major acres used to cultivate the alfalfa were established at that time. During the "Gold Rush" era (1850s), alfalfa made its way from Chile to the southwest of the United States.

2.5 California to Utah

Authorities have investigated and formulated that nearly a century ago, alfalfa was brought from Chile to California, from where it was introduced to eastward Utah. Utah spread alfalfa seeds in other North American areas. There is a misconception that California was not the source of the introduction of alfalfa into Utah, whereas there were many origins of its introduction into Utah as one type "winter hardy" arise from Britain and its neighboring islands.

2.6 Lucern: Alternative Name for Alfalfa

In 1850s and 60s, imported immigrants from Utah named Lucern also called Lucerne taken out of England, Wales, and Chile. Many Mormon emigrants to Utah (1890) came from Britain and European origin and used to call alfalfa Lucern.

Present leadership and chronicles gave evidence that Lucern was brought to Utah by eight immigrants:

- Three imported seeds were from England
- One immigrant from California
- One immigrant of Swiss origin
- Three immigrants not specified Lucern seed.

The evolutionary history of alfalfa shows that Britishers instead of the Spanish acted as the primary repository of alfalfa seed, from where these spread across the United States (to Chile to California to Utah). Hence, alfalfa from Britain and neighboring islands is the major ancestor of alfalfa development in the United States (Brough et al. 1977). And today the United States is the largest producer of alfalfa.

3 Traditional Breeding and Its Challenges

Traditional plant breeding incorporates choosing plants that possess traits of interest and crossing them to generate individuals of desired characteristics. Conventional plant breeding is a basic method to increase productivity with available genetic resources. After 150 years of research, Mendal's law revolutionized crop breeding. Conventional plant breeding is always key to developing unique featured plants with the following traits but inappropriate for the enhancement of genome and production value. For the betterment of crop quality and productivity, plant breeders utilized the combined knowledge of genetics that saves time and is also cost-effective (Ahmar et al. 2020).

However, conventional breeding methods have been used for thousands of years but still, they have some challenges.

3.1 Limited Genetic Diversity

Conventional breeding can only be done between plants or animals that can sexually mate with each other, limiting the introduction of various new traits (Ahmar et al. 2020).

3.2 Time-Consuming Approach

Development of new varieties through conventional breeding can take many years, often a decade, to achieve the desired results (Wanga et al. 2021).

3.3 Environmental Sensitivity

Traits developed through conventional breeding may not always be stable under different environmental conditions, making it challenging to breed plants that can withstand climate change (Wang et al. 2023).

3.4 Complex Traits

Breeding for traits controlled by multiple genes (quantitative traits) is more complex and less predictable compared to traits controlled by a single gene (qualitative traits). They are not easy to control through conventional breeding (Ahmar et al. 2020).

3.5 Disease and Pest Resistance

Developing resistance to diseases and pests is a continuous challenge, as pathogens and pests evolve (Wanga et al. 2021). Despite these challenges, conventional plant breeding remains a crucial tool in agriculture. Often, complemented by modern techniques such as genetic engineering and genomics to overcome some of its limitations.

4 Breeding Goals for Alfalfa

In the field of agriculture, to increase the productivity of the crop, farmers must focus on some breeding goals. In the case of alfalfa, resistance against stresses (biotic and abiotic) and pests, nutritional quality factors, yield quality of forage, and production of the hybrid crop are the most important agronomic goals. In breeding strategies, traditional breeding approaches are more time-consuming but they give maximum output. In the era of modernization of agriculture, we have to implement the combined approach of breeding for the production of alfalfa (Azad et al. 2023).

Pakistan, which is continuously facing economic fluctuations, needs to establish sustainable breeding goals for alfalfa production. Alfalfa ranked second among all other forage crops in Pakistan. The data was reported by Latif et al. (2023) in the year 2016–17: four main forage crops were reported as Berseem (*Trifolium alexandrinum*), Lucerne (*Medicago sativa*), Oat (*Avena sativa*), and Ryegrass (*Lolium* sp.), which provided 29,653 kg/ha. The specificity of this crop is that it can survive in extreme summers (53 °C) and winters (−24 °C), which makes it unique and versatile. The other thing that makes the alfalfa sustainable to the environment is its deep rooting system. The deep rooting system of alfalfa helps in the field rotation of crops, because it maintains soil nitrogen and promotes water infiltration in the soil. Therefore, it is necessary to maintain the breeding goals, quality of alfalfa, and its production for the betterment of an eco-sustainable environment (Fernandez et al. 2019).

Breeding strategies require some basic goals for alfalfa production, such as resistance against stresses (biotic and abiotic) and pests, nutritional quality factors yield quality of forage, and production of the hybrid crop is the most important agronomic goal.

4.1 Enhancement in Nutritional Value, Yield, and Resistance Against Stress

To find out the nutritional value and yield rate of alfalfa, we have to discuss them briefly.

4.1.1 Nutritional Value of Alfalfa

As we discussed earlier, alfalfa is a great source of food for livestock animals rich with proteins and lipids. The nutritional values of alfalfa mostly depend on its life cycle and reproductive stage. The development of chlorophyll content, flavonoids, proteins, phenolic content, and lignin content mostly decreases at its maturity stage as compared to other stages of the life cycle. The availability of degrading enzymes into cellulose is disturbed due to high lignification, which may interrupt the process of digestion in the animals. The use of conventional breeding and modern breeding techniques improves the lignin-lowering enzymes. The harvesting period has also influenced the nutritional quality of alfalfa (Azad et al. 2023).

4.1.2 Biomass Yield of Alfalfa

There are so many factors that are responsible for the negative influence on the yield of alfalfa. Cropping systems, crop durability, dry matter production efficiency, breeding cultivars, growth, drainage system, and fertilization are factors that are responsible for the biomass yield of alfalfa. The use of multi-omics in the field of agriculture may help in the productivity rate of alfalfa (Azad et al. 2023).

4.1.3 Resistance Against Stress

There are numerous stresses, including both biotic and abiotic factors, that severely affect alfalfa crops. These include salt, drought, cold, and heat stress, as well as pest attacks, all contributing to reduced growth and productivity. Azad et al. (2023) noted that approximately 197 alfalfa cultivars are sensitive to high temperatures (above 25 °C), while cold stress (below 10 °C) hinders growth and decreases photosynthetic content and yield. Moreover, heavy metal stress induces oxidative stress, potentially leading to cell death. To address these challenges and enhance alfalfa yield, nutritional value, and stress resistance, a combination of plant breeding, modern breeding, and multi-omics approaches (genomics, transcriptomics, proteomics, and metabolomics) are employed. The integration of these techniques aims to mitigate both external and internal changes in the forage.

5 Advances in Breeding Techniques

Breeding is a fundamental process that causes genetic variations in populations and ensures the continuation of species. Variations lead to adaptations and evolution that enable organisms to survive under changing environmental patterns. Progress in the field of research and technology revolutionizes the world with advancements in breeding techniques. In recent years, significant advancement has been seen in

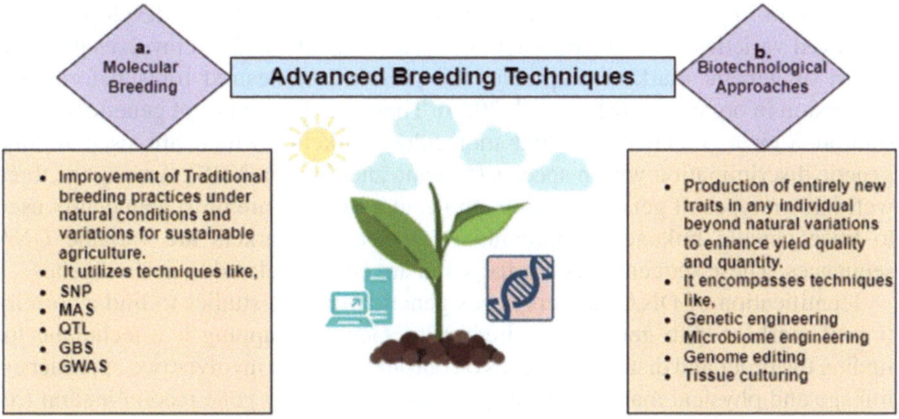

Fig. 1 Advanced breeding techniques. (**a**) *Molecular breeding*: It encompasses molecular biological tools for the improvement of desired traits under natural conditions by a combination of conventional breeding programs with advanced molecular approaches such as SNP (single nucleotide polymorphisms), MAS (marker-assisted selection), QTL (quantitative trait locus), GBS (genotyping-by-sequencing), GWAS (genome-wide association studies). (**b**) *Biotechnological approach*: A more advanced method than molecular breeding to produce novel traits to enhance crop yield under changing climatic conditions by employing techniques of genetic engineering, microbiome engineering, genome editing (CRISPR/Cas editing tool), and tissue culturing. These techniques can be used individually but their multiple combinations prove to be robust for sustainable agriculture to meet future challenges of food and climate

breeding techniques of fodder crops for livestock nutrition (Fig. 1). These advancements cover the following categories.

5.1 Molecular Breeding

It provides a cutting-edge approach for the utilization of molecular biology, genomics, and bioinformatics tools to improve the traits of plants and animals. In contrast to traditional time-consuming breeding approaches that involve observation of organisms over several generations, molecular breeding allows quick identification and selection of desired traits. The molecular breeding approach works at the DNA level to enable more precise breeding outcomes. Traditional breeding is an old practice with limited scope and is unable to cover many aspects of the production of new traits. While with a powerful set of tools (MAS and GS), molecular breeding revolutionized agriculture with novel traits, far beyond the scope of traditional breeding.

5.1.1 Marker-Assisted Selection

It is an approach of plant breeding, extensively utilized to improve crop yield, disease tolerance, and desirable traits. Researchers are finding valuable genes for the mitigation of drought tolerance, pathogens, and infections, aiming to enhance productivity (both quality and quantity). Marker-assisted selection facilitates to selection of organisms with genes of desired traits for faster development of improved varieties (Li and Brummer 2012; Aasim et al. 2022). It involves the utilization of genetic markers to predict the presence of desired traits before their expression in organisms (Shen et al. 2020). There are three types of genetic markers (morphological, biochemical, and molecular markers), which are used to find genetic discrimination within species (Sharma and Sharma 2018). Among the three well-known types of genetic markers, molecular markers are most abundantly used to trace genetic linkages of individuals. Molecular markers are variable DNA sequences. These sequences act as flags and are closely related to desired genes.

Identification of DNA markers needs genetic mapping studies to find a correlation of markers with genes of desired traits. Genetic mapping is a technique for finding the order and distance of genes on chromosomes. It involves two approaches: linkage and physical mapping. Linkage maps can identify gene recombination frequencies in terms of map units called centi-Morgan (cM) as mass functions. The process of linkage map development comprises three phases: establishing the map population, detection of polymorphisms, and analysis of genetic markers by utilizing computational tools (Map marker and Jion map, etc.). With the help of genetic maps, plant breeders identify and select those DNA markers that more closely related to genes of desirable traits. Markers are used to design primers to amplify specific genes of desired traits by polymerase chain reaction (PCR). The PCR product is used to screen a breeding population by using techniques of gel electrophoresis and DNA sequencing (Sharma and Sharma 2018). Based on screening, plants having alleles of desired traits could be selected for pollination. The selected plants transfer the desired traits to their offspring. Marker-assisted selection (MAS) can also extend to develop a new plant variety. For this purpose, breeders select plants of desired traits and linked markers by MAS over multiple generations.

This process concentrates genes of desired traits in resulting offspring. In this way, a new plant variety of enhanced characteristics is formed, which can be released for commercial use. The backcross marker-assisted procedure is broadly used in plant breeding programs (Sharma and Sharma 2018). Backcross involves a cross between offspring and recipient parent, which drags undesired genes along with genes of interest. Linkage drag may hinder plant improvement by utilizing unadapted germplasm due to the introduction of extra undesirable genes from a donor. MAS helps to overcome this challenge, as it targets specific genes of the desired trait like potato leafhopper is cured by using this approach since the late 1990s (Elden and Elgin 1987). Scientists have reported several molecular markers to fasten plant breeding programs with higher precision rates. De Mori and Cipriani (2023) described 44 markers that are commonly found in 15 fruit crops and are responsible for controlling various fruit traits. MAS has immense potential to improve traits of major fruit crops, as markers such as MDo.chr9.4 incorporate red skin color to apples, Cprfc1 for flesh color in peach, and LG3–13.146 for flesh color in cherry, papaya, and peach (De Mori and Cipriani 2023). MAS technique also incorporates the identification of molecular markers located on chromosome 11, which are associated with pathogen resistance in tomatoes (Orchard et al. 2023).

Marker-assisted selection (MAS) is a faster process than conventional breeding. Unlike conventional breeding, it eliminates the need to wait for multiple generations to observe a physical trait. MAS also allows early detection of traits (before physical expression), which otherwise cannot be observed directly. DNA-based selection aided with molecular markers offers a more targeted approach that helps to screen specific genetic variations associated with desired traits (Shen et al. 2020). Hence, it offers a more reliable approach without influences of environmental factors to develop more productive and resilient crops. However, molecular breeding with MAS has certain limitations, such as a large budget required to perform marker tests and analysis. Availability and reliability of DNA markers are also concerns, as markers of many traits are not identified yet. Linkage problems also exist, because markers may get separate from desired genes due to recombination events during reproduction (Tharun et al. 2024).

5.1.2 Genomic Selection

It is an advanced technique that is used to identify the potential of an individual for specific traits by analyzing its whole genome. It helps to select complex traits with influences of multiple genes, hence providing a more complex approach than marker-assisted selection (MAS).

Genomic selection involves establishing a representative crop population with a wide range of genetic diversity (Montesinos-Lopez et al. 2022). Under ideal conditions, plant individuals of the population possess known phenotypic data of desirable traits. These traits may be related to stress tolerance, yield, or any other phenotypic characteristics. The entire population is used to collect samples of representative plants. The sample may be taken from leaves, stems, roots, or floral parts, depending on the trait under study. After DNA extraction from all tissue samples, it is subjected to advanced DNA sequencing techniques for genotyping of plant individuals. This genotyping generates vast genomic data of the entire population to identify SNPs (single nucleotide polymorphisms). Phenotyping studies of representative crop populations should be made by recording the desired traits (yield, nutritional value, or disease tolerance) of all individuals. Phenotypic data possess information about desired traits, while genotypic data encompasses aspects of SNPs. These genotypic and phenotypic data are analyzed by statistical tools of algorithms to make a correlation between specific SNPs and the expression of desired traits (Merrick et al. 2022). This correlation helps to identify possible ways in which genetic variations (due to SNPs) affect the phenotypic expression of desired traits. The statistical models of correlation are entirely based on genotypic data and can be used to predict the genetic potential of untested plant individuals of the same species (Montesinos-Lopez et al. 2022). Based on genomic data, breeders select plant individuals of desired traits for pollination to produce new offspring. The offspring are further analyzed to a genomic and phenotypic extent for confirmation of predicted performance. Over multiple breeding cycles, desired genes concentrate to establish a new variety of superior phenotypic traits. However, genotypes can interact with the environment and produce different phenotypes, which hinder breeding program efficiencies. Moreover, the identification of association among genotypes with varied environments is a vital but time-consuming and high-cost process. A possible approach to resolve the issue is multi-environment models facilitated by pedigree and genetic markers (Merrick et al. 2022). Burgueno et al. (2012) used REML (restricted maximum likelihood) and dispersion parameter software to explore the expression of gene patterns in wheat under different environmental conditions. Marker-based models could predict genotype efficiency across environments and reduce extensive testing of conventional breeding (Hawkins and Yu 2018; Burgueno et al. 2012). Bhering et al. (2015) validated three methods (Bayesian-LASSO, RRBLUP, and GBLUP) of genomic selection to achieve organisms with superior traits. Among all three methods, RRBLUP is found to be more robust and cost-effective with a high precision rate for breeding programs. Genomic selection provides predictions about desired traits depending upon an individual's whole genome but its integration with other methods yields robust and precise results. One of the best approaches to genomic selection is the genomic estimated breeding value (GEBV) to find broad-spectrum genetic variability of individuals by considering all markers (Jazayeri and Torres 2017).

Genomic selection simultaneously provides a wide range of accurate genomic data about the entire population and provides insights into the influences of multiple genes on complex traits (Shen et al. 2020). It also conserves time and cost by eliminating the need for marker assays, which makes it a more advanced approach to molecular breeding over marker-assisted selection (MAS) (De Mori and Cipriani 2023). Genomic selection has the potential to increase production rates by reducing plant breeding cycles (Li et al. 2015) and by providing estimated breeding values (Hawkins and Yu 2018). However, limitations of genomic selection are also a concern, as it needs high expertise to establish a representative population and to develop statistical models. Genomes of some less researched plant species are not sequenced completely, which affects the reliability of data about desired traits. However, due to next-generation sequencing (NGS) and high-throughput technologies (HTT), genomic selection is evolving rapidly to overcome all these limitations in cost-effective manner and to provide sustainable solutions for future agriculture.

5.2 Biotechnological Approaches

Biotechnology is the unique combination of biology, engineering, and other sciences that utilize technology to manipulate living organisms for the production of commercially valuable products with a significant impact on human welfare. Biotechnological approaches in the form of traditional breeding practices have been in use for centuries to improve crops and livestock. However, advances in this discipline have taken place in recent years with the emergence of innovative techniques including genetic engineering, marker-assisted selection, omics technologies, and microbiome engineering (Ali et al. 2024). These techniques cover a wide range from agriculture to the human health care system by developing genetically modified organisms (GMOs), new medicines, gene therapies, bioremediation, and production of biofertilizers. Advances in biotechnological approaches are revolutionizing crop improvement practices in the following ways.

5.2.1 Improved Gene Discovery

It is a multifaceted approach with a combination of DNA sequencing, bioinformatics, and functional genomics. These combined approaches encompass high-throughput DNA sequencing, variant calling, annotations, transcriptomics analysis, genome-wide association studies (GWAS), and functional validation to discover new genes in crops with novel characteristics. Illumina is a widely used high-throughput DNA sequencing technique that efficiently produces massive amounts of DNA sequences from crop samples (Li and Durbin 2009). Variations exist in DNA sequences among varieties of a certain crop. These variations are often called SNPs (single nucleotide polymorphism) or Indels (insertion or deletions), which can act as potential markers for desired genes. These SNPs and Indels can be traced out by employing bioinformatics tools. Bioinformatics tools can compare identified variations with existing gene databases to predict their possible functions. Transcriptomics analysis provides an approach to analyze gene transcripts (mRNA molecules) in different tissues of a specific crop under controlled conditions such as stress (Satya et al. 2022). By comparing expression patterns of transcripts, genes associated with specific traits (stress) can be identified. Li and Durbin (2009) analyzed the transcriptome of alfalfa by using an RNA sequencing model to detect SNPs and Indels with the aim of yield improvement. Genome-wide association studies provide a statistical approach to create a link between genomic and phenotypic variations such as yield and stress tolerance in crop populations (Wu et al. 2022). This association helps to identify specific SNPs, which paved the way to pinpoint genes responsible for desired traits (Shen et al. 2020). Upon detection of the desired gene for a specific trait, it further subjected to functional validation by gene expression manipulation and reverse genetics. This combined approach enhances the reliability of the target gene and its association with the desired trait. Several GWAS models are utilized to identify gene loci, among which HapFM is the most efficient model for gene discovery (Wu et al. 2022).

5.2.2 Genome Editing with CRISPR

It is an advanced technique in genetic engineering for gene manipulation in living organisms including crops. CRISPR/Cas system is a defense system in bacteria against viruses and other foreign intruders. It works in conjunction with various Cas enzymes. The most commonly used Cas enzyme system is CRISPR/Cas9, which has endonuclease activity to cut double-stranded DNA and hence acts as molecular scissors in the genome editing process (Prado et al. 2024). CRISPR/Cas9 cannot work independently; instead, it needs instructions for the precise cutting of DNA molecules. These instructions are provided by a specialized RNA sequence known as guide RNA or simply gRNA (Fig. 2). Scientists design this gRNA with sequences complementary to target DNA. CRISPR/Cas9 and gRNA complex get entry into the host cell via a suitable delivery system, which may be nanoparticles, electroporation, bacteria, or viral vectors. After entering inside nucleus, gRNA binds to the

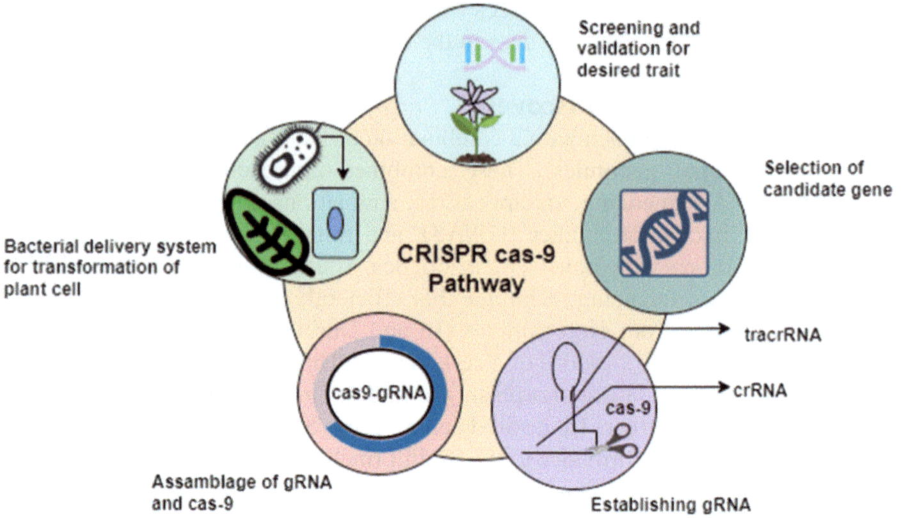

Fig. 2 CRISPR/Cas9 pathway. The pathway involves the selection of candidate genes to construct gRNA (which is established by a combination of tracrRNA and crRNA). The gRNA has a nucleotide sequence complementary to the target DNA, which in conjunction with Cas9 (sgRNA-Cas9) entered plant cells via the appropriate vector. Being complementary, gRNA precisely binds at the target DNA of the host plant cell, and the enzymatic activity of Cas-9 cleaves the host double-stranded DNA at the target site. In this way, part of the candidate gene could be deleted or inserted at the cleaved site of the target DNA to produce GM plants with desired traits. GM plant is further cloned to establish a field trial population for screening and validation of desired traits

complementary region of target DNA and the CRISPR enzyme utilizes its endonuclease activity to cut double-stranded DNA at a specific target point (Prado et al. 2024). In this way, nucleotide sequences related to desired characteristics can be inserted or deleted from the target DNA. After genome editing, PCR and sequencing techniques are used to confirm the success and efficiency of desired modifications on target DNA.

Genome editing with CRISPR/Cas9 allows modifications and development of crops with novel traits (Shen et al. 2020). Manipulation of phytoene desaturase (PDS) and PALM1 genes are well-studied examples of *Agrobacterium tumefaciens*-based CRISPR/Cas9 editing. PDS gene expression produces the PDS enzyme, which involves the carotenoid pathway in plants. PDS gene mutations are responsible for albinism, dwarfism lower reproduction. CRISPR/Cas9-mediated editing of the PDS gene cured albinism and dwarfism in alfalfa plants (Lyu 2020). Prado et al. (2024) reported that only 32 GM cultivars get approval for cultivation at the commercial level in 45 countries across the globe. CRISPR technology offers valuable insights for sustainable agriculture of the future through gene modifications of valuable crops, especially perennials. Although it is a modest technique of genetic engineering, it needs high expertise and careful designing to avoid unintended cuts on the genome due to similarity with target sequences. As ethical concerns exist regarding genome editing, CRISPR-based crops have not gained popularity and are restricted to the cultivation of tomato, soybean, canola, camelina, and maize at the commercial level (Prado et al. 2024).

5.2.3 Engineering for Stress Tolerance

It is the era of climatic changes, leading to harsh environmental conditions with extreme temperatures, which accelerates the processes of drought and salinity. Most crops are susceptible to these altered environmental conditions, which act as stressors for plants and are a major cause of yield and economic losses. There is a need to equip plants with novel characteristics to handle worsening environmental conditions. Engineering of crop genomes with stress-tolerant genes is a feasible approach under these circumstances. The major task for crop engineering is the identification of the target gene, most compatible with encountered stress and plant physiology (Wu et al. 2022). Several approaches including germplasm studies, gene expression profiling, and comparative genomics are used for the identification of target genes. Target genes responsible for stress management can be further manipulated to enhance stress tolerance ability by employing techniques such as modifications of regulatory region, overexpression, and genome editing by CRISPR (Prado et al. 2024). In this way, new crops with improved stress (heat, cold, drought salinity, etc.) tolerant ability can be produced to enhance yield under specific environmental challenges (Shen et al. 2020). However, the multigenic nature of stress tolerance and some unintended effects of GM plants on other crops may pose a threat to natural balance. So, manipulated crops should pass through functional validation and field testing before their integration into breeding programs at commercial levels. Regulatory guidelines should be undertaken strictly for the use of engineered crops to avoid potential risks.

5.2.4 Symbiotic Microbiome Engineering

In recent years, engineering of symbiotic microbes has grabbed attention to improve nutrient availability and uptake by plants for better health profiles (Fig. 3). NGS (next-generation sequencing) is a useful technique to identify the composition and function of symbiotic bacteria, fungi, and other microbes in the vicinity of plants. One example is *Sinorhizobium meliloti*, a well-known symbiotic bacterium associated with the genus *Medicago* plants. This symbiotic bacterium has immense potential to improve alfalfa yield (Ticha et al. 2020). *Bacillus* and *Pseudomonas* are mostly found in plant rhizosphere and have broader potential to reduce harmful microbes for plants. These microbes assist plants by incorporating nutrients or by modifying host physiology via chemical signaling. The microbial communities produce vast chemicals with antagonistic functions such as subtilin, pyoluteorin, and surfactin (Ticha et al. 2020). Metagenomics techniques are used to analyze genomes of microbiota for the identification of candidate genes of stress tolerance, pathogen resistance, or nutrient acquisition. The next step involves the isolation and culturing of microbes under controlled conditions to find their individual effects on plants (Gao et al. 2024). Microbes with beneficial effects for stress tolerance are supplied to plants by various strategies such as seed inoculation, foliar applications, or soil amendments including prebiotics and selective agricultural practices. These

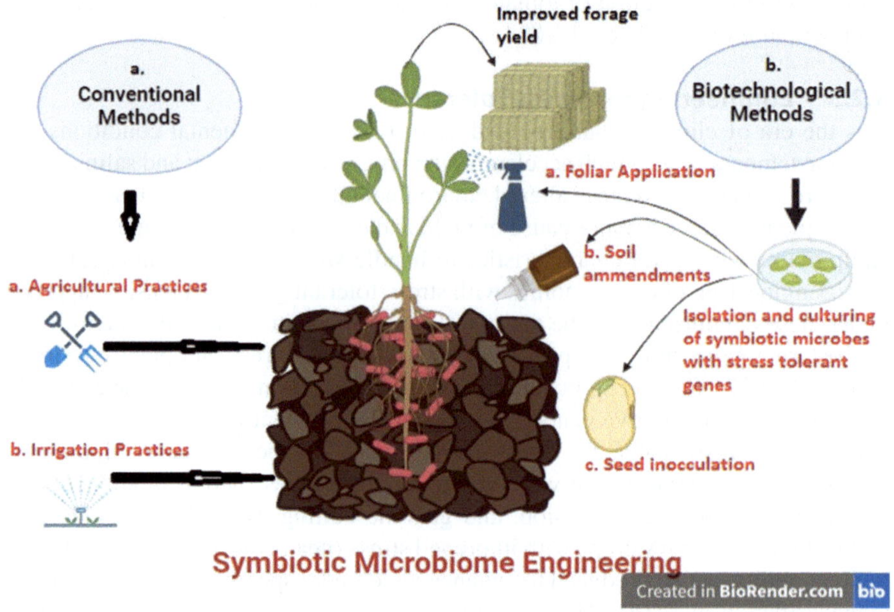

Fig. 3 Symbiotic microbiome engineering is a technique to improve nutrient uptake of plants by utilizing symbiotic microbes to enhance crop yield. It includes (**a**) *Conventional methods* such as conventional agricultural and irrigation practices for improvement of symbiotic microbial status and soil health and (**b**) *Biotechnological methods*, which encompass the isolation of symbiotic microbes from soil and then the culturing of microbes containing genes of interest for desired traits (stress-tolerant genes) on culture media. After establishing pure cultures, symbiotic microbes containing candidate genes subjected to plant inoculation either by foliar application, soil amendments, or seed inoculation to introduce candidate genes for stress mitigation and crop improvement

microbes produce certain chemicals that help to regulate plant hormones and protective molecules to tolerate stress conditions such as drought (Gao et al. 2024), salinity, and heat stress. Microbiome engineering is an evolving technique to increase crop yield under stressful conditions (Ticha et al. 2020). Despite benefits, there are concerns about microbiome complexity and the long-term stability of introduced microbes. Moreover, introduced microbes may compete with existing microbiomes to pose unintended effects on soil health and the ecosystem. These issues need to be addressed by proper monetizing, optimization, and regulatory guidelines for the use of specific microbes in agriculture. By utilizing the power of beneficial microbes, a resilient and productive agricultural system can be established to meet the food needs of future generations.

5.2.5 Enhancing Forage Quality

To eliminate the need for additional food supplements, biotechnological approaches in combination with traditional breeding are utilized to enhance forage quality by improving the digestibility and nutritional value of fodder crops. Species with better traits can be selected either by traditional breeding or marker-assisted selection (MAS) (Shen et al. 2020). Traditional practices, to improve forage quality, include proper maintenance of crops by irrigation, fertilization, harvesting at optimal stage of maturity, and proper hay and silage preservation. Biotechnological approaches include: (1) *genetic engineering* of rumen microbes in livestock to enhance digestibility of cellulose and production of proteins (Brooker et al. 1989); (2) *biofortification* to enhance vitamins, minerals, or proteins in fodder crops by genetic modifications (Talla et al. 2023); and (3) *microbiome engineering* to enhance the availability and uptake of nutrients by plants. Integration of these technologies provides a holistic approach for significant improvement of forage quality to enhance animal health, productivity, and a sustainable livestock production system. However, certain unintended risks of gene manipulations are associated with biotechnological approaches that can be addressed or minimized by strict regulatory measures. Public awareness, safety concerns, and economic feasibility should be considered to get more benefits from advances in biotechnological approaches.

5.2.6 Advanced Bio-imaging Techniques

Advanced bio-imaging techniques are profoundly used to observe metabolic pathways in plants. Plants need macro- and micronutrients from the soil for growth and development. Soil deprivation of these nutrients may lead to stunted growth and reduced plant yield. Foliar application is an alternative and efficient source to provide vital nutrients to plants under nutrient-deficient conditions. However, it is difficult to assess the pathway adopted by nutrients as a result of foliar applications and their quantity available for plant use. Advances in bio-imaging techniques (such as LA-ICP Mass spectrometry) made it possible not only to detect the amount used by plants but also the underlying pathways adopted by these nutrients to enter plant cells (Arsic et al. 2020). The omics techniques integrated with mass spectrometry imaging (MSI) provide a robust approach to analyzing the actual operating status of crops in the presence of stressors (both biotic and abiotic).

Plants have the potential to activate defensive pathways under stressful conditions such as the production of certain hormones and metabolites to mitigate salt, water, temperature, and pathogen attacks by altering cellular activities and membrane potentials, which can be examined by MSI (Lu et al. 2024). In the last decade, biosensors have become popular among farmers due to their smaller size and precision rate. A biosensor is a combination of biomolecules (enzymes, antibodies, and nucleic acids) and detectors to identify certain chemical substances. Biosensors are used to detect metabolic activities, signal recognition, and manufacturing of nanoparticles. These are also utilized in analytical tools to detect and read signals in the form of images, which provides a robust way to understand underlying cellular mechanisms (McLamore et al. 2021). The latest microscopic techniques (LSFM and airy can microscopy) help study the molecular basis and developmental stages of plants by providing data about genetically engineered lines of valuable crops, which contribute to biotechnological advances in crop breeding programs (Ticha et al. 2020). Table 1 shows a comparison between molecular and biotechnological approaches for alfalfa breeding.

Table 1 Comparison between molecular breeding and biotechnological approaches

Attributes	Molecular breeding	Biotechnological approaches	Reference
Root of variation	Natural conditions	Any condition	Singh et al. (2020)
Spectrum	Narrow (limited to natural variations)	Broad (covering any individual)	Pérez et al. (2013)
Genetic tools	GWAS QTL MAS	Genome editing Genetic engineering Tissue culturing	Susmitha et al. (2023)
Speed	Slow	Rapid	Pérez et al. (2013)
Accuracy	Comparatively low	High	Pérez et al. (2013)
Regulative obstacles	Low flexibility	Highly flexible	Singh et al. (2020)

6 Genomic Tools and Techniques Utilized in Alfalfa Research

Alfalfa (*Medicago sativa* L.) is a profound fodder crop cultivated in temperate and subtropical areas across the globe but its genome is not fully sequenced. Limited information about reference genomes is a barrier in alfalfa breeding programs. Alfalfa is an autotetraploid (four copies of each chromosome) species, which needs specialized techniques and software for analysis to address the complexity existing in chromosomes. Chromosomal complexity is a hindrance in historical progress and improvement of genomic traits for advanced breeding programs in alfalfa. Several genomic tools and techniques can be used to research the alfalfa plant. Selection of the right technique is based on research goals, budget, and available resources (Fig. 4).

6.1 Single Nucleotide Polymorphism (SNP)

Two major approaches can be used to identify SNPs in alfalfa: (1) *Re-sequencing with a reference genome*: it needs a pre-sequence genome of alfalfa species that acts as a reference to analyze target DNA for re-sequencing. High-throughput sequencing techniques (RNA-seq, Illumina, and PacBio sequencing) are used to create short reads of target alfalfa DNA (Shen et al. 2020). These reads can be compared with reference alfalfa genome by bioinformatics tools to identify mismatches (SNPs). (2) *Genotyping-by-sequencing (GBS)*: it is commonly used in cases when reference genome for individuals is not available. Restriction enzymes can be used for selective digestion of the genome followed by desired ligation and sequencing by high-throughput techniques (Li et al. 2015). Sequences of different individuals can be analyzed by bioinformatics tools to identify SNPs. Sanger sequencing or PCR techniques can be used for further validation of SNPs. After validation, several

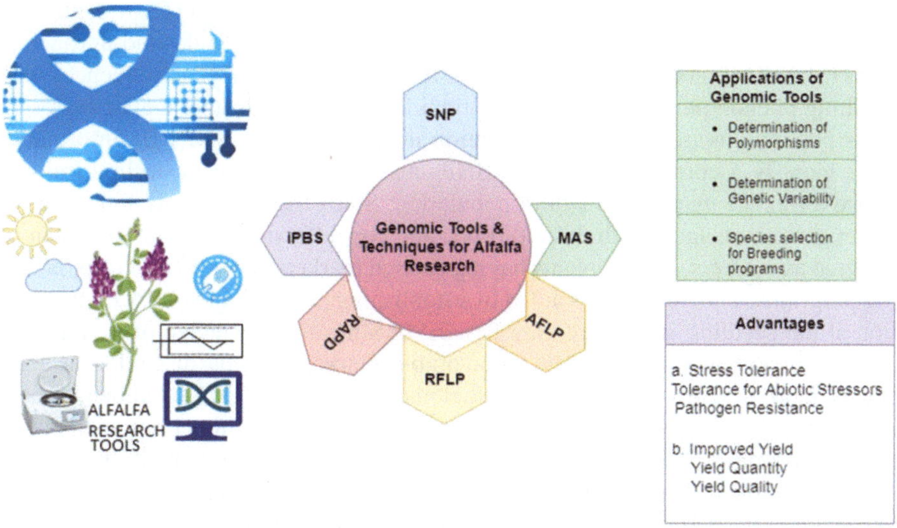

Fig. 4 Genomic tools for alfalfa research: several modern genomic tools and techniques individually or in combination are used for alfalfa (*Medicago sativa* L.) breeding programs. These techniques include SNP (single nucleotide polymorphisms), MAS (marker-assisted selection), AFLP (amplified fragment length polymorphism), RFLP (restriction fragment length polymorphism), RAPD (random amplified polymorphic DNA), iPBS (inter primer binding site), which helps to detect polymorphisms, genomic variations, pedigree analysis, and evolutionary links among alfalfa genotypes by employing bioinformatics tools. Data validated by bioinformatics tools is used to select alfalfa plants with better traits of stress tolerance (heat, cold, drought, salinity, and pathogens) to improve forage yield (quality and quantity)

approaches (SNP, KASPar, or Illumina Infinium SNP arrays) can be used for SNP genotyping. SNP arrays can perform genotyping of several samples simultaneously. This array utilizes high-throughput techniques and predesigned probes for SNP genotyping. KASPar assays utilize primer-based PCR amplification of SNP alleles. Illumina Infinium SNP arrays are applicable commercially for rapid genotyping by utilizing preselected SNPs for alfalfa. Han et al. (2014) utilized Illumina arrays to validate aluminum tolerance and regrowth capability on a group of alfalfa plants. This array significantly detected more than 3000 SNPs to create a detailed genetic map of alfalfa. Bioinformatics tools can also generate links between SNPs and desired traits to trace out genetic diversity and genomic prediction accuracy among alfalfa plants (Li et al. 2015). Lyu (2020) utilized Canu and ALLHiC algorithmic programs to create a reference genome of alfalfa containing 32 scaffolds without overlaps.

6.2 Marker-Assisted Selection (MAS)

MAS is a marker-based approach to utilize DNA sequences (SNPs) for the selection of plants with desired traits. The technique involves the creation of linkage maps by utilizing phenotypically well-characterized alfalfa population. QTL (quantitative trait locus) technique can identify chromosomal loci linked to desired agronomic traits (Shen et al. 2020; Altaf et al. 2024). It helps to associate gene sequences with resulting phenotypes (De Mori and Cipriani 2023). Chromosomal loci of representative populations are used for identification and genotyping of SNPs. Bioinformatics tools identify a correlation between SNPs and the expression of desired traits to create a linkage map of the alfalfa population. SNPs that are closely linked to the target gene, associated with desired traits, and cost-effective for genotyping can be selected as markers. Alfalfa plants having desired markers can be selected for breeding programs. This selection is aided by establishing parental lines or by early selection at the seedling stage. Seedling-based MAS has the advantage of saving time and resources, as seedlings with unfavorable markers can be discarded at early stages. In alfalfa, broad chromosomal regions are located by QTLs. Efficient breeding programs need the identification of underlying genes within chromosomal loci (Li and Brummer 2012). QTLs are associated with certain traits in alfalfa, such as aerial morphology, nitrogen nutrition, and flowering time (Julier et al. 2007). RCT1 gene cloned from *M. truncatula* could potentially be used to cure anthracnose (fungal disease) in alfalfa (Yang et al. 2008). After proper validation and improvements, if needed, established alfalfa varieties with desired traits can be introduced for commercial breeding programs. The creation of a valid linkage map is crucial to pinpointing the location of genes (QTL mapping) and isolating genes by map-based cloning. The flowering time of plants is controlled by genes and environment and it can be predicted by establishing linkage maps. A linkage map is developed to identify the flowering time in the alfalfa parental line by SNP markers, QTLs, and RNA-seq analysis (Scossa et al. 2021). This linkage map will act as a reference for future breeding plans of the alfalfa crop. SRAP (sequence-related amplified polymorphism) acts as a marker tool to analyze the DNA of various organisms. SRAP functioning relies upon PCR amplification of DNA by utilizing two primers. The banding pattern of amplified DNA determines polymorphisms. These markers are used to determine genetic variability, the selection of species for breeding programs, and the production of new cultivars with favorable features (Rhouma et al. 2017). Although it is a powerful technique, comprehensive understanding of target genes, desired traits, and continuous refinement of the breeding program is needed for its successful implication at the commercial level.

6.3 Amplified Fragment Length Polymorphism (AFLP)

AFLP is used to identify genetic diversity within a population. It allows detection of a large number of polymorphisms from a relatively small amount of DNA. Two restriction enzymes are used for the digestion of extracted DNA of the alfalfa population. One restriction enzyme can act as a frequent cutter, and it cuts DNA fragments of 4–6 nucleotide bases. While other restriction enzyme acts as a rare cutter to cut less frequent and relatively longer fragments of 6–8 nucleotides. The whole process digests DNA into fragments of variable lengths. Then ligation is processed by adding an adapter sequence of a few nucleotides in digested DNA fragments. Primer-based amplification is used to amplify adapter and selective nucleotide sequences to generate a comprehensive fingerprint. Bands with varied lengths of DNA fragments are separated by using polyacrylamide gel electrophoresis. UV-illuminator is used to visualize banding patterns of DNA. Banding patterns represent the presence or absence of certain bands. Comparison of the presence or absence of different bands among population individuals can be made by bioinformatics software to indicate polymorphisms (Wang et al. 2012). Bioinformatics tools can also construct dendrograms to depict relationships between wild and cultivated alfalfa populations. However, AFLP only depicts the presence or absence of DNA fragments instead of underlying alleles. Minor variations in reaction conditions also affect the efficacy of the process. A representative population is needed to interpret banding patterns. Despite limitations, AFLP is a good choice to identify parental lines and genetic diversity for breeding programs, to study germplasm collections, and to identify genetic relationships among wild and cultivated plants. Abdel-Tawab et al. (2011) adopted the AFLP technique to recognize genes for salt stress tolerance in alfalfa cultivars. Their study comprised cultivating salinity-resistant and salinity-sensitive alfalfa plants under saline conditions. By comparing the genetic makeup of both categories, they find salinity-resistant genes among alfalfa species. The findings paved the way to develop resistant alfalfa genotypes with improved yield under challenging saline ecosystems.

6.4 Restriction Fragment Length Polymorphism (RFLP)

RFLP can be used to detect polymorphisms by exploiting restriction enzymes. The selection of restriction enzyme is based on the type of gene and expected level of polymorphism. Restriction enzymes digest genomic DNA into short DNA sequences at specific recognition sites. Agarose gel electrophoresis is used to separate DNA bands based on fragment lengths, as short sequences move faster across the gel. By employing the Southern blotting technique, DNA fragments are transferred on a nylon membrane. A labeled probe (specialized short single-stranded DNA fragment) is used to hybridize target DNA sequence (Labombarda et al. 2000). Probe binds to specific target DNA sequence due to the presence of complimentary nucleotides (Pupilli et al. 2000). Hybridized labeled probe can be recognized by autoradiography or chemiluminescence techniques (Jiang and Pu 2021). The location of the probe pinpoints the position of the target gene on the DNA fragment. Analysis of the banding pattern determines variations in target genes among individuals of the population (Pupilli et al. 2000). RFLP is a reliable technique due to the specificity of labeled probes. This technique is applicable to identify SNPs, alleles associated with specific traits (disease resistance and stress tolerance), and genetic relationships of different alfalfa plants or germplasm lines (Williams et al. 1993). Before the advancement in genomic approaches, RFLP was commonly used for genetic mapping, disease diagnosis, and genetic diversity. However, processes of Southern blotting, probe labeling, and hybridization make RFLP a time-consuming and labor-intensive approach. Moreover, RFLP has a limited scope of analysis, as it provides access to detect polymorphisms associated only with the probed region. To overcome these limitations, RFLP is mostly replaced by other efficient methods such as SNP genotyping.

6.5 Random Amplified Polymorphic DNA (RAPD)

RAPD is a PCR-based technique to detect genetic diversity among alfalfa plants. It can detect a large number of polymorphisms with minimal prior genomic knowledge about alfalfa (Taški-Ajduković et al. 2014). The technique involves PCR-based amplification of isolated DNA by using a single random primer (6–10 nucleotide-long sequence). Random primer is not gene-specific; instead, it is designed to bind randomly on the genome to produce DNA fragments of different sizes. Agarose gel electrophoresis is used to separate DNA fragments in the form of bands. Bands can be visualized by UV-transilluminator and further analyzed by bioinformatics tools to identify genetic polymorphisms between different alfalfa samples (Ertus et al. 2014). It is a simple and inexpensive technique to access genetic diversity among alfalfa germplasm collections, e.g., *M. sativa*, *M. arborea*, and *M. falcate* (Komal et al. 2014). Ahsyee et al. (2013) utilized the RAPD approach to find the genetic relationship among nine alfalfa varieties. RAPD helps to determine hereditary relationships among alfalfa ecotypes, which aids in selecting favorable genotypes for sustainable cultivation practices (Ertus et al. 2014). Like RFLP, RAPD can only identify the presence or absence of DNA fragments instead of underlying alleles, and the efficiency of the process is affected by minor changes in reaction conditions (Ertus et al. 2014). Random amplification of genomic DNA may also lead to complex banding patterns, which may be difficult to interpret for the identification of polymorphisms (Ahsyee et al. 2013). Although RAPD provides an easy approach to analyze initial genetic diversity in alfalfa research, now it is largely replaced by advanced techniques of SNP genotyping and AFLP.

6.6 iPBS Retrotransposon Markers

iPBS (inter primer binding site) is a retrotransposon marker-based technique that provides a cost-effective approach for organisms' initial genetic diversity assessment. Retrotransposons are self-replicating genetic elements that can insert their copies at new locations of the genome. Retrotransposons are extensively found in the genomes of plants and contribute to genetic diversity among them (Alzohairy et al. 2014; Uçer et al. 2024). Retrotransposons have PBS (primer binding site) region to which the reverse transcriptase enzyme binds and initiates its replication. iPBS retrotransposon system utilizes two primers (18–20 nucleotide sequence) that bind at PBS sequences of retrotransposons. Bioinformatics tools or PBS sequences of other plant species can be used to design primers for alfalfa retrotransposons. Designed iPBS primers amplify retrotransposons between PBS sites during PCR amplification process. The agarose gel electrophoresis technique is used to separate bands of amplified retrotransposons depending upon size and distance traveled on the gel (Eren et al. 2023). Visual and statistical analysis by UV-transilluminator and bioinformatics tools, respectively, of banding patterns reflects the diversity of retrotransposon insertions within the genome of the alfalfa population. Bioinformatics software can score presence or absence of bands in plant samples, determine genetic diversity by using genetic diversity indices (Nei's genetic distance or Shannon's diversity index), and construct dendrograms to find genetic relationships among alfalfa plants (Rhouma et al. 2017). The conserved nature of PBS sequence across plant genomes broadens the applicability of iPBS technique to various plant species. Readily available primers and PCR techniques make it cost-effective and reliable approach to identifying polymorphisms. iPBS banding patterns can be used to identify subpopulations within large germplasm collections of alfalfa (Eren et al. 2023). Due to the specificity of iPBS markers, this system can be integrated into plant breeding programs as marker-assisted selection. iPBS marker system has certain limitations, as it only focuses on the retrotransposon region of the genome, so it provides limited information instead of actual genes. Like RAPD and AFLP, iPBS system only detects the presence or absence of bands and does not provide any data about alleles.

7 Genomic Selection (GS) Strategies in Alfalfa Improvement

7.1 Introduction

Genomic selection (GS) has emerged as a revolutionary breeding tool in plant breeding, offering significant potential for enhancing crop improvement programs. In alfalfa (*Medicago sativa*), a perennial forage crop, GS provides an opportunity to accelerate the breeding cycle and improve traits such as yield, quality, and stress tolerance (Annicchiarico et al. 2021; Hawkins and Yu 2018). The key GS strategies are discussed in the following sections.

7.1.1 Training Phase

In genomic selection, the training phase is a critical phase that plays an important role in accurate genomic predictions. In this phase, we develop models by using training populations to predict which alfalfa plant will have the best traits (Paolo et al. 2022; Annicchiarico et al. 2015).

1. *Training population design*: To ensure the best traits in alfalfa, the training population must be phenotypically diverse and genetically varied. In alfalfa, this often involves selecting a wide range of germplasm lines and breeding populations.
2. *Genotyping*: High-density genotyping technologies, such as SNP arrays or genotyping-by-sequencing (GBS), are used to obtain genome-wide marker data from

the training population. This data forms the basis of the genomic prediction models.
3. *Phenotyping*: Precise and consistent phenotyping of key traits (e.g., yield, forage quality, stress tolerance) is conducted. High-throughput phenotyping technologies, including remote sensing and digital imaging, are increasingly being used to enhance accuracy and efficiency.
4. *Model training*: Statistical models (e.g., GBLUP, Bayesian methods) are developed to associate genotype data with phenotypic traits. These models are then validated and refined to ensure high predictive accuracy.

7.1.2 Breeding Phase

The breeding phase applies the genomic prediction models developed in the training phase to select and advance individuals with the desired traits (Lebedev et al. 2020; Resende Jr et al. 2012).

1. *Selection of breeding candidates*: Using genomic estimated breeding values (GEBVs), individuals with superior genetic potential for target traits are selected. This enables early and accurate selection without extensive phenotypic trials.
2. *Crossing and population development*: Selected individuals are crossed to develop new breeding populations. Genomic selection allows for the identification of optimal parent combinations, enhancing genetic gains.
3. *Validation and field trials*: The performance of selected lines is validated through multi-environment field trials to assess their stability and adaptability. This step is crucial to account for genotype-by-environment interactions.
4. *Cycle advancement*: The breeding cycle is accelerated by reducing the time required for phenotypic evaluation. This leads to faster release of improved alfalfa varieties with enhanced traits.

7.2 Genome Editing Technologies

Genome editing technologies, such as CRISPR/Cas9, are transforming plant breeding by enabling precise modifications to the genome (Gao 2021; Zhang et al. 2018).

1. *CRISPR/Cas9*: This technology allows for targeted edits at specific genomic loci, facilitating the introduction of desirable traits or the removal of deleterious alleles. In alfalfa, CRISPR/Cas9 has been used to improve traits such as disease resistance and stress tolerance.
2. *Base editing and prime editing*: These advanced genome editing techniques offer even greater precision, enabling the modification of single nucleotides or small DNA segments without introducing double-strand breaks.
3. *Regulatory considerations*: The adoption of genome editing in alfalfa breeding must navigate regulatory frameworks to ensure safety and acceptance. This includes compliance with biosafety regulations and addressing public concerns.

7.3 Integration of Genomic Data

Integrating genomic data with phenotypic and environmental data enhances the accuracy and utility of genomic selection (Fang et al. 2024; Scossa et al. 2021; Crossa et al. 2016).

1. *Multi-omics approaches*: Combining genomics with transcriptomics, proteomics, and metabolomics provides a comprehensive understanding of the genetic basis of traits. This holistic approach can improve the predictive power of GS models.
2. *Environmental data*: Incorporating environmental variables (e.g., soil type, climate conditions) into GS models helps to account for genotype-by-environment interactions, improving the robustness of predictions.
3. *Data management*: Effective data management systems are crucial for handling large datasets. This includes databases for storing and retrieving genomic, phenotypic, and environmental data.

7.4 Bioinformatics Tools

Bioinformatics tools are essential for analyzing and interpreting genomic data in alfalfa breeding (Chen et al. 2020; Perez and de los Campos 2014).

1. *Genomic prediction software*: Tools such as GBLUP, BayesB, and rrBLUP are commonly used for developing and validating genomic prediction models. These tools handle large datasets and complex trait architectures.
2. *Sequence analysis*: Software for sequence alignment (e.g., BWA, Bowtie), variant calling (e.g., GATK, SAMtools), and genome assembly (e.g., SPAdes, Canu) are critical for processing high-throughput sequencing data.
3. *Data integration platforms*: Platforms such as Galaxy, Taverna, and iPlant Collaborative provide integrated environments for combining and analyzing multi-omics data, facilitating comprehensive genomic studies.

8 Challenges and Future Directions in Alfalfa Breeding and Genomics

8.1 Challenges

Despite the advancements in genomic selection and breeding, several challenges remain in alfalfa improvement (Yang et al. 2022).

1. *High costs and infrastructure needs*: Implementing genomic selection requires significant investment in genotyping and phenotyping technologies. Small breeding programs may struggle with these costs.

2. *Phenotyping bottlenecks*: While genotyping has become faster and cheaper, phenotyping (measuring plant traits) remains labor-intensive and expensive. Accurate and high-throughput phenotyping is crucial for building reliable prediction models.
3. *Genotype-by-environment interactions*: Alfalfa plants can perform differently in various environments. This variability makes it challenging to predict which genotypes will perform best across diverse conditions. Multi-environment trials are needed, but they are time-consuming and costly.
4. *Complex trait architecture*: Many important traits in alfalfa, such as yield and stress tolerance, are controlled by multiple genes. Understanding and predicting these complex traits require sophisticated models and large datasets.
5. *Regulatory and public acceptance*: Genomic and gene editing technologies face regulatory hurdles and public skepticism. Ensuring the safety and acceptance of these technologies is critical for their widespread adoption.

8.2 Future Directions

The future of alfalfa breeding and genomics looks promising with several exciting directions (Fang et al. 2024).

1. *Advanced genomic technologies*: New technologies such as CRISPR/Cas9 and other gene editing tools will enable precise modifications to improve alfalfa traits. As these technologies become more refined, their application in breeding will likely increase.
2. *Integrated breeding approaches*: Combining genomic selection with traditional breeding methods can optimize the strengths of both approaches. This integration can lead to faster and more accurate development of superior alfalfa varieties.
3. *High-throughput phenotyping*: Advances in phenotyping technologies, such as drones and remote sensors, will provide rapid and accurate measurements of plant traits. These tools will help overcome current phenotyping bottlenecks.
4. *Big data and machine learning*: Leveraging big data and machine learning can enhance the analysis of complex traits and improve prediction models. These technologies will enable breeders to make more informed decisions based on vast amounts of genomic and phenotypic data.
5. *Environmental adaptation*: Climate is becoming a key threat to agriculture (Sarıkaya and Tatar 2023), and developing alfalfa varieties that can adapt to changing environmental conditions is crucial. This includes breeding for resilience to climate change, improving water and nutrient use efficiency, and enhancing resistance to diseases and pests.
6. *Collaborative breeding programs*: Collaboration among researchers, breeders, and industry stakeholders will be essential. Sharing data, resources, and expertise will accelerate the development and adoption of improved alfalfa varieties.

9 Ethical and Regulatory Considerations

The application of genomic technologies in alfalfa breeding presents several ethical and regulatory challenges. The use of genetic engineering and gene editing techniques, such as CRISPR/Cas9, to introduce or modify traits in alfalfa must be carefully regulated to ensure both safety and public acceptance. Concerns have been raised about the potential environmental impacts of genetically modified alfalfa, such as gene flow to wild relatives and unintended effects on ecosystems. Therefore, regulatory frameworks are needed to address these issues and to ensure that genetically modified alfalfa varieties are rigorously tested for safety and efficacy before they are commercially released (Hrbáčková et al. 2020).

In addition, ethical issues related to the ownership and control of genetic resources, as well as the potential socioeconomic impacts on smallholder farmers, need to be addressed. It is crucial to ensure equitable access to the benefits of genomic technologies and to build public trust through transparent and inclusive decision-making processes. This approach is essential for the sustainable development of alfalfa crop improvement.

In conclusion, while genomic technologies offer significant potential for advancing alfalfa crop improvement, it is vital to address the technical, methodological, ethical, and regulatory challenges associated with their use. Ongoing research and collaboration among scientists, policymakers, and stakeholders will be key to overcoming these challenges and achieving sustainable progress in alfalfa breeding (Zhou and Luo 2024).

10 Conclusion

The chapter highlights the significant potential of genomic technologies to advance alfalfa crop improvement, particularly through genomic selection, gene editing (such as CRISPR/Cas9), and the integration of multi-omics data. These advancements could lead to improved traits such as higher yield, better stress tolerance, and enhanced nutritional value. However, the successful implementation of these technologies faces challenges including high costs, the complexity of traits, genotype-by-environment interactions, and regulatory hurdles. Ethical considerations, particularly regarding the environmental impact and socioeconomic effects on smallholder farmers, are also crucial. The conclusion emphasizes the need for ongoing research, collaboration, and transparent decision-making processes to address these challenges and to ensure sustainable development in alfalfa breeding. In summary, while there is great promise in using genomic tools for improving alfalfa, careful attention to the associated challenges and ethical issues is essential for achieving long-term success.

Author Contributions

Aina Inam: Write-up.
Sadia Afzal: Write-up.
Lalarukh Azhar: Write-up.
Nadia Ghaffar: Literature review and write-up.
Sumera Javad: Draft analysis and content analysis.
Ajit Singh: Literature review and proofreading.

References

Abdel-Tawab FM, Fahmy EM, El-Nahrawy MA, Sharawy WM, Sayed MR (2011) Detection of molecular markers associated with salt tolerance in alfalfa (Medicago sativa L.). Egyp J Genet Cytol 40(1)

Aasim M, Katirci R, Baloch FS, Mustafa Z, Bakhsh A, Nadeem MA et al (2022) Innovation in the breeding of common bean through a combined approach of in vitro regeneration and machine learning algorithms. Front Genet 13:897696

Ahmar S, Gill RA, Jung KH, Faheem A, Qasim MU, Mubeen M, Zhou W (2020) Conventional and molecular techniques from simple breeding to speed breeding in crop plants: recent advances and future outlook. Int J Mol Sci 21(7):2590

Ahsyee SR, Al-Sloge O, Ćalić I, Branković G, Zorić M, Momirović U, Šurlan-Momirović G (2013) Genetic diversity of alfalfa domesticated varietal populations from Libyan genbank revealed by RAPD markers. Arch Biol Sci 65(2):595–602

Ali A, Shahbaz M, Ölmez F, Fatima N, Umar UUD, Ali MA et al (2024) RNA interference: a promising biotechnological approach to combat plant pathogens, mechanism and future prospects. World J Microbiol Biotechnol 40(11):339

Altaf MT, Tatar M, Ali A, Liaqat W, Mortazvi P, Kayihan C et al (2024) Advancements in QTL mapping and GWAS application in plant improvement. Turk J Bot 48(7):376–426

Alzohairy AM, Gyulai G, Ramadan MF, Edris S, Sabir JS, Jansen RK, Bahieldin A (2014) Retrotransposon-based molecular markers for assessment of genomic diversity. Funct Plant Biol 41(8):781–789

Annicchiarico P, Nazzicari N, Li X, Wei Y, Pecetti L, Brummer EC (2015) Accuracy of genomic selection for alfalfa biomass yield in different reference populations. BMC Genomics 16(1):1020

Annicchiarico P, Nazzicari N, Pecetti L (2021) Genomic selection for higher yield and quality in alfalfa. In: Yu LX, Kole C (eds) The alfalfa genome. Springer, Cham, pp 191–210

Arsic M, Le Tougaard S, Persson DP, Martens HJ, Doolette CL, Lombi E, Husted S (2020) Bioimaging techniques reveal foliar phosphate uptake pathways and leaf phosphorus status. Plant Physiol 183(4):1472–1483

Azad MAK, Barwal SK, Moni A (2023) Exploring impact of integrated breeding strategies in enhancing yield, nutritional quality, and stress tolerance in alfalfa. Plant Trends 1(1):1–15

Bhering LL, Junqueira VS, Peixoto LA, Cruz CD, Laviola BG (2015) Comparison of methods used to identify superior individuals in genomic selection in plant breeding. Genet Mol Res 14(3):10888–10896

Brooker JD, Lockington RA, Attwood GT, Langridge P, Nield JK, Langridge U (1989) Engineering ruminal flora for improved protein quality. In: The biology of wool and hair. Springer, Dordrecht, pp 425–440

Brough RC, Robison LR, Jackson RH (1977) The historical diffusion of alfalfa. J Agron Educ 6(1):13–19

Burgueno J, de los Campos G, Weigel K, Crossa J (2012) Genomic prediction of breeding values when modeling genotype× environment interaction using pedigree and dense molecular markers. Crop Sci 52(2):707–719

Chen C, Chen H, Zhang Y, Thomas HR, Frank MH, He Y, Xia R (2020) TBtools: an integrative toolkit developed for interactive analyses of big biological data. Mol Plant 13:1194–1202

Crossa J, Jarquín D, Franco J, Pérez-Rodríguez P, Burgueño J, Saint-Pierre C, Vikram P, Sansaloni C, Petroli C, Akdemir D, Sneller C (2016) Genomic prediction of gene bank wheat landraces. G3 Genes Genom Genet 7(7):2333–2345

De Mori G, Cipriani G (2023) Marker-assisted selection in breeding for fruit trait improvement: a review. Int J Mol Sci 24(10):8984

Elden TC, Elgin JH (1987) Recurrent seedling and individual-plant selection for potato leafhopper (Homoptera: Cicadellidae) resistance in alfalfa. J Econ Entomol 80(3):690–695

Eren B, Keskin B, Demirel F, Demirel S, Türkoğlu A, Yilmaz A, Haliloğlu K (2023) Assessment of genetic diversity and population structure in local alfalfa genotypes using iPBS molecular markers. Genet Res Crop Evol 70(2):617–628

Ertus MM, Sabanci CO, Şensoy S (2014) The determination of molecular diversity among some alfalfa (Medicago sativa L.) ecotypes using RAPD markers. Yuzuncu Yıl Univ J Agr Sci 24(1):7–15

Fang L, Tao L, Mingyu L, XueMing D, Yuling H, Congzhuo X, Siqi L, Jia Z, Xiaojuan H, Qiang Z, Dong L, Zhipeng L (2024) MODMS: a multi-omics database for facilitating biological studies on alfalfa (*Medicago sativa* L.). Hort Res 11(1):245

Fernandez AL, Sheaffer CC, Tautges NE, Putnam DH, Hunter MC (2019) Alfalfa, wildlife, and the environment, 2nd edn. National Alfalfa and Forage Alliance

Gao C (2021) Genome editing in crops: from bench to field. Nat Rev Genet 22(7):415–432

Gao H, Huang Z, Chen W, Xing A, Zhao S, Wan W, Li H (2024) Mild to moderate drought stress reinforces the role of functional microbiome in promoting growth of a dominant forage species (*Neopallasia pectinata*) in desert steppe. Front Microbiol 15:1371208

Han Y, Khu DM, Li X, Farmer A, Martinez JM, Brummer EC, Monteros MJ (2014) High density array for SNP genotyping and mapping in tetraploid alfalfa. In: Quantitative traits breeding for multifunctional grasslands and turf. Springer, Cham, pp 255–259

Hawkins C, Yu LX (2018) Recent progress in alfalfa (*Medicago sativa* L.) genomics and genomic selection. Crop J 6(6):565–575

Hrbáčková M, Dvořák P, Takáč T, Tichá M, Luptovčiak I, Šamajová O, Šamaj J (2020) Biotechnological perspectives of omics and genetic engineering methods in alfalfa. Front Plant Sci 11:592

Jazayeri SM, Torres ROV (2017) Genomic and transcriptomic approaches toward plant selection. J Sci Res 2(8):54–64

Jiang Y, Pu K (2021) Molecular probes for autofluorescence-free optical imaging. Chem Rev 121(21):13086–13131

Julier B, Huguet T, Chardon F, Ayadi R, Pierre JB, Prosperi JM, Huyghe C (2007) Identification of quantitative trait loci influencing aerial morphogenesis in the model legume *Medicago truncatula*. Theor Appl Genet 114(8):1391–1406

Komal T, Mustafa M, Ali Z, Kazi AG (2014) Heavy metal induced adaptation strategies and repair mechanisms in plants. J Endocytobiosis Cell Res 25:33–41

Krishna TPA, Veeramuthu D, Maharajan T, Soosaimanickam M (2023) The era of plant breeding: conventional breeding to genomics-assisted breeding for crop improvement. Curr Genomics 24(1):24

Labombarda P, Pupilli F, Arcioni S (2000) Optimal population size for RFLP-assisted cultivar identification in alfalfa (*Medicago sativa* L.). Agronomie 20(2):233–240

Latif A, Sun Y, Noman A (2023) Herbaceous alfalfa plant as a multipurpose crop and predominant forage specie in Pakistan. Front Sustain Food Syst 7:1126151

Lebedev VG, Lebedeva TN, Chernodubov AI, Shestibratov KA (2020) Genomic selection for forest tree improvement: methods, achievements and perspectives. Forests 11(11):1190

Li X, Brummer EC (2012) Applied genetics and genomics in alfalfa breeding. Agronomy 2(1):40–61

Li H, Durbin R (2009) Fast and accurate short read alignment with Burrows–Wheeler transform. Bioinformatics 25(14):1754–1760

Li X, Wei Y, Acharya A, Hansen JL, Crawford JL, Viands DR, Brummer EC (2015) Genomic prediction of biomass yield in two selection cycles of a tetraploid alfalfa breeding population. Plant Genome 8(2):2014–2012
Lu W, Wang J, Xie Y, Wu S, Li L, Xu H (2024) Mass spectrometry imaging: an emerging technology in plant molecular physiology. Phytochem Rev 23(2):551–577
Lyu J (2020) Transform alfalfa research. Nat Plants 6(6):596
McLamore ES, Alocilja E, Gomes C, Gunasekaran S, Jenkins D, Datta SP, Zhou A (2021) FEAST of biosensors: food, environmental and agricultural sensing technologies (FEAST) in North America. Biosens Bioelectron 178:113011
Merrick LF, Herr AW, Sandhu KS, Lozada DN, Carter AH (2022) Optimizing plant breeding programs for genomic selection. Agronomy 12(3):714
Montesinos-Lopez OA, Montesinos-Lopez A, Acosta R, Varshney RK, Bentley A, Crossa J (2022) Using an incomplete block design to allocate lines to environments improves sparse genome-based prediction in plant breeding. Plant Genome 15(1):e20194
Orchard CJ, Kressin J, Chompookam S, Chuapong J, Onmanee N, Van Leeuwen K, Francis DM (2023) Marker-assisted selection to combine alleles for four disease resistance genes of tomato collocated on chromosome 11. HortScience 58(5):495–501
Paolo A, Nelson N, Taoufik H, Meriem L, Monica C, Daniel B, Cristina MM, Charles EB, Luciano P (2022) Alfalfa genomic selection for different stress-prone growing regions. Plant Genome 15:e20264
Perez P, de los Campos G (2014) Genome-wide regression and prediction with the BGLR statistical package. Genetics 198(2):483–495
Pérez RM, Vives V, Zandalinas SI, López-Climent MF, Muñoz V (2013) Gómez-Cadenas a (2013) biotechnological approaches to study plant responses to stress. Biomed Res Int 1:654120
Prado GS, Rocha DC, Santos LND, Contiliani DF, Nobile PM, Martinati-Schenk JC, Souza AAD (2024) CRISPR technology towards genome editing of the perennial and semi-perennial crops citrus, coffee and sugarcane. Front Plant Sci 14:1331258
Pupilli F, Labombarda P, Scotti C, Arcioni S (2000) RFLP analysis allows for the identification of alfalfa ecotypes. Plant Breed 119(3):271–276
Resende MFR Jr, Munoz P, Resende MD, Garrick DJ, Fernando RL, Davis JM, Jokela EJ, Martin TA, Peter GF, Kirst M (2012) Accuracy of genomic selection methods in a standard data set of loblolly pine (*Pinus taeda* L.). Genetics 190(4):1503–1510
Rhouma HB, Taski-Ajdukovic K, Zitouna N, Sdouga D, Milic D, Trifi-Farah N (2017) Assessment of the genetic variation in alfalfa genotypes using SRAP markers for breeding purposes. Chil J Agric Res 77(4):332–339
Sarıkaya MF, Tatar M (2023) Global climate change and crop production. Uluslararası Sivas Bilim ve Teknoloji Üniversitesi Dergisi 2(2):65–78
Satya P, Bhattacharjee S, Sarkar D, Roy S, Sharma L, Mandal NA (2022) Transcriptomics in plant. In: Plant genomics for sustainable agriculture. Springer Nature, Singapore, pp 99–127
Scossa F, Alseekh S, Fernie AR (2021) Integrating multi-omics data for crop improvement. J Plant Physiol 257:153352
Sharma S, Sharma A (2018) Molecular markers based plant breeding. Adv Res 16(1):1–15
Shen C, Du H, Chen Z, Lu H, Zhu F, Chen H, Wang T (2020) The chromosome-level genome sequence of the autotetraploid alfalfa and resequencing of core germplasms provide genomic resources for alfalfa research. Mol Plant 13(9):1250–1261
Shi S, Nan L, Smith KF (2017) The current status, problems, and prospects of alfalfa (*Medicago sativa* L.) breeding in China. Agronomy 7(1):1
Singh B, Das A, Parihar AK, Bhagawati B, Singh D, Pathak KN, Dwivedi K, Das N, Keshari N, Midha RL, Kumar R (2020) Delineation of genotype-by-environment interactions for identification and validation of resistant genotypes in mungbean to root-knot nematode (Meloidogyne incognita) using GGE biplot. Sci Rep 10(1):4108
Susmitha P, Kumar P, Yadav P, Sahoo S, Kaur G, Pandey MK, Singh V, Tseng TM, Gangurde SS (2023) Genome-wide association study as a powerful tool for dissecting competitive traits in legumes. Front Plant Sci 14:1123631

Talla VVR, Sandal SS, Walia P, Vidya A (2023) Quality enhancement in forage crops of cereals and legumes. Int J Plant Soil Sci 35(16):166–179

Taški-Ajduković K, Nagl N, Milić D, Katić S, Zorić M (2014) Genetic variation and relationship of alfalfa populations and their progenies based on RAPD markers. Open Life Sci 9(8):768–776

Tharun PS, Talekar N, Akkati VR (2024) Applications of molecular marker implementation for enhanced oilseed breeding through marker-assisted selection (Mas). J Exp Agric Int 46(5):865–876

Ticha M, Illésová P, Hrbáčková M, Basheer J, Novák D, Hlaváčková K, Šamaj J (2020) Tissue culture, genetic transformation, interaction with beneficial microbes, and modern bio-imaging techniques in alfalfa research. Crit Rev Biotechnol 40(8):1265–1280

Uçer VA, Aglar E, Mortazavi P, Qureshi SA, Ali A, Tatar M et al (2024) Exploring genetic diversity of Turkish fig (Ficus carica L.) germplasm using inter-Primer Binding Site (iPBS) retrotransposon markers. Genet Resour Crop Evol 72:5487–5498

Wang Y, Bi B, Yuan QH, Li XL, Gao JM (2012) Association of AFLP and SCAR markers with common leafspot resistance in autotetraploid alfalfa (*Medicago sativa*). Genet Mol Res 11(1):606–616

Wang X, Miao J, Kang W, Shi S (2023) Exogenous application of salicylic acid improves freezing stress tolerance in alfalfa. Front Plant Sci 14:1091077

Wanga MA, Shimelis H, Mashilo J, Laing MD (2021) Opportunities and challenges of speed breeding: a review. Plant Breed 140(2):185–194

Williams CE, Wielgus SM, Haberlach GT, Guenther C, Kim-Lee H, Helgeson JP (1993) RFLP analysis of chromosomal segregation in progeny from an interspecific hexaploid somatic hybrid between *Solanum brevidens* and *Solanum tuberosum*. Genetics 135(4):1167–1173

Wu X, Jiang W, Fragoso C, Huang J, Zhou G, Zhao H, Dellaporta S (2022) Prioritized candidate causal haplotype blocks in plant genome-wide association studies. PLoS Genet 18(10):e1010437

Yang S, Gao M, Xu C, Gao J, Deshpande S, Lin S, Zhu H (2008) Alfalfa benefits from *Medicago truncatula*: the RCT1 gene from *M. truncatula* confers broad-spectrum resistance to anthracnose in alfalfa. Proc Natl Acad Sci U S A 105(34):12164–12169

Yang B, Yao Z, Zhenfei G (2022) Research progress and prospect of alfalfa resistance to pathogens and pests. Plants 11(15):2008

Zhang Y, Massel K, Godwin ID, Gao C (2018) Applications and potential of genome editing in crop improvement. Genome Biol 19(1):210

Zhou X, Luo D (2024) The impact of genomic studies on alfalfa crop improvement. Legume Genom Genet 15:71–81

Recent Advancements in Breeding and Genomics of Red Clover

Sakshi Lingamurti Chinta and Vaishnavi Dilip Umbarkar

Abstract

The red clover (*Trifolium pratense* L.) is a flowering herbaceous plant that belongs to the Fabaceae family. Native of these plants are Europe, Western Asia, and North West Africa. It is a short-lived perennial plant. It is grown as fodder and cover crops. This plant has many applications in agricultural, food, and medical fields. In agriculture it is valued for its nitrogen fixation ability, improves soil fertility, and serves as a weed suppressor. The flowers of this plant are used to make jelly and tisanes. Red clover has anti-inflammatory and anti-dermatosis properties. It contains a chemical product called isoflavones, which can be used in the treatment of menopause, whooping cough, cancer, arthritis, and asthma. The aim of breeding red clover is high yield as it has many applications in various fields, improving persistence, resistance to pests and diseases on crops, resistance to biotic and abiotic factors, and for its quality characteristics such as protein quality and stability. The key step in crop breeding is selection and conventional breeding is based on phenotypic selection. Genome selection is a new breeding method where a large number of DNA markers are used to predict the genetic value of an individual or population. It helps to identify better properties such as harvest yield, protein quality, persistence, and disease resistance of the plant. Genomics also helps to improve crop breeding. Genomics focuses on the structure, function, mapping, and editing of the genome. Nowadays many new technologies are used for genomic analysis of plants. New generation sequencing is used for sequencing the red clover genome and assembly of the genome is done by using the de novo assembly method. The structure of the genome is identified using fluorescence in situ

S. L. Chinta (✉)
Department of Microbiology, HPT Art's and RYK Science, Nashik, Maharashtra, India

V. D. Umbarkar
Department of Microbiology, Sanjivani Art's, Commerce and Science College, Kopargaon, Maharashtra, India

F. S. Baloch et al. (eds.), *Forage Crops in the Omics Era*,
https://doi.org/10.1007/978-981-95-9249-4_9

hybridization and DNA markers are used for comparison of the genome with other model plant genomes. This new technology helps to improve the plant varieties and provides a better understanding of the plants.

Keywords

Red clover · Breeding · Genome selection · Genomics

1 Introduction

Red clover (*Trifolium pratense* L.) is a widely cultivated crop of the Fabaceae family for animal feed in the immoral area. The term *Trifolium* is derived from Latin, with three from "*tres*" and leaf from "*folium*." This shows the most distinctive feature of the genus, the trifoliate leaves; nevertheless, four- or even five-leaf clovers can be seen. The four-leaf clover occurs at a frequency of 1:6000, but the five-leaf clover occurs in 1 in 10,000 natural populations. Another distinctive trait of *Trifolium Pratense* L. is the white V-shaped mark on its leaves. These triangular marks serve as distinguishing markers for plants in the *Trifolium* genus (Petrauskas et al. 2023). Red clover is distinguished by its purple-red flowers. Compared to cultivated red clover, wild red clover grows low to the ground, has longer stems and smaller leaves, and has more flower heads. However, compared to wild red clover, produced red clover is taller, has bigger leaves, and yields fewer flower heads. Historical accounts indicate that the red clover was first grown in Spain in the thirteenth century. Red clover farming then spread to both southern and northern European nations (Osterman 2024). In Turkey, red clover was domesticated about 1000 years ago (Vleugels 2013). *Trifolium pratense* L., or red clover, is a prominent forage legume in temperate climates worldwide. This cultivar is frequently found in pastures and meadows in Europe, Asia, and Africa, according to Możdżeń et al. (2020). It provides important feed for ruminants and pigs (Mehmat et al. 2021).

Red clover is a perennial plant that only lasts 2–3 years. In terms of soil reactivity, it may grow in a broad pH range (5–8.5). For the development of clover leaves, a pH range of 6–7.6 is advised. Depending on how many plant nutrients are present in the soil, clover can also withstand low pH levels. Sufficient input micronutrients, such as potassium (K) and phosphate (P), are necessary for red clover development in agricultural soils. Phosphorus is a vital plant nutrient, especially for root and nodule development in legumes. The lifespan of the clover is impacted by the addition of phosphorus to the soil, particularly during the first year of establishment, according to data from soil research. Red clover can tolerate poor drainage, although for optimal growth, well-drained soils are ideal. Red clover (*Trifolium pratense L.*) grows best in clay loam, well-drained clay, and loamy soils with high water-holding capacity. Red clover can thrive at temperatures up to 5 °C. It grows in areas with annual rainfall above 350 mm. The optimum amount of precipitation is 550 mm or more. Red clover grows quite well in this climate zone (Mehmat et al. 2021). Timothy (*Phelum pratense* L.), meadow fescue (*Festuca pratensis* Huds.), tall fescue (*F. Arundinacea* Schreb), and perennial ryegrass are among the grasses that are

Table 1 Ideal conditions for the growth of red clover plants (Mehmat et al. 2021)

Parameters	Ideal condition
Climate	Cold and humid
pH	6–7.6
Temperature	18 °C
Micronutrients	Phosphorus and potassium
Precipitation	550 mm
Soil	Loamy, clayey, silty-loamy, and well-drained soils

typically cultivated alongside clover. Italian ryegrass (*Lolium italicum* L.) is also the most suitable species when grown in combination with red clover (Mehmat et al. 2021). Ideal growing conditions for the cultivation of this crop are given in Table 1.

2 Applications of Red Clover

Red clover is a nutrient-dense crop that has many applications in agriculture, medicine, and industry. It is produced not only for its quality feed properties, particularly when it is used as a cover crop, but also for providing various ecological services (Osterman 2024).

2.1 Agricultural Use

For ruminants, pigs, and poultry, red clover is a great source of forage that can be used as dry fodder, silage, straw, or green fodder. It is also a nutrient-rich plant that supplies feed on the farm all-year round. To increase fertility, red clover can be grown as a green manure crop. Because of their ability for biological nitrogen fixation (BNF), forage legumes play a significant role in preserving soil fertility. They also enhance animal output by offering a high-protein feed source. Regarding the sustainability of agriculture, this promotes the usage of legumes (Mehmat et al. 2021). Red clover reduces soil erosion from wind and rain and covers the ground all winter as it is a perennial crop. Complementary and annual crops that minimize soil compaction are therefore gentler on soil organisms. The increased root mass of a perennial crop increases soil carbon levels, which helps soil microbiota and other crops in crop rotation. The most well-known ecological benefit of clover is its capacity to work in symbiosis with rhizobacteria to improve the amount of nitrogen that is accessible in the soil. Meadow clover was a vital crop in crop rotation in the 1900s during the development of contemporary fertilizers. Red clover root system-associated rhizobacteria transform atmospheric nitrogen (N2) into soil-available nitrogen. When the bacteria infect the roots, the plant develops nodules. All legumes can form such symbiotic relationships; however, red clover is the most effective compared to other clovers. The increase in available nitrogen caused by symbiosis with rhizobacteria increases the protein content of clover (Osterman 2024). Red clover is used as a cover crop to protect soil, maintain soil health, and mitigate risks in situations of biotic and abiotic stress. In addition, many properties such as the ability to fix free nitrogen from the soil, suppression of weeds and flowering, a shape that attracts a large number of pollinators, increase the value of clover as a cover crop. In sustainable agriculture, growing green manure crops improves soil fertility, lowers the need for chemical pesticides and fertilizers, and boosts productivity and product quality. In many places across the world, including Korea, northeast Iowa, and Estonia, red clover is a fairly common green manure crop. Red clover enhances soil structure with its BNF properties and reduces greenhouse gas emissions by using fewer synthetic fertilizers. Red clover promotes carbon sequestration. The addition of red clover raises the soil's organic carbon content and adds substantial quantities of organic matter to both the soil's aboveground and subsurface components. Growing red clover is significantly cheaper than other fodder crops (Mehmat et al. 2021).

2.2 Medicinal Use

Red clover is a wild plant from the legume family. Red clover is consumed by animals, including livestock. Due to its numerous applications, red clover is a plant that is good for people. Coumestrol, lignans, phenolic glycosides, salicylates, coumarins, sitosterols, carcinogenic glycosides, saponins, carbohydrates, minerals, vitamins, starch, fatty acids, and volatile oils are among the substances found in the plant. Calcium, iron, magnesium, selenium, manganese, chromium, phosphorus, potassium, zinc, molybdenum, beta-carotene, and vitamins B, C, and E are among the several minerals found in red clover seeds. Red clover's antispasmodic, estrogenic, and expectorant qualities make it beneficial for people. Red clover has medicinal applications such as essential oils and medicinal ointments, teas, pills, and so on (Costin et al. 2021). In the medical field, it is used to treat psoriasis, eczema, cancer, whooping cough, and respiratory issues. According to doctors, red clover "purifies" the blood by acting as a diuretic (helps the body get rid of excess fluids), an expectorant (helps clear the lungs of mucus), promotes blood circulation, and helps cleanse the liver. Red clover contains isoflavones, which are compounds of plant origin that cause estrogenic effects in the body. Isoflavones have shown promise in treating a variety of menopause-related disorders, including hot flashes, cardiovascular health, and osteoporosis. However, as researchers become more aware of the risks associated with estrogen use, some concerns have been raised about the safety of isoflavones. Additionally, evidence that red clover reduces menopausal symptoms such as hot flashes is conflicting (Mount Sinai Health System n.d., *Red Clover*). Red clover plays an important role in traditional medicine by controlling and curing various diseases. Red clover contains isoflavones as well as phenolic and polyphenolic compounds that are used to treat coughs, muscle pain and menopausal symptoms, asthma, joint disease, wound healing, cancer prevention, and reproductive problems. In addition to traditional uses, clover and its derivatives have been extensively studied for their anticancer effects (Akbaribazm et al. 2021). Red clover has been used in traditional medicine for centuries and has attracted interest since the 1950s. Red clover is increasingly popular as a source of bioactive chemicals and a basis for herbal medicines and nutritional supplements. Red clover flowers are historically consumed in

Italy, while nectar from the flowers is consumed in Slovakia (Costin et al. 2021). Red clover is used in Turkey as a sedative, analgesic, and antiseptic. In Pakistan, it is used to treat meningitis, fever, and mouth and throat injuries. Red clover is used to make a tea that is used to treat diarrhea and stomach problems in several regions of Europe. Red clover is also used by Americans to treat respiratory, neurological, and reproductive ailments in addition to skin blemishes (Akbaribazm et al. 2021). Phytoestrogens are recommended instead of synthetic estrogenic chemicals for treating hormonal problems with hormone replacement therapy (HRT). In women, low estrogen can cause symptoms including hot flashes, restlessness, and vaginal dryness. Red clover contains estrogenic compounds that can mimic the release of sex hormones in the female body. However, women who take red clover supplements after menopause may experience less severe and frequent menopausal symptoms. Accordingly, isoflavones reduced the amount of bone loss associated with uric acid menopause without producing any changes in osteogenic markers such as alkaline phosphatase and osteocalcin. The effects of crude red clover extract and identified components on cardiovascular disease have been investigated in various studies. The use of industrial drug treatments containing active red clover components, as well as biochanin A and formononetin, accelerated vascular vasoconstriction, sac, and pulmonary artery pace. Promensil and Rimostil are commercial red clover tablets containing isoflavones (Akbaribazm et al. 2021).

2.3 Industrial Use

Red clover has a high content of proteins, vitamins, and minerals, which is why it is used as animal feed. It can be used fresh in animal feed or dried and stored. In addition to being incredibly vital as animal feed, it also has a beneficial effect on ecological systems. Red clover, which is abundant in secondary metabolites, finds application in both conventional and modern medicine, including the pharmaceutical sector (Habschied et al. 2024). Despite being mostly utilized as animal feed, it is becoming more and more well-known as a major source of bioactive compounds in the food, medicine, and cosmetic industries. The leaves and petals of red clover can be consumed either raw or cooked, in traditional forms such as tea or syrup, or as part of a balanced diet in salads. Red clover sprouts, with their high vitamin and mineral content, are considered a superfood. Red clover is also well known for its high isoflavone content. These isoflavones are structurally similar to 17β-estradiol, classifying them as phytoestrogens that exhibit a mild estrogenic effect in animals. Red clover isoflavones have the potential to serve as a natural estrogen substitute, particularly for menopausal women, once the precise chemicals causing this biological impact have been identified. As a result, the pharmaceutical sector now uses red clover as a raw material to make nutritious supplements. A variety of red clover nutritional supplements are now available on the market with the primary goal of alleviating menopausal symptoms. Furthermore, red clover flower extracts, which can be ingested as a beverage or added to food, may help with weight gain, cholesterol reduction, and osteoporosis prevention (Mikulić et al. 2023). The importance of flavonoids and other phenolic compounds as preventive dietary elements with estrogenic, antibacterial, anti-inflammatory, anti-allergenic, and anticancer properties has become a growing topic of study. These chemicals are used in nutraceuticals, cosmetics, medicine, food additives, and agri-food trade (Kazlauskaite et al. 2022). In the food industry, red clover is used to make jellies, honey, tisanes, and essay dishes (Foodb n.d.). As a resource-rich red clover, it is used in commercial production.

3 Recent Advances in Red Clover Breeding

Plant breeding is the study of modifying plant traits to produce desired traits. This technique involves merging parent plants to create the next generation with the best traits. The genetic potential of plants is increased through plant breeding. Breeding is the process of creating a multigenerational genetically diverse population and new combinations of plants with certain desirable characteristics (NAPB n.d.). Simply put, plant breeding is the process of integrating agricultural genetics into a plant to create a desired plant. Plant breeding is a cyclical process of selecting desired traits from parents and their offspring. To create variation, plant breeders use various techniques such as mutagenesis, artificial crossing, and chromosomal manipulation. The process of creating plant varieties helps breeders select desired traits. Over the past 150 years, plant breeders have succeeded in shortening the life cycle of plants. A shorter life cycle allows the integration of another main crop into the crop calendar (Acquaah 2015). Over the last 25 years, global red clover seed output has averaged ten million kg each year. Enhancing persistence, biomass yield, seed yield, and other characteristics is the main objective of breeding. By choosing for particular tolerance to disease, harsh environmental factors such as cold or drought, as well as grass combinations, breeding techniques increase persistence. Another important breeding goal is to increase the production of red clover seeds. According to seed production data for Oregon, USA, from 1944 to 2017, the seed yield of diploid red clover grew at an annual rate of 7.1%. In addition, tetraploid red clover produces fewer seeds, which is an area of concern for breeders. Another attribute of breeding is the improvement of feed quality. Biomass production, durability, disease and insect resistance, biotic and abiotic stress tolerance, and feed quality enhancement, including protein quality and stability, are key objectives of plant breeding (Riday 2023). As the population grows, so does the demand for food, which leads to the development of husbandry practices. However, to overcome this difficulty, the food supply must increase by 71–100%. This goal can be easily achieved through crop breeding and plant breeding technologies. Selection was the initial stage of plant breeding before Mendel's theory, and the identification of DNA and RNA is based on phenotypic traits such as plant size, shape, color, and height. After Mendel's theory and the discovery of DNA and RNA, plant breeding moved into the molecular era. Today, breeding is based on genetic traits (Lamichhane and Thapa 2022). Over the last 20 years, modern breeding technologies have been integrated with conventional husbandry practices. Several different traits are being investigated, such as disease resistance, improved nutrition, and sensitivity to soil and microbial communities. To achieve these goals, conventional breeding is integrated with other scientific disciplines for a comprehensive study of the genome. The genetic potential of plants is being increased through the application of new technologies including high-throughput phenotyping (HTP), marker-assisted selection, and genomic selection. These new procedures lead to modern breeding. Modern breeding is an effective and advanced strategy that is based on scientific research and knowledge. They are using these new techniques to select plant genetic traits that may be beneficial for plant improvement. This recent progress in plant breeding is helping to achieve the goal of a global dietary supplement (Acquaah 2015).

4 Main Breeding Goals of Red Clover

4.1 Persistence

The main breeding goal of clover is to improve endurance. The ability of a plant to endure in the field for 3–4 years is known as persistence. Signs of clover persistence vary with plant community and mowing frequency. Since meadow clover often has a lower level of persistence than other companion plants, improving it is a key breeding goal. Increased red clover persistence would improve sustainability by lowering the consumption of fossil fuels and synthetic fertilizers, minimizing nutrient loss during plowing and sowing, and increasing the net income of farmers through higher yields, better forage quality, and cost-effective operations. The reason for the limited persistence of clover is the short-lived perineum without vegetative reproduction, which limits its ability to survive the winter. It is also susceptible to nematodes and fungal root-infecting pathogens and faces competition from grasses that are more common in seedbeds with medium to high levels of nitrogen fertilization. Persistence is closely linked to winter survival because red clover populations in seed stands typically decrease throughout the winter. Winter survival may be directly attributed to the capacity to tolerate a variety of biotic and abiotic winter stressors, including water-saturated soil, frost, low-temperature pathogenic fungus, ice cover, and lack of light. On the other hand, circumstances in the late autumn or early spring growing season can be important. This condition can lead to the overall weakening of the plant due to pathogens, defoliation, or competition. In red clover, study on different genetic materials has indicated a trade-off between shoot biomass potential and its capacity to survive the winter. Several other additional characteristics, such as a specific disease resistance mechanism, vegetative growth habit, and other architectural features, are likely related to disease resistance, competitiveness, and ability to regrow and survive repeated defoliation (Ergon and Bakken 2022). The lack of persistence caused by widespread plant mortality is the primary issue limiting meadow clover globally. Clover survival is influenced by several factors, including biotic (diseases, pests, competition, and the plant's morphophysiology), abiotic (temperature, pH, and soil fertility), and management (frequency and severity of grazing or mowing). These elements combine to form a complex, and the meaning of each varies according to the setting. Forage species now confront new difficulties. They have to survive and make different types of biomasses and also deal with the negative effects of climate change such as low water availability and environmental temperature. In the past, meadow clover was considered a biennial plant, but modern cultivars persist for 3–4 years. Selecting for survival after exposing plants to their surroundings is the most straightforward way to breed for durability. It could take years to find better genotypes using this time-consuming method (Inostroza et al. 2020).

Since clover is a short-lived perennial, the main goal of red clover breeding is to increase stamina. Klimenko et al. (2010) used marker-assisted selection in clover to identify potential quantitative trait loci (QTL) associated with plant persistence, a complex trait that is associated with resistance to diseases such as *Sclerotinia trifoliorum* and *Fusarium* (*F. oxysporum*, *F. solani*, and *F. roseum*), as well as abiotic stress from low temperatures (winter hardiness). Using a simple sequence repeat (SSR), restriction fragment length polymorphism (RFLP), random amplified polymorphic DNA (RAPD), and sequence tagged site (STS) markers, two cross-derived complete sib mapping populations were used to identify QTL. A total of 10 and 23 potential QTL regions for plant persistence were found in each full-sib family. These DNA markers are sufficient for monitoring the inheritance of the main agronomic traits (Řepková and Nedělník 2014).

4.2 Biomass Yield

Biomass yield is the ratio of biomass produced to the amount of substrate used, such as nutrients and energy, during plant growth. Biomass yield is a highly quantitative trait that is influenced by both genotype and environment. To improve yield, observation over several harvests and years is key (Riday 2009). Yield can be improved by breeding. Meadow clover is key to a sustainable livestock production system due to its ability to fix 150 kg N ha/year, which can then be used for grass. They provide a huge amount of high-quality forage with a crude protein content of 18–19%, which results in high voluntary intake and better animal performance. Additionally, recent studies have demonstrated that red clover contains greater levels of the enzyme polyphenol oxidase (PPO) than other forage legumes. PPO has a beneficial effect on nitrogen utilization in ruminants and ruminant products (meat and milk). Increasing the biomass of perennial forages remains a critical aspect of the profitable pastoral industry and is therefore a major goal of the forage breeding project (Marshall et al. 2017). It increases the amount of protein provided by clover and grass, which is grown during the four harvest years. Persistence breeding is intimately linked to biomass yield, hence breeding for persistence involves breeding for biomass yield. Research has been published showing that stands in later years generate more biomass through breeding than stands from earlier years. This observation indicates that biomass yield depends on greater persistence. Increased persistence is an important factor in increasing biomass yield (Riday 2023).

Table 2 Forage yield and crude protein ratio of red clover in varied soil and climate conditions (Mehmat et al. 2021)

Precipitation regime (mm)	Soil structure	Forage yield (t DM/ha)	Crude protein (%)
195.5	Loamy, pH = 7.4	8.42	–
296.0	Clay-loam, pH = 7.75	6.41	17.9
450	Loamy, pH = 7.6	11.33	–
836–1230	Alluvium—colluvium's brown soil, brown acid soil	8.9–9.9	15.7
338	Chernozem soil	8.1–12.2	14.9–18.7
–	pH = 5.3	4.8–7.2	–
–	Endocalcari–Endohypogleyie, cambisol, pH = 7.2–7.3	16.6–17.8	
374.5–669.4	Alluvial soil, pH = 4.8	6.3	–
374.5–680.3	Alluvial soil, pH = 4.8	10.4	–
888.5	Clay-loam, pH = 6.9	9.10	–
287.1–504.3	Eutric cambisol, silt clay-loam	19.29	16.3–18.5
744–066	pH = 6.6	14.8–15.0	–
439.7–1011.5	Clay-loam, pH = 7.35–7.91	13.29	18.9
–	Clay, pH = 6.2	5.6–11.7	18.7–19.9
1037.2–1379.8	Andisol, family typic hapludands loam, pH = 6.21	6.2–15.9	–

Effective nodulation can improve forage yield by 63%. The yield is connected not only with nodulation but also with the activity of the nitrogenous substance (Jalůvka et al. 2014). The amount of dried biomass that is obtained after trimming and removing the moisture present in the new forage biomass is known as forage dry matter yield (Gebremedhin et al. 2019). In forage breeding efforts, dry matter yields are frequently assessed and impacted by environmental, cultural, and genotypic factors. Both yield and quality vary with environmental conditions, with climate appearing to be the most important environmental factor determining red clover production. The most important aspect of the climate is humidity, and red clover grows in a place with a greater distribution of humidity during the growing season. Meadow clover has been found to produce greater yields in mountainous areas with medium altitude (400–700 m) than in low- and high-altitude areas (Mehmat et al. 2021). Yield and protein contents of this crop under various soil and climatic conditions are provided in Table 2.

The yield and distribution of dry matter production during the growing season are compared between the novel candivars and the regular species (Jalůvka et al. 2014). Breeding for high dry matter yield in forage crops is expected to be enhanced by the introduction of genomic selection. These new techniques enable rapid, non-destructive, inexpensive, and high-throughput phenotypic growth, development, and yield of plants in the field (Gebremedhin et al. 2019).

4.3 Forage Quality

One important perennial legume that yields high-quality fodder is red clover. Improving the quality of legumes for fodder is one of the primary objectives of breeding initiatives. Red clover is a high-value fodder for grazing animals and, unlike other legumes, has a reduced proportion of protein breakdown during ensiling. Due to the poor and insufficient supply of vegetable proteins in the nutrition of animals and humans, the goal of several breeding projects is to increase the quality of legumes. Qualitative traits of perennial legumes are extremely complex and their expression is influenced by several factors, including genetic factors, environmental factors, and applied management (frequency of cutting, height of cutting, and maturity of plants at cutting). The most valuable forage quality parameters include crude protein (CP), crude fiber content, neutral detergent fiber (NDF), acid detergent fiber (ADF), digestible dry matter (DDM), dry matter intake (DMI), relative feed value (RFV), total digestible nutrients (TDN), net energy lactation (nel), net energy maintenance (nem), net energy gain (neg), and digestibility of forage and from leaves to stem (estimation depends on different stages) (improvement of the nutritional value of forage). Forage quality is influenced by many independent factors, including maturity, crop type, harvest and storage, environmental conditions, soil fertility, and diversity. Meadow clover varies in quality according to the degree of ploidy and earliness. In 1989, Makarenko and Pribytkoy reported that protein content decreased from early to late populations. Tetraploid red clover has more protein, water-soluble carbohydrates (WSC), potassium, and phosphorus, but less than five diploid red clovers. Studies of CP, PDIN (protein supplementation under rumen nitrogen limitation), PDIE (protein supplementation under rumen energy limitation), ME (metabolizable energy), and nel levels revealed that tetraploid clover produced better forage quality than diploid. The environment has a significant effect on the nutritional content of the forage. As noted by Buxton (1996), plant environment influences the impact of forage maturity, but results in interannual, seasonal, and regional variation in the quality of forage harvested at the same level of maturity. The observed seasonal fluctuations in forage quality mean that the assessment of the stability of forage quality includes intersession stability within 1 year and inter-year stability (Drobná and Jančovič 2006). Plant development, especially the transition from vegetative to reproductive growth, has a great influence on biomass yield and forage quality. In general, vegetative plants produce more biomass and feed with a higher crude protein (CP) content and lower lignin content than reproductive plants. This suggests that genetically modifying genes that influence the transition from the vegetative to the reproductive phase could be a beneficial method for improving red clover. *Mir156* is one such gene, a microRNA that acts as a master regulator of the vegetative phase transition in land plants. Genetic tinkering can improve feed (Zheng et al. 2016).

4.4 Disease Resistance

Red clover mortality is most often associated with the decline of the crown and roots. *Fusarium* ssp. root rot often occurs in these conditions. Winter injury and mechanical damage to the crown increase the occurrence of pathogens. Cultural techniques for dealing with *Fusarium* include management of the stand for optimal growth in the fall and prevention of mechanical damage and plant damage. Many breeding strategies have attempted to increase cultivar resistance. However, selection for *Fusarium* resistance in red clover, the diversity of species and isolates infecting red clover, and the complicated genetic resistance and interaction of fungi with the environment are problematic, *Sclerotinia trifoliorum* Eriks is another important root pathogen of red clover. *Sclerotinia* is the primary fungal disease affecting red clover in Europe, Japan, and South America. Infection resistance was high in the population. RAPD markers are associated with resistance and disease. Klimenko et al. (2010) recently mapped QTLs associated with *Sclerotinia* and *Fusarium* resistance, as well as QTLs associated with winter hardiness. *Kabatiella caulivora* is a foliar fungal disease that occurs in cooler climates. *Colletotrichum trifoli* B. and E. are found in warmer regions; cultivars resistant to these were developed in the 1950s in North America. Breeding for clover rust and leaf spot has not yet begun. It is only bean yellow mosaic virus (BYMV). Considered agronomically relevant and resistance breeding has been effective for this virus. Breeding for resistance was done in nematode pests: root nematode, root-damaging nematode, clover nematode, and stem nematode. The necessity for later stem nematode breeding in Europe was carried out during the twentieth century, resulting in most commercial cultivars being resistant to this pest. Other pests that attack red clover include pea aphid, root borer, clover borer, red or clover borer, clover root curculio, stem borer, and potato rootworm (Riday 2010). Assisted segment markers and linkage maps are used in clover breeding. Simple sequence repeat (SSR) markers are available. For example, trait determination based on the heritability of SSR markers has been successfully applied in family selection studies (Vleugels 2013).

5 Breeding Techniques

5.1 Genomic Selection

Genomic selection (GS) is an advanced method of plant breeding that predicts how well-untested individuals will perform for particular qualities using a genome-wide genetic marker (Varshney 2021). The way a person's genotype influences the population mean is known as breeding value. An estimated genetic breeding value is produced by quantifying breeding value using genetic markers. A GS model may be based on many markers with a minor contribution throughout the genome, or on a limited number of markers that greatly influence the phenotype (Wang et al. 2018). Genes related to breeding value can be found using genome-wide association studies (GWAS) (Osterman 2024). Genomic selection is a relatively new plant breeding technique that uses a large number of DNA markers to predict the genetic value of an individual or group. This enables early identification of promising candidates with superior traits such as crop yield, protein and fiber quality, disease resistance, and resilience (Varshney 2021; Wang et al. 2018). Genomic selection is particularly useful for breeding desirable traits that are controlled by multiple genes. Genomic selection increases the power of selection while reducing the need for field trials and shortening the time needed for the breeding cycle and breeding program. Genomic prediction is used when many markers scattered throughout the genome are used for Genomic Estimated Breeding Value (GEBV). GEBVs allow us to determine which individuals will perform better and are suitable as parents in hybridization or for subsequent generational development of a breeding program because their molecular marker profiles are identical to those of other training material plants that have been shown to perform better in specific environments (Bhat et al. 2016). It is possible to predict the performance of genomic selection of untested populations by mapping the genetic influence on attributes. The effect of genome-wide genetic markers that capture all quality trait loci (QTLs) that influence any trait variation is added to GEBV determination (Anand et al. 2023). QTLs are areas of the genome that affect quantitative traits genetically. Finding their precise placement in the genome is challenging and time-consuming. To detect and measure the impact of a QTL without pinpointing its precise location, molecular markers such as SNPs (single nucleotide polymorphism) and their associations with characteristics are utilized (Osterman 2024). The primary goal of GS is not to find QTL but rather to generate predictions about future progeny performance based on DNA samples collected in the present. Breed values can be predicted using penalized regression and Bayesian methods. However, the quality of the prediction can be assessed by comparing the estimated and observed breeding values. In genomic selections, two distinct datasets—training and validation populations—are used to provide these estimated and true breeding values. Breeding values in a validation population can be predicted using an already genotyped and phenotyped population. Genome-wide selection successfully mapped small-effect genes with numerous QTLs. These genes are important in plant breeding because they determine most of the commercially and agronomically important traits. With the increasing cost of crop phenotyping, genomic selection can be an effective method for improving quantitative traits that are difficult or costly to recognize using phenotypic expression (Lamichhane and Thapa 2022).

For a genomic selection of red clover, a late and medium-late flowering diploid and a medium-late flowering tetraploid were selected for field trials. Additions genotyped by the GBS (genotyping-by-sequencing) method. GBS is an extremely efficient next-generation DNA sequencing for simultaneous identification of new SNP markers and genotyping of the studied genetic material. DNA was extracted from the plants followed by genotyping based on GBS. Genomic data are used for genetic diversity analysis to assess the overall genetic variability in the red clover gene pool. Phenotypic traits of the accession including forage yield, disease resistance, persistence, and forage quality were studied. Phenotypic and genotypic data used for

GWAS using statistical models that integrate parental and population structure to reveal gene and DNA markers associated with target traits. A core set of genetic markers reflecting genetic variability across the red clover genome, including those associated with several desirable traits, will be used to construct and validate genome-wide prediction techniques. Different genomic best linear unbiased predictions (GBLUP) and other approaches will be used to estimate the GEBV for each target trait (Geleta et al. 2019).

6 Genomics of Red Clover

Genomics is a branch of biology that focuses on the study of all the DNA of an organism. Genomics includes the identification, characterization, function, mapping, and editing of the genome (Genomic.gov n.d.-a, n.d.-b). Red clover (*T. pratense* L.) is a forage legume with a diploid ($2n = 2x = 14$) and tetraploid ($2n = 4x = 28$) genome Additionally, the species is self-incompatible, which leads to a very variable genome. Due to the domesticated gene pool's low allelic variety and physical limitations that make it challenging to find new alleles, many aims are challenging to accomplish. Since clovers are completely incompatible with each other, they must cross-pollinate to remain a species, leading to a diversified population structure (Mehmat et al. 2021). Numerous exceptional agronomic qualities have developed as a result of its broad genetic diversity. The genomic structure of red clover and critical genetic information remain unknown, severely limiting breeding and agricultural development. The original purpose of fluorescence in situ hybridization was to study the structure of the genome. SSR and SNP markers were then used to create a linkage genomic map of red clover (Yan et al. 2022). Red clover's genetic makeup is yet unknown. The latest assortment of genome assemblies ranges in size from 348.8 to 440 MB and consists of seven haploid chromosomes. Based on the National Centre for Biotechnology Information's most recent report (NCBI 2021), 246,086 nucleotides, 60,487 protein groups, and 169 genes with various functions were reported. Most studies have assessed the genetic diversity of small populations using markers such as SNP (single nucleotide polymorphism), VNTR (variable number tandem repeat), AFLP (amplified fragment length polymorphism), RAPD (random amplification polymorphic DNA), and SSR–microsatellite polymorphism (single sequence). Modern genetic and genomic techniques can enhance breeding and backcrossing research's identification and selection procedure. By using genetic markers, it is possible to increase agronomic and qualitative attributes, find new alleles for tolerance to biotic and abiotic stress, and so on (Mehmat et al. 2021). Fluorescence in situ hybridization (FISH) was used to examine the genomic genome of clover. The 836.6-cm-long consensus high-density linkage map, which includes 204 RFLPs, 181 AFLPs, and 1414 SSRs, was created recently. Next-generation sequencing (NGS) is constantly evolving along with new sequencing technologies; however, Illumina's sequencing technology remains the benchmark for high-throughput and massively parallel sequencing. Illumina reads are shorter and have a higher error rate than Sanger sequencing but can provide greater genome coverage at an inexpensive and evolving cost. De novo assembly analysis was used to examine the red clover genome assembly (Istvánek et al. 2014).

7 Genomic Assembly of Red Clover

Genome assembly is the process of reassembling a large wide variety of tiny DNA sequences to reconstruct the unique chromosomes from which the DNA was produced. It is a critical step after doing next-generation sequencing, PacBio SMRT sequencing, or nanopore sequencing. There are two types of genome assembly: mapping to a new line-up and a reference genome.

De novo assembly is the system of building a brand-new genome from scratch without using reference genomic statistics. A reference genome or reference assembly is a computerized database of nucleic acid sequences containing a species' gene set (CD Genomics n.d.). A high-quality chromosome-level genome was built utilizing Illumina, PacBio HiFi, and Hi-C data to expedite genetic and red clover genome studies as well as future breeding (Baker 2012). The genome was assembled using de novo assembly. With $n50 = 55$ MB, the assembly had 96 scaffolds (~423 MB). A genome size of 456 Mb was projected by K-mer analysis, with heterozygosity accounting for 1.57% and repetitive sequences making up 54%. The red clover genome was then put together using PacBio's third-generation HiFi sequencing (TGS) technology after the genome size was estimated using standard next-generation sequencing (NGS) data assembly techniques. TGS produces ultra-long reads, GC preference, and genome assembly by integrating high-quality HiFi reads, compensating for NGS limitation in assembly applications and removing the need for PCR amplification. This increases the accuracy of assembly results and makes sequencing more efficient and assembly process. A total of 1,764,862 N50 18-length HiFi players were produced kbp. There are 194 contigs (423 Mb) in the assembled genome, with the longest contig measuring 34 Mbps and the N50 size 13 Mbps. The constructed genome has an average GC content of 33.6%. The integrity and quality of the assembly were then assessed by comparison of Illumina values with assembled contigs. Both were long terminal repeats (LTRs) transposable elements and micro invertible repeatable transposable elements (MITE). For prediction, structural features were used; these components accounted for 52.81% and 30.01%, or the assembled genome contained 7605 individual repeats. Thirteen types of 16,227 ncRNAs were also discovered. Similarly, RNA sequencing and de novo prediction methods were used to yield 46,089 transcripts and 44,588 high-confidence gene models following the exclusion of gene models with early stop codons and frameshifts. The seven chromosomes have an unequal distribution of the gene models. Transcripts and genes had average lengths of 1730 and 3620 bp, respectively. There was an average of one transcript per gene, and there were an average of five exons per transcript. CDs, exons, and introns have average lengths of 1585, 348, and 495 bp, respectively (Yan et al. 2022).

References

Acquaah G (2015) Conventional plant breeding principles and techniques. In: Advances in plant breeding strategies: breeding, biotechnology and molecular tools. Springer, Cham, pp 115–158

Akbaribazm M, Khazaei F, Naseri L, Pazhouhi M, Zamanian M, Khazaei M (2021) Pharmacological and therapeutic properties of the red clover (Trifolium pratense L.): an overview of the new findings. J Tradit Chin Med 41(4):642–649

Anand A, Subramanian M, Kar D (2023) Breeding techniques to dispense higher genetic gains. Front Plant Sci 13:1076094

Baker M (2012) *De novo* genome assembly: what every biologist should know. Nat Methods 9:333–337

Bhat J, Ali S, Salgotra R, Mir Z, Dutta S, Jadon V, Tyagi A, Mushtaq M, Jain N, Singh P, Singh G, Prabhu K (2016) Genomic selection in the era of next generation sequencing for complex traits in plant breeding. Front Genet 7:221

Buxton DR (1996) Quality-related characteristics of forages as influenced by plant environment and agronomic factors. Animal Feed Science and Technology 59(1–3):37–49

CD Genomics (n.d.) An overview of genome assembly. https://www.cd-genomics.com/an-overview-of-genome-assembly.html

Costin D, Muntea L, Butaş A, Mariş R (2021) Medicinal and therapeutic uses of red clover (Trifolium pratense L.) – review. Hop Med Plants 29:73–81

Drobná J, Jančovič J (2006) Estimation of red clover (Trifolium pratense L.) forage quality parameters depending on the variety, cut, and growing year. Plant Soil Environ 52(10):468–475. Using sensor technologies. Agronomy, 9(2),65

Ergon S, Bakken A (2022) Breeding for intercropping: the case of red clover persistence in grasslands. Euphytica 218(7):98

Foodb (n.d.) Showing food red clover. https://foodb.ca/foods/FOOD00930

Gebremedhin A, Badenhorst P, Wang J, Spangenberg G, Smith K (2019) Prospects for measurement of dry matter yield in forage breeding programs. Agronomy 9(2):65

Geleta M, Gustafsson C, Nadeau E, Ortiz R, Parsons D, Andersson A, Ohlund L (2019) Genomic selection in red clover (Trifolium pratense) – a research project funded by SLU Grogrund – Center for Breeding Food Crops. https://pub.epsilon.slu.se/id/eprint/17089/contents

Genomic.gov (n.d.-a) Genomics. https://www.genome.gov/genetics-glossary/genomics

Genomic.gov (n.d.-b) Fluorescence in situ hybridization. https://www.genome.gov/genetics-glossary/Fluorescence-In-Situ-Hybridization

Habschied K, Babić K, Horvat D, Arapović M, Puljić L, Kajić N, Krstanović V, Mastanjević K (2024) Red clover (Trifolium pratense) as a source of phytoestrogens in beer. Fermentation 10(8):423

Inostroza L, Ortega-Klose F, Vásquez C, Wilckens R (2020) Changes in root architecture and aboveground traits of red clover cultivars driven by breeding to improve persistence. Agronomy 10(12):1896

Istvánek J, Jaros M, Krenek A, Řepková J (2014) Genome assembly and annotation for red clover (Trifolium pratense; Fabaceae). Am J Bot 101(2):327–337

Jalůvka L, Dostál V, Meyer V, Bayle B, Lapage F, Chloupek O (2014) Comparison of breeding methods for forage yield in red clover. Acta Univ Agric Silvic Mendel Brun 57(2):45–52

Klimenko I, Razgulayeva N, Gau M, Okumura K, Nakaya A, Tabata S, Kozlov NN, Isobe S (2010) Mapping candidate QTLs related to plant persistency in red clover (Trifolium pratense L.). Theor Appl Genet 120(6):1253–1263

Kazlauskaite JA, Ivanauskas L, Marksa M, Bernatoniene J (2022) The effect of traditional and cyclodextrin-assisted extraction methods on Trifolium pratense L. (red clover) extracts antioxidant potential. Antioxidants 11(2):435

Lamichhane S, Thapa S (2022) Advances from conventional to modern plant breeding methodologies. Plant Breed Biotechnol 10(1):1–14

Marshall A, Collins R, Vale J, Lowe M (2017) Improved persistence of red clover (Trifolium pratense L.) increases the protein supplied by red clover/grass swards grown over four harvest years. Eur J Agron 89:38–45

Mehmat A, Harun B, Sehim A (2021) Red clover (Trifolium pratense l.). In: Legumes processing and potential. Iksad Publications, Ankara, p 268

Mikulić M, Krstonošić MA, Kladar N, Vasiljević S, Katanski S, Mamlić Z, Rakić D, Cvejić J (2023) Phytochemical composition of different red clover genotypes based on plant part and genetic traits. Foods 13(1):103

Mount Sinai Health System (n.d.) Red clover. https://www.mountsinai.org/health-library/herb/red-clover

Możdżeń K, Wójcik M, Kowalczyk K (2020) Allelopathic interaction between two common meadow plants: Dactylis glomerata L. and Trifolium pratense L. Biologia 75(5):653–663

National Center for Biotechnology Information (2021) Database resources of the National Center for Biotechnology Information. Nucleic Acids Res 49(D1):D10–D17

National Association of Plant Breeders (NAPB) (n.d.) What is plant breeding? https://www.plantbreeding.org/about-us/what-is-plant-breeding/?acsSsoAuthCheck=1

Osterman J (2024) Advancing red clover breeding through genomic selection methods. Doctoral thesis, Acta Universitatis Agriculture Science, Faculty of Landscape Architecture, Horticulture and Crop Production Science

Petrauskas G, Norkevičienė E, Baistruk-Hlodan L (2023) Genetic differentiation of red clover (Trifolium pratense L.) cultivars and their wild relatives. Agriculture 13(5):1008

Řepková J, Nedělník J (2014) Modern methods for genetic improvement of Trifolium pratense. Czech J Genet Plant Breed 50(2):92–99

Riday H (2009) Correlations between visual biomass scores and forage yield in space-planted red clover (Trifolium pratense L.) breeding nurseries. Euphytica 170(3):339–345

Riday H (2010) Progress made in improving red clover (Trifolium pratense L.) through breeding. Int J Plant Breed 4:22–29

Riday H (2023) Current red clover breeding research

Varshney RK (2021) The plant genome special issue: advances in genomic selection and application of machine learning in genomic prediction for crop improvement. Plant Genome 14(3):e20178. https://doi.org/10.1002/tpg2.20178

Vleugels T (2013) Breeding for resistance to clover rot (Sclerotinia spp.) in red clover breeding (Trifolium pratense). PhD thesis, Faculty of Bioscience Engineering, Ghent University

Wang X, Xu Y, Hu Z, Xu C (2018) Genomic selection methods for crop improvement: current status and prospects. Crop J 6(4):330–340

Yan Z, Sang L, Ma Y, He Y, Sun J, Ma L, Li S, Miao F, Zhang Z, Huang J, Wang Z, Yang G (2022) A de novo assembled high-quality chromosome-scale Trifolium pratense genome and fine-scale phylogenetic analysis. BMC Plant Biol 22(1):332

Zheng Q, Liu J, Goff B, Dinkins R, Zh H (2016) Genetic manipulation of miR156 for improvement of biomass production and forage quality in red clover. Crop Sci 56(3):1199–1205

Recent Advancements in Breeding and Genomics of White Clover

Vaishnavi Dilip Umbarkar, Sakshi Lingamurti Chinta, and Dipali Shantaram Gosavi

Abstract

White clover (*Trifolium repens* L.) belongs to the Fabaceae family, which also includes several well-studied forage legumes. Allotetraploid white clover, a cool-season perennial legume widely used as livestock forage, is an important candidateformarker-assistedbreeding.ThisspeciesisalsousedasanEco–Eco-evolutionary model to study variations in chemical defense within the species. It has a well-known variation for cyanogensis (release of HCN after tissue damage), with cyanogenic plants being more common in warmer areas around the world. A genetic linkage map of white clover was created using simple sequence repeat (SSR) markers derived from the DNA sequences of several *Trifolieae* species, including white clover. By incorporating foreign genes into superior plant material of white clover, scientists have conducted the first-ever field test of genetically modified white clover. A new breeding method has been developed for cross-pollinated crops such as white clover that overcomes challenges by identifying second-generation (F2) plants that have two identical copies of specific introduced genes RNA-seq and qRT-PCR experiments confirmed that genes are more highly expressed during early or intermittent responses to cold stress, indicating that they likely play important roles in determining white clover's cold tolerance.

Keywords

White clover · Breeding · Marker-assisted breeding · Cyanogensis · Cold tolerance · *Trifolium repens* L

V. D. Umbarkar (✉) · D. S. Gosavi
Department of Microbiology, Sanjivani Art's, Commerce and Science College, Kopargaon, Maharashtra, India

S. L. Chinta
Department of Microbiology, HPT Art's and RYK Science, Nashik, Maharashtra, India

F. S. Baloch et al. (eds.), *Forage Crops in the Omics Era*,
https://doi.org/10.1007/978-981-95-9249-4_10

1 Introduction

White clover (*Trifolium repens* L.) is a species of flowering plant that belongs to the large family *Leguminosae* (commonly known as Fabaceae). The *Trifolium* genus, which includes white clover, is a broad group of plants comprising roughly 255 species (Ellison et al. 2006). White clover has spontaneously expanded to many places of the world with warm temperatures and plenty of rain (more than 750 mm/year) (Jahufer et al. 2002). It grows nicely in temperate regions (not very hot or cold), subtropical climates (warm, humid), and Northern and Southern Hemispheres (Ellison et al. 2006). White clover is a resilient and multipurpose plant that thrives in a variety of circumstances. White clover is also a highly variable species, which means it can appear differently depending on the location and conditions. It may have different leaf forms and sizes, flower colors, and sizes. This means that white clover may be adapted to a wide range of settings and applications, making it a highly flexible and beneficial plant. White clover is a versatile plant that thrives under diverse situations (Lluga-Rizani et al. 2021).

Its productivity is determined by growing it alongside grass and calculating the output from repeated harvests throughout the year (Woodfield and Caradus 1994; Connolly et al. 2001). However, between 1991 and 2009, the usage of synthetic nitrogen fertilizers on dairy farms increased six times, reaching an average of 120 kg/ha. This intensification has led to the overuse of nitrogen fertilizers, generating worries about nitrate pollution in groundwater and streams (Weith et al. 2022).

2 Application of *Trifolium repens* L. in Traditional Medicine

In Turkey, three distinct species of *Trifolium* (*T. pratense*, *T. arvense*, and *T. repens*) are used to form a medical mixture that relieves pain (analgesic), prevents infection (antiseptic), promotes relaxation (sedative), helps with coughing (expectorant), and improves overall health (tonic) (Sabudak et al. 2008). In Iran, the above-ground sections of the white clover plant (*T. repens*) are used to cure jaundice in newborns (a condition that causes the skin to turn yellow) and skin disorders (Ahmad and Zeb 2020).

They also utilized a tea produced from the white clover plant (*T. repens*) to treat diarrhea. In Pakistan, both white clovers (*T. repens*) are common plants used to treat a variety of health conditions, including fever, feeling feverish, meningitis, pneumonia, and sore throat. In Pakistan's mountainous regions, people utilize the entire white clover plant (*T. repens*) to prepare extracts or drinks that assist treat a variety of health conditions such as stomach pain, joint difficulties, coughs, colds, and fever. They also use it as an eye wash to relieve irritations (Ruckle et al. 2018). White clover's long history of usage in traditional medicine implies that it has unique chemicals that can aid with a variety of health issues (Ahmad and Zeb 2020). In short, white clover is a useful plant with possible health advantages. When they go from flower to bloom, they mistakenly transfer pollen, allowing white clover plants to proliferate and generate seeds (Burghardt et al. 2001).

Furthermore, plants have unique features such as helping animals digest their food more effectively, Promoting the formation of beneficial bacteria in the stomach, enhancing digestive system efficiency, regulating hormone balance, and so on. Plants, for example, contribute to the development of a healthy rumen (stomach) environment in cows, which is essential for digestion and overall health (Kolodziejczyk-Czepas 2012).

These plants possess unique characteristics that make them beneficial for preventing soil erosion (keeping dirt from washing away) and enhancing soil quality (making it healthier). Furthermore, several plants have been employed in traditional medicine in numerous parts of the world for their therapeutic components, which include Quercetin (found in white clover). These chemicals, known as secondary metabolites, have potential health advantages and are utilized to treat a variety of diseases in traditional medicine (Sato et al. 2005).

White clover is also an important herb in traditional medicine, which is used to treat a variety of diseases, including rheumatic diseases (joint pain, inflammation), parasite infestations (deworming), wore throat, A fever, pneumonia (a lung infection), and meningitis (an infection of the brain and spinal cord membranes). White clover is an adaptable plant that provides benefits for both animal feeding and human health (Weith et al. 2022).

3 Benefits of *Trifolium* Species

3.1 Harnessing the Benefits of *Trifolium* Species for Sustainable Land Management in New Zealand's Pastoral Sector

White clover is an important legume in New Zealand's agricultural business. However, with wilting, chopping, and acid additions, these issues have been resolved, making it a viable option for hay and silage production (Egan et al. 2021).

White clover is useful in agriculture because it fixes nitrogen from the air into the soil, minimizing the demand for synthetic fertilizers. It promotes the growth of companion grasses without the need for synthetic fertilizers, which can fix up to 600–700 kg of nitrogen per hectare per year. Several factors influence white clover output, including pests and illnesses, droughts, grazing management, and changes in farming laws. One important concern is that white clover does not survive well in drought situations. As global temperatures rise, white clover's performance diminishes, particularly during prolonged droughts. The optimal temperature for growing is 20–24 °C. Inadequate temperatures lead to decreased production (Caradus et al. 2023).

3.2 Environmental Benefits

Pastoral agriculture, particularly involving grazing animals such as cows and sheep, has been characterized as contributing to environmental issues such as air pollution from methane emissions and water pollution from nitrogen runoff. One concern was that nitrogen fixation by white clover could increase nitrous oxide (N_2O) emissions, a powerful greenhouse gas. Research indicates that surplus nitrogen and soil carbon have a greater impact on N_2O emissions than nitrogen fixation alone (Woodfield et al. 1998).

Recent advances in molecular biology have revealed a mechanism to "turn on" the production of condensed tannins in white clover leaves by utilizing a specific gene. Laboratory investigations have demonstrated that these tannins can protect proteins in the rumen before releasing them into the animal's stomach, improving digestion and lowering methane emissions by up to 15% in early testing. Genetically modifying organisms is prohibited in many sectors, but this breakthrough has the potential to improve environmental effects, animal health, and productivity in pasture-based farming systems (Caradus et al. 2023).

4 Advances in Breeding Techniques

In the early 1900s, scientists in New Zealand began cultivating *Trifolium* species (such as white clover) to improve farming. They understood how vital these plants were to agriculture, so they dedicated a lot of time and resources to breeding them. *Trifolium* species play a major role in New Zealand's grasslands (Egan et al. 2021).

The modernization of white clover began in New Zealand in the 1870s (Egan et al. 2021). In 2018, researchers compared two selection strategies: AHS (among half-sibling family selection): choosing among families and AWHS (among and within half-sibling family selection): choosing both among and within families. AWHS family selection was the most effective technique, particularly during early selection cycles. In the first cycle selection, AWHS had a 2.5% genetic gain compared to 1% for AHS (Egan et al. 2021). White clover yield and percentage clover in the sward increased by 6% every period of 10 years through genetic gain. Scientists are employing modern genetic tools and techniques to improve forage crops, such as white clover, to make them more durable and suitable for animal consumption (Brock et al. 1988).

Medium-leaved clovers, on the other hand, perform well in a variety of grazing conditions, except in cases of extremely close and continuous grazing. Large- and medium-leaved clovers are often combined and used in dairy pastures to balance productivity and adaptability. Small-leaved clovers are low-growing with numerous leaves and thin bran (Egan et al. 2021).

In New Zealand, white clover was first divided into four types (Brock et al. 1988):

Type 1: White clover is a highly productive, medium-leaved perennial that thrives in ancient, fertile pastures. It has a natural toxin and is an excellent choice for grazing.

Type 2: White clover, commonly known as "New Zealand Wild White No. 2," is a perennial with smaller, thicker foliage than Type 1, although it is still a durable and adaptable type.

Type 3: White clover, often known as "Common New Zealand White," is a medium-leafed clover variety that develops moderately in its first year but struggles to live and thrive in subsequent years.

Type 4: White clover is a low-growing, nonpersistent kind with small leaves and limited productivity, belonging to the annual type of plant.

This strategy worked properly, leading to the generation of novel cultivars such as "Grasslands Demand" and "Grasslands Sustain" (Brock et al. 1988). van den Bosch et al. (1993) discovered that selecting desirable root characteristics in white clover resulted in lower yields (production) and reduced persistence (plants did not survive as well). Breeders also observed that introduced white clover types from outside New Zealand performed well in the first year, but fell behind local New Zealand varieties in productivity and persistence in future years (Egan et al. 2021).

4.1 Genetic Linkage Map of *Trifolium repens*

Researchers used a technique known as linkage analysis to generate a genetic map of white clover. They examined 179 individual plants from white clover germplasm and sought genetic markers (such as little flags) on their DNA. They identified genetic variants at 415 distinct loci on DNA using 343 SSR primer pairs; some of these tools originated from white clover, red clover, and other related plants (Barrett et al. 2004; Jones et al. 2003), 78 points were identified only in one parent plant (GA43), 247 in other parent plant (SRVR), and 90 in both. Scientists examined the DNA of white clover and discovered eight pairs of similar chromosomes (homologous pairs). Each pair had two similar chromosomes (homologs). Genetic markers were grouped into 25 clusters (linkage groups), and each cluster had 3–38 markers, the lengths of the eight pairs of chromosomes ranged from 82 to 155 units (cM). Markers were not evenly distributed across the chromosomes, with some having as few as 12 markers and others having as many as 41 markers (Caradus et al. 2023). Caradus et al. (2023) also discovered that a consensus map was created using 22% of polymorphic genetic markers from both parent plants, which is significantly lower than prior work on white clover. These primer pairs were mapped to two comparable chromosomes (homologs) in a different study. In the current work, these primer pairs were mapped to two similar chromosomes, but on separate groups.

A total of 43 special markers (TRSSRs) were designed using SSRs (short, repeating DNA sequences) discovered in a unique library of white clover DNA clones. Eight-six percent of genetic markers (SSRs) worked well within the current study population. These markers can be easily transferred and used in different white clover populations (Faville et al. 2004). Ryegrass and other plant species such as perennial have a lower transferability rate of markers between populations (61%). White clover species share similar genetic makeup (conserved genome), but not with distant species (Caradus et al. 2023).

4.2 RNA Sequencing Techniques

Crops frequently confront harsh conditions such as cold, dryness, and salty soil (Knight and Knight 2012), which can severely limit their growth and spread. Cold stress in particular is a major issue. Scientists have identified two types of genes that help crops survive in freezing temperatures. The first group includes genes that produce antioxidants, transporters, and other compounds that protect plants from cold damage. The second group consists of genes that regulate the plant's internal process such as how it responds to environmental signals and controls its genetic functions. Transcription factors (TFs) from families such as AP2/ERF, WRKY, MYB, and NAC are key in helping plants react to cold stress. Additionally, protein kinases play an essential role in how plants adjust to cold by activating various signaling pathways. Examples include calcium-dependent protein kinases/CPKs, MAPK cascades, and receptor-like kinases (RLKs) (Chen et al. 2021).

White clover grows in milder areas and is susceptible to cold stress, particularly in locations with hard winters (Li et al. 2024). The advent of sequencing technology, such as RNA sequencing, has significantly improved their ability to measure gene expression in. This technique allows scientists to quickly and accurately identify which genes are active and how they are expressed, deepening our understanding of plant biology plants (Zhu et al. 2024). It is widely used in genetics research to analyze microarray and RNA sequencing data. Essentially, this method is a powerful tool for exploring the genetic makeup of various organisms, including plants, animals, and humans. This approach has been crucial in advancing our knowledge of genetics and biology (Yu et al. 2020; Li et al. 2021).

To investigate how white clover responds to cold stress, researchers analyzed gene activity over time using RNA sequencing. To ensure accuracy, they experimented three times at each time point. They used modern sequencing technology to produce cDNA libraries (which are copies of the plant's RNA) and generated over 657 million data points (or reads) to better understand how cold affects the plant's genes (Zhang et al. 2022).

4.3 Quantitative Reverse Transcription PCR (qRT-PCR)

Quantitative reverse transcription PCR (qRT-PCR) is a common technique for measuring gene activity in plants. To reliably evaluate gene activity levels, scientists require a trustworthy reference gene that is continuously expressed under a variety of situations. These reference genes, sometimes known as "housekeeping genes," must be carefully selected for each plant species to ensure correct findings. This approach is faster and more sensitive than others, such as northern blotting, semi-quantitative PCR, and ribonuclease protection assays, which take longer and are less exact. One of the most difficult tasks is establishing correct normalization to account for variances in RNA concentration and quality. Even these housekeeping genes can vary in expression levels based on the tissue type or experimental settings. This diversity could be related to changes in how these genes are controlled or how their mRNA is degraded. As a result, a gene that is stable in one set of conditions may not be as dependable in another, emphasizing the need to carefully select reference genes for each study (Narancio et al. 2018).

5 Future Direction

To improve white clover (*Trifolium*), scientists can employ novel technologies and breeding targets such as landscape genomics, investigating how the environment influences species evolution. Disease and pest resistance strengthens the plant against threats, controlling plant reproduction through male sterility, enhancing breeder tools for informed decision-making, climate change adaptation, getting the plant ready for shifting conditions. Methane output features reduce plant's environmental impact isoflavones investigating the plant's possible health advantages. Landscape genomics is an emerging area that helps scientists to understand how the environment affects animal behavior and ecosystem services. Using this strategy in breeding programs allows scientists to improve their understanding of plant behavior and maximize ecological services and make informed breeding decisions, which further characterize the plant's genetic material (germplasm).

Historically, breeding strategies for white clover (*Trifolium*) focused mostly on resistance to diseases and pests. However, in recent years, the attention has switched toward production-related characteristics (e.g., yield, growth rate). This shift in focus raises concerns because climate change is anticipated to raise temperatures, making pests and illnesses more prevalent (Sarıkaya and Tatar 2023). White clover is more susceptible to these dangers without ongoing breeding for resistance. Scientists are looking into novel chemical methods for making male white clover plants sterile (unable to produce viable pollen). This could transform hybrid clover seed production in the future. Sterile male plants facilitate hybrid seed production. Cross-pollinated species, such as white clover, are often improved through population-wide breeding programs. By using sterile male plants, breeders may focus on enhancing individual parent plants, making the process more efficient.

To speed up white clover breeding initiatives, scientists should enhance techniques for analyzing plant properties (phonemics) and genes (genomics), add more genes to the white clover genome, and improve measurement methods for plant characteristics, particularly root absorption of water, nutrients, and carbon. There is now a measurement gap for structural qualities such as root properties, which is a bottleneck in the improvement of white clover. To address this, scientists should create high-throughput phenotyping systems that can ensure plant attributes quickly and accurately, assist breeders in understanding and utilizing genetic variation in stored white clover germplasm, and maximize the use of existing germplasm to improve white clover varieties.

Future research into white clover will focus on developing features to assist farmers adapt to climate change and promote sustainability. Key areas include drought tolerance and nutrient efficiency, lower methane emissions (greenhouse gas), and improve root systems for greater water absorption. White clover's shallow roots make it susceptible to drought, therefore studying natural variations in root features can help it adapt to changing conditions (Zhang et al. 2007).

Using molecular markers (Nadeem et al. 2018) and recent genomic approaches to select advantageous features in plants will remain crucial. As farming increasingly relies on legume-based pastures such as white clover, the requirement for chemical fertilizers will decrease, resulting in less flow of nutrients into rivers and decreased N_2O outpouring. White clover can be genetically modified to contain features that reduce methane emissions and nitrogen losses. For example, causing white clover to develop condensed tannins in its leaves can assist achieve these environmental aims while also minimizing bloating to boost the health of animals by enhancing flesh, milk, and fiber output.

The key difficulty will be to balance concerns about the risks of genetic alteration against the possible benefits to the environment, animal health, and farming productivity. Improving white clover performance with microbial technology could promote sustainable farming by lowering the use of synthetic fertilizers and pesticides (Caradus et al. 2023).

References

Ahmad S, Zeb A (2020) Phytochemical profile and pharmacological properties of *Trifolium repens*. J Basic Clin Physiol Pharmacol 32(1). https://doi.org/10.1515/jbcpp-2020-0015

Barrett P, Healy-Farrell L, March JS (2004) Cognitive-behavioral family treatment of childhood obsessive-compulsive disorder: a controlled trial. J Am Acad Child Adolesc Psychiatry 43(1):46–62. https://doi.org/10.1097/00004583-200401000-00014

Brock JL, Hay MJ, Thomas VJ, Sedcole JR (1988) Morphology of white clover (*Trifolium repens* L.) plants in pastures under intensive sheep grazing. J Agric Sci 111(2):273–283

Burghardt F, Proksch P, Fiedler K (2001) Flavonoid sequestration by the common blue butterfly Polyommatus icarus: quantitative intraspecific variation in relation to larval hostplant, sex and body size. Biochem Syst Ecol 29(9):875–889. https://doi.org/10.1016/s0305-1978(01)00036-9

Caradus JR, Roldan M, Voisey C, Woodfield DR (2023) White clover (Trifolium repens L.) benefits in grazed pastures and potential improvements. In: Production and utilization of legumes – progress and prospects. IntechOpen. https://doi.org/10.5772/intechopen.109625

Chen X, Ding Y, Yang Y, Song C, Wang B, Yang S, Guo Y, Gong Z (2021) Protein kinases in plant responses to drought, salt, and cold stress. J Integr Plant Biol 63(1):53–78. https://doi.org/10.1111/jipb.13061

Connolly V, Crops Research Centre, Teagasc, & DLF-TRIFOLIUM (2001) Breeding improved varieties of white clover

Egan LM, Hofmann RW, Ghamkhar K, Hoyos-Villegas V (2021) Prospects for *Trifolium* improvement through germplasm characterization and pre-breeding in New Zealand and beyond. Front Plant Sci 12:653191. https://doi.org/10.3389/fpls.2021.653191

Ellison NW, Liston A, Steiner JJ, Williams WM, Taylor NL (2006) Molecular phylogenetics of the clover genus (Trifolium—Leguminosae). Mol Phylogenet Evol 39(3):688–705. https://doi.org/10.1016/j.ympev.2006.01.004

Faville MJ, Vecchies AC, Schreiber M, Drayton MC, Hughes LJ, Jones ES, Guthridge KM, Smith KF, Sawbridge T, Spangenberg GC, Bryan GT (2004) Functionally associated molecular genetic marker map construction in perennial ryegrass (*Lolium perenne* L.). Theor Appl Genet 110(1):12–32

Jahufer MZZ, Cooper M, Ayres JF, Bray RA (2002) Identification of research to improve the efficiency of breeding strategies for white clover in Australia – a review. Aust J Agric Res 53(3):239. https://doi.org/10.1071/ar01110

Jones ES, Hughes LJ, Drayton MC, Abberton MT, Michaelson-Yeates TP, Bowen C, Forster JW (2003) An SSR and AFLP molecular marker-based genetic map of white clover (Trifolium repens L.). Plant Sci 165(3):531–539. https://doi.org/10.1016/s0168-9452(03)00212-7

Knight MR, Knight H (2012) Low-temperature perception leading to gene expression and cold tolerance in higher plants. New Phytol 195(4):737–751. https://doi.org/10.1111/j.1469-8137.2012.04239.x

Kolodziejczyk-Czepas J (2012) *Trifolium* species-derived substances and extracts—biological activity and prospects for medicinal applications. J Ethnopharmacol 143(1):14–23. https://doi.org/10.1016/j.jep.2012.06.048

Li L, Qiao Z, Cai J, Gu X, Liang Y, Chen N, Li M, Guo X, Miao J, Wei K (2021) Mineral element contents and gene expression in Sophora tonkinensis during florescence. Biotechnol Biotechnol Equip 35(1):1593–1602. https://doi.org/10.1080/13102818.2021.1988707

Li M, Chen X, Huang W, Wu K, Bai Y, Guo D et al (2024) Comprehensive identification of the β-amylase (BAM) gene family in response to cold stress in white clover. Plants 13(2):154

Lluga-Rizani K, Šoljan D, Berisha N, Kurteshi K, Letaj K (2021) Morphological variability of *Trifolium repens L.* (Fabaceae). Hacquetia 20(2):281–290. https://doi.org/10.2478/hacq-2021-0009

Nadeem MA, Nawaz MA, Shahid MQ, Doğan Y, Comertpay G, Yıldız M et al (2018) DNA molecular markers in plant breeding: current status and recent advancements in genomic selection and genome editing. Biotechnol Biotechnol Equip 32(2):261–285. https://doi.org/10.1080/13102818.2017.1400401

Narancio R, John U, Mason J, Spangenberg G (2018) Selection of optimal reference genes for quantitative RT-PCR transcript abundance analysis in white clover (Trifolium repens L.). Funct Plant Biol 45(7):737. https://doi.org/10.1071/fp17304

Ruckle ME, Bernasconi L, Kölliker R, Zeeman SC, Studer B (2018) Genetic diversity of diurnal carbohydrate accumulation in white clover (Trifolium repens L.). Agronomy 8(4):47. https://doi.org/10.3390/agronomy8040047

Sabudak T, Isik E, Oksuz S (2008) Two new compounds from Trifolium resupinatum var. microcephalum. J Asian Nat Prod Res 10(11):1017–1021. https://doi.org/10.1080/10286020802278038

Sarıkaya MF, Tatar M (2023) Global climate change and crop production. Uluslararası Sivas Bilim ve Teknoloji Üniversitesi Dergisi 2(2):65–78

Sato S, Isobe S, Asamizu E, Ohmido N, Kataoka R, Nakamura Y, Kaneko T, Sakurai N, Okumura K, Klimenko I, Sasamoto S, Wada T, Watanabe A, Kohara M, Fujishiro T, Tabata S (2005) Comprehensive structural analysis of the genome of red clover (*Trifolium pratense L.*). DNA Res 12(5):301–364. https://doi.org/10.1093/dnares/dsi018

van den Bosch J, Black IK, Cousins GR, Woodfield DR (1993) Enhanced drought tolerance in white clover. Proc N Z Grassl Assoc 55:97–101

Weith SK, Jahufer MZZ, Hofmann RW, Anderson CB, Luo D, Ehoche OG, Cousins G, Jones EE, Ballard RA, Griffiths AG (2022) Quantitative genetic analysis reveals potential to breed for improved white clover growth in symbiosis with nitrogen-fixing rhizobium bacteria. Front Plant Sci 13:953400. https://doi.org/10.3389/fpls.2022.953400

Woodfield DR, Caradus JR (1994) Genetic improvement in white clover representing six decades of plant breeding. Crop Sci 34(5):1205–1213. https://doi.org/10.2135/cropsci1994.0011183x003400050011x

Woodfield D, McNabb W, Kennedy L, Cousins G, Caradus J (1998) Floral and foliar tannin content in white clover. In: Proceedings of the Trifolium conference, vol 15, p 19

Yu Y, Xin F, Dong L, Ge L, Zhai C, Shen X (2020) Weighted gene coexpression network analysis identifies critical genes in different subtypes of acute myeloid leukaemia. Biotechnol Biotechnol Equip 34(1):925–936. https://doi.org/10.1080/13102818.2020.1811767

Zhang Y, Sledge MK, Bouton JH (2007) Genome mapping of white clover (*Trifolium repens L.*) and comparative analysis within the *Trifolieae* using cross-species SSR markers. Theor Appl Genet 114(8):1367–1378. https://doi.org/10.1007/s00122-007-0523-3

Zhang X, Yang H, Li M, Chen C, Bai Y, Guo D, Guo C, Shu Y (2022) Time-course RNA-seq analysis provides an improved understanding of genetic regulation in response to cold stress from white clover (Trifolium repens L.). Biotechnol Biotechnol Equip 36(1):745–752. https://doi.org/10.1080/13102818.2022.2108339

Zhu XT, Sanz-Jimenez P, Ning XT, Ul Qamar MT, Chen LL (2024) Direct RNA sequencing in plants: practical applications and future perspectives. Plant Commun 5(11):101064. https://doi.org/10.1016/j.xplc.2024.101064

Minor Legume Species as Genetic Resources for Breeding the Forage Crops

Anuradha Chetukuri, Swapnil Shamrao Baraskar, Niranjan Ravindra Thakur, Vinod Anand Ubarhande, Patlolla Ramakanth Reddy, and Venkata Chanda Sameer Kumar

Abstract

Minor legume species represent important genetic assets for the enhancement of forage crop breeding. Despite being relatively underutilized, these species exhibit important traits that may contribute to the enhancement of feed quality and agricultural productivity. While more than 1500 legume species are available for use in cattle feed, only approximately 60 are widely cultivated as forage crops. Forage legumes have been in use for over 11,000 years, where species were originally used as human food before being adapted for animal feed. Currently, genomic resources together with molecular breeding methods are being employed to improve the genetic potential of forage legumes. Gene banks such as ICARDA and CIAT maintain large collections of fodder legumes that include underutilized species. Genetic resources in minor legumes are mostly undefined because of a lack of genomic information. Establishment of genomic resources and mapping population development among these under-explored legumes will streamline identification of desirable traits and their utilization in breeding programs. Forage legumes enhance sustainable cattle farming by improving the soil's fertility, reducing erosion, and promoting nutrient cycling. Addressing the fodder needs and enhancing forage production systems in the future requires utilization of major diversities in minor legume species. This may be achieved by incorporating less-known legume species into breeding programs to further enhance the nutritional value and acceptability of the fodder, leading to better cattle health and productivity. Moreover, such genetic diversity within these minor legumes acts as an insurance policy against climate change, diseases, and pest attacks because of the underutilized alleles in these crops.

A. Chetukuri (✉) · S. S. Baraskar · V. A. Ubarhande · P. Ramakanth Reddy · V. C. S. Kumar
Professor Jayashankar Telangana Agricultural University (PJTAU), Hyderabad, Telangana, India

N. R. Thakur
Vasantrao Naik Marathwada Krishi Vidyapeeth (VNMKV), Parbhani, Maharashtra, India

F. S. Baloch et al. (eds.), *Forage Crops in the Omics Era*,
https://doi.org/10.1007/978-981-95-9249-4_11

Keywords

Legumes · Genomic resources · Sustainable · Productivity · Alleles

1 Introduction

The immense global population, now reaching 8 billion people, has greatly challenged the supply of natural resources and agricultural systems. This explosive growth in the population has led to a greater demand for food, water, and land; the consequence is the increasing number of more intensive agricultural methods applied to exert pressure on ecosystems. It likewise poses an immediate threat to livestock and agricultural practices through changes in the seasons of growth, disturbance of weather patterns, and increase in extreme events such as heat waves, flooding, and drought. Climate variation threatens food production due to a reduction in crop yield and deterioration in soil health (Menesch et al. 2023; Jabbar et al. 2023; Sarıkaya and Tatar 2023; Allen et al. 2011). Livestock are susceptible to adverse conditions inasmuch as heat stress, reduced availability of water, and shifts in disease patterns impinge directly on their well-being, performance, and health. Increased temperatures, along with changes in rainfall patterns, may lead to a decline in pasture conditions and lower forage quality and quantity—the very staple that supplies the nutritional needs of cattle. Moreover, diseases and pests encouraged by the rise in temperature would threaten agricultural produce and livestock, respectively; hence, these conditions enhance the possibility of food insecurity. The conjuring needs from a growing population coupled with climate change demand urgent actions that would make agricultural systems resilient enough to meet these challenges. Climate change adaptation needs sustainable agriculture practices, improved water resource management, and innovative breeding in crops and livestock. If climate change is left unchecked and the rise in population is not aligned with sustainable management, then truly, the future of global food security will be in jeopardy and might lead to disastrous impacts on agriculture and livestock, affecting billions of people dependent on those resources.

Forages are a critical component in the nutritional requirements of livestock, forming the most basic part of the diet for ruminant animals such as cattle, sheep, and goats. Forages include numerous types of grasses and legumes that are mostly consumed directly by the animal or harvested and fed to them. Three main categories define forages: grasses, including species like bermudagrass, timothy, and ryegrass; legumes, including alfalfa, clover, and vetch; and fodder crops. The most common type is grasses, which supply fiber and energy, but legumes are valued for their high protein content essential to animal growth, as well as milk and overall health. Alfalfa and clover are just two of the forages from legumes that are critical in animal nutrition. With more protein, vitamins, and minerals compared to grasses, they not only provide better nutritional quality but also increase the palatability and digestibility of the diet. As a result, the productivity of livestock improves and feed conversion is boosted. In addition, due to the symbiotic relationship between legumes with their root-nodulating Rhizobium bacteria, they have the astonishing capability to fix atmospheric nitrogen. This fixing of nitrogen enriches the soil naturally, therefore reducing dependence on synthetic fertilizers and enabling sustainable agricultural practices. In inclusion within pasture and forage systems, legume forages make significant contributions to soil structure, biodiversity, and overall ecosystem health. Legume forages are also indispensable in modern agriculture because they offer a well-rounded diet that can meet livestock nutritional requirements while contributing positively toward environmental sustainability. This, in turn, contributes to increased animal production and ecological resilience.

Hence, encouraging the use of minor legume forages should be considered a vital step in adapting agriculture to resilient, sustainable diversification. While the larger legumes have much attention for their nutritional and ecological benefits, the smaller legume species like sainfoin, birdsfoot trefoil, and fenugreek do show properties which allow big changes in fodder systems. Under these conditions, these minor legumes often exhibit a finer fit to specific conditions compared with more common legumes in terms of poor soils, dry areas, and low-fertility plains. The diversification of a wider range of legume species into forage systems would allow farmers to create agricultural landscapes that are more resilient to climate change of any type of limitation. Such diversification of minor legumes boosts productivity and stability in the productivity of fodder systems. Legumes can be grown in companion or rotation with one another based on their diverse growth patterns, nutrition profile, and root systems. This brings in a different kind of organic matter and establishes varied soil microorganisms, which not only increase overall quality from a nutritional perspective but also augment soil health. Furthermore, a greater variety of pollinators and beneficial insects may be supported by the different flowering dates of minor legumes (Cole et al. 2022), which further contributes to ecosystem services like pest control and pollination. Besides, the promotion of minor legume forages will reduce dependency on the small range of dominant species. This lowers the risks of possible pests, diseases, or market fluctuations that might adversely affect the primary legumes. This surely will contribute to the development of more sustainable agriculture, since such diversification encourages the use of locally adapted species that require fewer inputs and are also better able to resist or tolerate local challenges. In general, the inclusion of minor legume forages in agricultural systems increases the environmental sustainability of farming, contributes to improved biodiversity, and helps in developing more resilient and productive agriculture toward global challenges such as climate change.

2 Characteristics and Uses of Minor Legume Species

2.1 Nutritional Status of Minor Legume Species

Nutrition is a critical part of health and development. Better nutrition leads to better health for infants, children, and mothers; sharper minds; stronger immune systems; safer pregnancy and childbirth; reduced risk of non-communicable diseases like diabetes and heart disease; and a longer life. SDG Target 2.2 by 2025, end all forms of malnutrition, including achieving by 2025 the internationally agreed targets on stunting and wasting in children under 5 years of age, and address nutritional needs of adolescent girls, pregnant, and lactating women and older persons. Mineral micronutrient deficiencies in humans are widespread globally, especially among women and children. However, micronutrients are those nutrients that are needed by the human body in small amounts, yet more than 2 billion people worldwide are estimated to have a deficiency of critical vitamins and minerals, with particularly important deficiencies of vitamin A, iodine, iron, and zinc, which have adverse effects on human health. In 2011, about 6.9 million children under the age of 5 died worldwide because of undernutrition, which caused them to be more vulnerable to diseases.

Legumes play a crucial role in improving food security worldwide, addressing nutrition needs, and promoting sustainable agriculture. Foods like beans, lentils, and peas are rich in protein, fiber, vitamins, and minerals, making them very important to human nutrition. As such, legumes are an important part of enhancing the lives of many communities with food insecurity, as they provide affordable, high-quality protein, and other vital nutrients. It can help solve malnutrition issues in countries that rely heavily on crops and where animal protein may be scarce. Prakash et al. (2001) screened seeds of 104 leguminous species belonging to 17 genera (Table 1) and reported that the variation of fiber contents was 4.1–8.9%, carbohydrate 18.4–49.2%, ash 1.8–7.2%, TIA 48.7–87.5 mg/g, oil 1.3–19.8% and protein 11.0–51.6%. The protein content (41–45%) in *Acacia mellifera* (41.6%), *Albizzia lebbek* (43.6%), *Bauhinia triandra* (42.7%), *Lathyrus odoratus* (42.8%), *Parkinsonia aculeata* (41.6%), *Psophocarpus tetragonolobus* (41.9%), *Sesbania paludosa* (41.2%), and *S. sesban* (43.8%) was in close proximity to soybean (42.8%), whereas *Bauhinia retusa* (51.6%), *B. variegata* (46.5%), *Delonix elata* (48.7%), and *Gliricidia maculata* (46.3%) showed higher percentages of protein than soybean. This high protein content makes them valuable for addressing protein deficiencies in diets, particularly in resource-poor settings. While there is great praise for the amino acid profiles in common legumes, underutilized legumes also provide essential amino acids; some varieties are considered to be inadequate in methionine and cysteine amino acids. Nevertheless, all these legumes can be adequately mixed with cereals in order to achieve a full amino acid profile. For example, Bambara groundnut contains a high proportion of lysine and therefore improves the amino acid profile of grains (Yao et al. 2015). Underutilized legumes are rich in dietary fiber, which aids digestion and helps in blood sugar level control. They are also sources of essential micronutrients such as iron, zinc, magnesium, and folate-often at concentrations equal to or exceeding those found in more commonly consumed legumes. The pigeon pea is an example of a legume noted for its high folate content, crucial for cellular activity and tissue growth. Most underutilized legumes contain antioxidants and bioactive compounds that help maintain different conditions of the body. Studies on *Canavalia gladiata*, or sword bean, have determined that antioxidant properties may be used to reduce oxidative stress and inflammation (Hwang et al. 2023). This has become increasingly recognized as an integral part of general health promotion and prevention of chronic diseases. Besides the nutritional benefits derived from them, underutilized legumes are faced with problems arising from hard-to-cook phenomena and antinutritional factors inhibiting digestion or

Table 1 Nutritional profile of minor legume crops

Taxon	Moisture (%)	Fiber (%)	Carbohydrate (%)	Ash (%)	Trypsin inhibitor activity (mg/g)	Protein (%)
Acacia auriculiformis	6.9	7.5	49.2	4.3	78.5	31.8
A. mellifera	5.8	6.8	38.3	3.6	67.9	41.6
A. richardiana	7.3	7.8	46.1	5.3	72.8	35.3
Albizzia lebbek	8.4	8.5	35.2	4.1	78.4	43.6
Bauhinia retusa	6.1	6.7	18.4	5.4	87.5	51.6
B. triandra	8.7	7.3	23.6	7.2	76.8	42.7
B. variegata	9.1	6.2	21.3	6.3	82.6	46.5
Caesalpinia pulcherrima	9.6	5.2	47.4	1.8	57.9	37.5
Delonix elata	6.2	8.1	26.2	3.8	68.3	48.7
Erythrina fusca	4.9	8.9	33.5	6.8	48.7	36.8
Gliricidia maculata	7.8	6.7	24.1	5.4	58.4	46.3
Indigofera enneaphylla	8.3	5.8	36.5	4.2	66.3	38.3
Lathyrus odoratus	8.9	6.2	35.8	3.8	58.7	42.8
Parkinsonia aculeata	6.5	4.5	43.7	2.6	49.8	41.6
Parkia biglandulosa	5.3	5.7	26.9	4.3	64.8	35.8
Pithecelobium dulce	6.5	5.3	36.2	4.8	54.7	35.5
Psophocarpus tetragonolobus	7.6	6.4	21.5	5.1	86.6	41.9
Sesbania bispinosa	8.8	4.8	47.8	3.6	65.8	39.5
S. paludosa	9.5	4.3	46.5	3.5	68.7	41.2
S. sesban	7.9	4.1	48.3	2.9	72.5	43.8

Source: Prakash et al. (2001)

absorption of some nutrients. Beans and lentils have had a good share of research work in addition to development aspects geared at enhancing the cooking aspects of such grains to improve their utilization. However, soaking, boiling, and fermentation processing methods have been shown to reduce antinutritional factors in underutilized legumes, increasing their digestibility manifold.

2.2 Use of Minor Legume Species in Forage Production

Among other things, minor legume species are attracting increased interest for forage production due to their nutritional advantages, environmental benefits, and potential role in enhancing the sustainability of agriculture. Inclusion of these species into forage systems can largely raise the general nutritional quality of the diet offered to the grazing animals. One of the most important benefits of using legumes in forage production is their ability to fix atmospheric nitrogen through symbiotic relationships with bacteria of the genus Rhizobium. This improves soil fertility, reduces the need to apply synthetic nitrogen fertilizers, and contributes to sustainable agriculture. For instance, legumes can increase nitrogen soil levels for the benefit of subsequent crops and improve the general soil condition. By reducing the need for synthetic fertilizers and improving soil carbon sequestration, legumes have the potential to decrease greenhouse gas emissions resulting from agricultural activities (Boddey et al. 2020). Legumes enhance soil health by improving soil organic matter content and reducing soil erosion. Their root structure works to stabilize soil and improve its structure, particularly in marginal areas (Marshall et al. 2016). This attribute is essential for maintaining the natural resource base in agricultural landscapes. The incorporation of minor legumes into forage systems can support biodiversity. Producing a diverse range of legumes with grasses can build more resilient ecosystems that, in turn, support a broader range of soil organisms and wildlife. Such diversity could promote more efficient pest and disease management with reduced dependence on chemical inputs. Minor legumes have the potential to exhibit niche partitioning and facilitation with other plant species, thus enabling resource utilization more efficiently in diverse communities (Rehling et al. 2021). This complementarity can lead to positive biodiversity–function relationships, where increasing legume diversity enhances overall ecosystem productivity and stability

3 Genetic Diversity and Breeding of Minor Legume Forage Species

3.1 Genetic Diversity of Minor Legume Species and Its Significance in Breeding

Genetic diversity in minor legume species forms the backbone of agricultural sustainability and resilience. This diversity enhances their breeding potential and contributes to food security and adaptability to changing climatic conditions. Genetic diversity makes it possible for the selection of these important traits, allowing legumes to tolerate different abiotic stresses such as drought, salinity, and extreme temperatures. This adaptability can help ensure productivity under weather and climate change. For instance, certain underutilized small legumes exhibit tolerance to poor conditions, which could be accessed through breeding programs for the development of more resilient cultivars. In addition, genetic variability in minor legumes may contribute to their resistance against pests and diseases. Breeding programs incorporating a diverse range of germplasm can develop varieties with enhanced resistance to common pathogens, thereby minimizing dependency on chemical pesticides while promoting sustainable agriculture. In addition, small legumes often have unique nutritional profiles further amendable through breeding (Samal et al.

2023). Genetic diversity provides breeders with the opportunity to select for improved protein content, essential amino acids, and micronutrients, contributing to better food quality and nutritional security. Legumes are crucial in agro-ecosystems, especially in terms of nitrogen fixation, which improves soil fertility. Greater genetic resources can enhance these ecosystem services and contribute to soil health and more sustainable agriculture.

The wild relatives and traditional landraces represent the huge reservoirs of genetic diversity for minor legumes. This is partly because, besides being a source of variation, the wild relatives are invariably carriers of desirable traits which can be used toward adaptation and resilience but which may have been lost in modern cultivars. For example, several landraces of chickpea and lentil have been recognized as genetic resources useful for breeding programs with yield and stress tolerance improvement (Singh et al. 2022b; Teso et al. 2022). International gene banks, such as those maintained at CIAT and ILRI, maintain extensive collections of accessions of legumes. Gene banks are key to the conservation of genetic diversity and assistance to breeding programs. The development of core collections enables exploitation of genetic resources by narrowing it down to workable numbers with maximum representation of variability. Improved phenotypic and genotypic knowledge of the accessions conserved in gene banks can stimulate their use in breeding programs. This will enable breeders to identify those characters that are highly useful for the purpose of varietal improvement.

Many modern legume varieties are derived from a limited number of parental lines, leading to a narrow genetic base (Gayacharan et al. 2023). This can make them less adaptable and resilient to diseases and environmental stresses. For instance, there are only a few common ancestors shared by a surprisingly large number of chickpea cultivars, hence compromising the gene pool for breeding. Most breeding programs have focused on short-term gains in yield at the cost of long-term genetic diversity. The immediate focus might be causing an eventual loss of important genetic traits valuable for future adaptation and sustainability. This situation is exacerbated by the predisposition toward using elite germplasm rather than exploring "exotic" germplasm. Compared with the major crops, the minor legumes receive fewer resources and attention in research; this can be a constraint to the study and utilization of genetic diversity (Nayak et al. 2022). This lack of investment can limit the development of innovative breeding programs that leverage the full genetic potential of these species.

3.2 Prospects of Conventional Breeding for Minor Legume Species

Traditional methods of improvement in minor legume species have been plagued by many setbacks, though promising advantages are being accomplished through germplasm evaluation, genetic diversity assessment, and the identification of promising accessions. Generally, breeders have been concerned with the maintenance of crop genetic resources, germplasm evaluation, and genetic diversity assessment in minor legumes. This has resulted in the identification of promising genotypes possessing a desirable attribute. Germplasm evaluation studies provided information on diversity, heritability, correlations between traits, and environmental influences that may guide breeding. For instance, in moth bean, accessions such as IC36245, IC36555, IC36667, IC36577, and IC36604 were identified as promising through germplasm evaluation (Kanishka et al. 2023). Handling problems of small flowers that drop off during crossing have made hybridization programs challenging in some minor legumes, like moth bean. Consequently, there have been few genetic studies. Variety development programs for minor legumes probably started in the late 1960s or early 1970s. Systematic evaluation of germplasm has identified useful accessions for various traits, which could be used in minor legume-breeding programs with a view to enhancing the level of resistance to both biotic and abiotic stresses along with improving grain yields.

Conventional improvement of minor legume species has several major constraints that need to be overcome before such crops can be further improved. Most minor legumes possess a very narrow genetic base due to the reason that they are underutilized crops. This narrow genetic diversity restricts the available traits for selection and makes it hard to achieve significant improvements in yield, disease resistance, and other agronomic traits. This situation is further aggravated by the self-pollinating habit of many legume species, resulting in lower genetic diversity within the population. Many minor legumes suffer from reproductive difficulties such as self-incompatibility, low fertility, and embryo abortion that complicate hybridization. Such challenges might cause crossing with other varieties to succeed at low rates, thereby restricting new breeding lines. Reduced seed set can also occur when a high percentage of flowers drop before setting fruit in some species (Singh et al. 2022a). Conventional breeding is generally a time-consuming and tedious process. New variety development through hybridization and selection usually requires several years in conventional breeding, which again is a serious drawback when considering the progress made on cereals. Breeding progress in legumes is slow, with relatively low genetic gain rates compared to major crops. Most legumes are sensitive to environmental variables including temperature, photoperiod, and genotype × environment interactions (Bhandari et al. 2017; Iannucci et al. 2008; Papastylianou et al. 2021). These sensitivities have the potential to impact flowering synchrony and maturity, complicating the process of breeding for improvement. In addition to this, slower growth rates in early stages of development translate to higher competition from weeds, further complicating cultivation (Frenda et al. 2013). Most minor legumes have anti-nutritional factors that can limit their use both in human diets and as animal feed. All these challenges need to be taken into consideration by breeding programs to enable the improvement of the nutritional quality of these legumes without compromising agronomic performance.

3.3 Potential Modern Breeding Techniques for Trait Improvement in Minor Legume Species

Modern biotechnology has revolutionized the improvement of crops through the attainment of new height in terms of productivity, sustainability, and resilience, particularly in agriculture. Such tools utilize genetic engineering, precision breeding techniques, and state-of-the-art analysis for novel crop varieties. Gene editing using CRISPR technology has also been applied in minor legumes, showcasing its potential for crop improvement. The editing process is complicated through challenges in transformation efficiency and genotype dependence. A promising way involves using the hairy root transformation system, which allows for fast assessment of the efficiency of sgRNA prior to stable transformation for new ways of enhancing overall efficiency in CRISPR applications in this species (De Koning et al. 2023). A significant study demonstrated CRISPR/Cas9-mediated DNA-free editing of the *4CL* and *RVE7* genes in chickpea protoplasts. This research aimed to understand drought response mechanisms and highlighted the potential for achieving polygenic adaptive traits such as drought tolerance through CRISPR technology. The study emphasized the importance of DNA-free editing to avoid foreign DNA integration, which is a concern in traditional methods (Badhan et al. 2021).

4 Challenges and Opportunities in the Utilization of Minor Legume Species

4.1 Technical Challenges in Breeding Minor Legume Species

The low use of minor legumes for breeding can be attributed to numerous factors which hinder the potential of these legumes such as narrow genetic base, genetic erosion, reproductive barriers, and limited germplasm availability (Sharma et al. 2013; Robertson and Bounejmate 1999; Samal et al. 2023). Most small legumes are characterized by a narrow genetic base due to their self-pollinating nature, hence limited genetic diversity. A narrow gene pool can hardly allow for the possibility of selection for desirable features, such as yield, resistance to diseases, and adaptability to changing environmental conditions. Such breeding programs result in varieties selected mostly with poor performance in terms of withstanding biotic and abiotic stresses. This forms a critical concern for underutilized legume species in terms of genetic erosion. Since only a few dominant species are widespread, minor legumes are genetically becoming increasingly limited and, therefore, increasingly vulnerable to diseases and other environmental changes. Continued reliance on the same or few cultivars risks the erosion of essential genetic traits that might well be needed in future breeding. In addition, these minor legumes have specific reproductive issues inhibitive of hybridization, such as embryo abortion and low seed set. These complex reproductive barriers can drastically reduce the success rate of conventional breeding methods and make the attainment of required genetic gain very difficult (Samal et al. 2023). Breeding programs are dependent on diverse available germplasm. Unfortunately, most minor legumes have less genetic resources due to less interest in their improvement compared to major legumes like soybean and chickpea. This actually diminishes the possibilities of trait improvements in minor legumes due to lack of diverse genetic materials.

4.2 Socio-Economic Factors Influencing the Adoption of Minor Legume Species in Forage Production

Socio-economic factors provide the most significant determinants in adopting the minor legume species in forage production. In fact, there exists an interaction of a number of socio-economic variables that condition the adoption of minor legume species in forage production, including household characteristics, financial resources, education, land tenure, and access to inputs. Addressing these through targeted interventions can boost the adoption rates of these critical forage legumes, which would ultimately pay off dividends in improved livestock production and sustainability of smallholder farming systems. Some of the key points that could be identified from recent studies pertain to aspects like awareness and adoption rates. A high level of awareness among farmers with respect to forage legumes is paramount. In a study undertaken in Zimbabwe, for instance, 94.7% of the households were aware of forage legume technologies, while 87.9% adopted them. The most adopted species included *Mucuna pruriens* and *Leucaena leucocephala* (Mapiye et al. 2006). However, the intensity of adoption varied, and many farmers were partial adopters since most of them experience constraints of various kinds. Household size, land size, and number of dairy cattle significantly influence the adoption of forage legumes (Washaya et al. 2024). Such technologies are bound to be adopted by larger sized households with more resources. Access to financial resources is of essence. In most cases, the binding constraints of farmers involve lack of capital, inhibiting them from investing in any new technology. The farmer's educational background and experience in farming also hold a key role. Generally, the younger, educated farmers would be more open to the adoption of new technologies because they tend to be better managers and willing to experiment. Land tenure issues inhibit the possibility of long-term investments in forage legumes. Farmers would not want to invest in legumes if doing so were to take away land from food crop production, often the first priority for ensuring household food security (Duncan et al. 2019). Good quality seeds and fertilizers, extension services support, are also very important. Inadequate inputs often result in low yields and persistence of legumes; this has been discouraging their adoption. One of the major constraints to the realization of forage legume adoption among smallholder farmers is a lack of fencing material. For instance, it has been reported that about 18.6% of households indicated that lack of fencing was one of the principal reasons for not adopting forage and browse legumes. This problem is particularly serious among the communal farmers, of whom approximately 91.5% reported that lack of fencing material was a major problem (Mapiye et al. 2006).

4.3 Opportunities for Public–Private Partnerships and Institutional Support

The way forward in improving minor legume species is developing strong PPPs with institutionalized support mechanisms. Collaboration, resource sharing, and supportive policy go a long way in enhancing productivity and the sustainability of the cropping of legumes for food security and agricultural resilience. Innovative breeding techniques, coupled with involvement with the community, will further ensure effectiveness and benefits to all concerned in these endeavors. The use of PPPs now allows public research institutions to share resources with private seed companies. Such collaboration might even develop improved legume varieties with wider resistance to climate change, which can also be more adaptable to the diverse agro-ecological conditions. Breeding will also be accelerated by pooling expertise and funding from both sectors, thereby enhancing genetic diversity in legume crops (Rubiales et al. 2021). Genomic selection is one such technique that offers more precise selection opportunities by markers distributed all over the genome which are further quantified to calculate the genomic estimated breeding value that aids in the selection of superior genotypes (Krishna et al. 2024). PPPs can, therefore, improve the inclusion of new breeding approaches like GS. These methods make the selection for identifiable features of interest such as yield and stress tolerance more efficient, which is necessary in minor legume species that may otherwise not get ample attention in conventional breeding initiatives. Applications of GS can decrease time and costs related to variety development drastically. Because minor legumes are relatively less profitable when compared with other major crops, they usually require upstream public funding in support of their research and development. PPPs may attract private investment in showcasing potential economic returns related to legume varieties. In addition, public funding can cover the breeding programs and seed production and distribution costs that are too expensive for private companies (Galluzzi et al. 2020).

5 Conservation and Utilization of Minor Legume Species

5.1 In Situ and Ex Situ Conservation Methods

The genetic diversity represents the most important feature in its conservation effort. These genetic resources are highly diverse and include legume forages from landraces to wild relatives and modern cultivars; all these constitute a useful pool for the improvement of crops. Biotechnological tools developed for marker-assisted selection will identify novel genes and alleles that may be transferred into commercially important varieties. This maintains desirable characteristics and will adapt the legume forages to changing conditions. The genetic diversity of the legume forage species is their key to long-term survival and adaptation. It allows adaptation to changing environments, resistance against pests and diseases, and maintenance of ecosystem functions. Thus, conserving genetic diversity represents a sustainable strategy that will ensure the persistence of these valuable legume forage resources (Tripathi et al. 2020). Genetic erosion in legume forages is enhanced by habitat loss, overgrazing, and the introduction of non-native species. Replacement of traditional landraces with modern cultivars in farming systems that have recently shifted from subsistence to market-oriented production contributes to loss of genetic diversity. In-situ conservation, defined by protection of legume forage species in their natural habitats is extremely important for maintaining genetic diversity. This method will maintain complicated interactions between species and the environment. Ex-situ conservation by germplasm collections provides a complement for in-situ efforts through the safeguarding of genetic diversity in seed banks and field gene banks (Salgotra and Chauhan 2023).

In-situ conservation of legume forages is an important strategy toward maintaining the genetic diversity and ecological integrity within the natural habitats of the legume species. A new approach is initiated in Switzerland with regard to in-situ conservation and is focused on the wild relatives of forage crops. This initiative establishes the integration of approximately 1566 hectares into an auction-based policy for the purposes of conserving the populations. This points out one of the most overlooked components of more classic conservation strategies: intraspecific genetic diversity (Kägi et al. 2023). In-situ conservation is, therefore, important in the Mediterranean region due to the rich biodiversity the region holds and the ecological role the species play. It was indicated that the diversity of legume species is different across the various sites where specific areas such as Samta and Rayyan resulted in higher richness when compared to other sites (Ajlouni et al. 2010). This variability underscores the importance of targeted conservation efforts that consider local ecological conditions and species interactions.

Ex-situ conservation of forage legumes is a highly important strategy in contributing to maintaining genetic diversity and, consequently, assures that such vital resources are available for future agricultural needs. It involves the exploration and collection, evaluation and characterization, conservation and utilization (Fig. 1; Thakur et al. 2024). Gene banks are storage facilities for genetic materials that provide the means with which to conserve PGRs crucial for food security and sustainable agriculture. It is estimated that approximately 1750 gene banks are available globally, and about 130 of these maintain more than 100,000 accessions each. ICAR-NBPGR serves as one of the major repository conserving over 60,000 accessions of legumes (Table 2). Major gene banks include the National Bureau of Plant Genetic Resources, ICAR-NBPGR of India, National Centre for Genetic Resources Preservation of the USA, and the International Agricultural Research Centres of the Consultative Group on International Agricultural Research, CGIAR.

5.2 Role of Gene Banks and Seed Banks in Preserving Genetic Diversity

The conservation of genetic diversity, especially in legume species that are so very crucial for global food security and ecological sustainability, requires facilities such

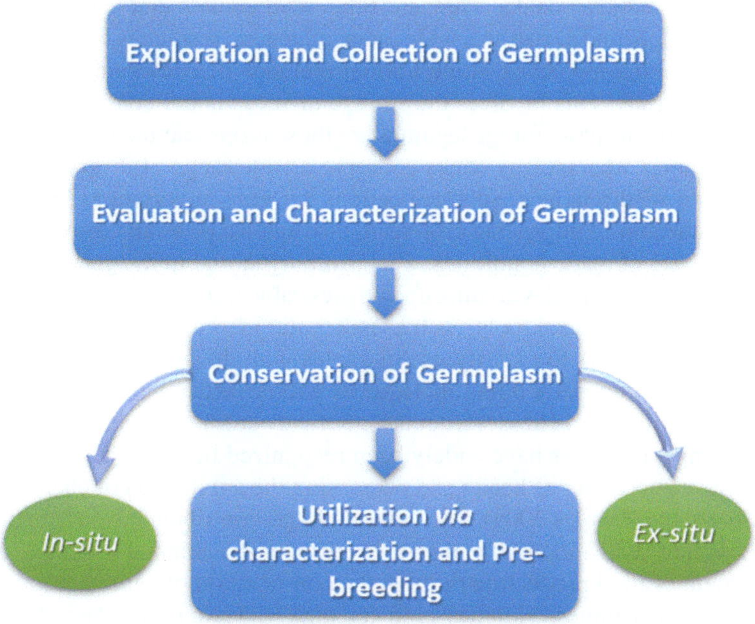

Fig. 1 Generalized workflow gene bank activities

Table 2 Details of accessions of legumes stored in Indian National Genebank

S. no.	Genus	Number of accessions	Number of species	Indigenous collection	Exotic collection
1	*Cajanus*	11,940	7	11,576	364
2	*Canavalia*	132	4	128	4
3	*Cicer*	15,053	10	11,908	3145
4	*Cyamopsis*	4308	1	4270	38
5	*Lablab*	1305	4	1298	7
6	*Lathyrus*	2747	10	2557	190
7	*Lens*	2612	7	1852	760
8	*Macrotyloma*	3157	2	3146	11
9	*Mucuna*	55	2	49	6
10	*Phaseolus*	4154	3	1730	2424
11	*Pisum*	4680	1	3528	1152
12	*Psophocarpus*	220	1	84	136
13	*Rhynchosia*	21	10	21	0
14	*Vicia*	786	4	411	375
15	*Vigna*	15,113	32	13,199	1914
	Total	*66,283*	*98*	*55,757*	*10,526*

Source: Gore et al. (2022)

as gene banks and seed banks. They are supportive centers that help in the collection, conservation, and access to a wide range of genetic resources of legumes. Gene banks use ex situ approaches as the normal methods of conserving legume seeds and other materials. This involves the storage of seeds under controlled conditions for longevity. In this case, forage legumes are those crops that include beans, lentils, and peas, among others. Gene banks conserve a wide range of diversity, including landraces and wild relatives. Such collections are very important to breeding programs with the aim of improving crop resilience against pests and diseases, in addition to changing climatic conditions. This diversity stored in the gene banks forms the basis of the new varieties coming up with desirable traits (Aubry 2023). Applying advanced biotechnological tools, such as molecular markers and genomic sequencing, allows researchers to identify useful alleles with the aim of transferring those into cultivated varieties. By doing so, legumes will be further enhancing their adaptability to environmental stresses as contributors toward food security in the face of climate change. Legumes have widely been recognized because they are able to fix soil nitrogen, improving its health and decreasing the application of chemical fertilizers (Crews and Peoples 2004). The gene bank has contributed to sustainable agriculture, which will hopefully result in less environmental damage and an improvement of food production through the conservation of genetic resources of a wide variety of legumes. Varieties bred for high yields and tolerance for climate variability, pests, and diseases can benefit from the seeds conserved in such banks.

Recent technological advances are enhancing the potentials of gene banks in effectively conserving legume species. One such project is the INCREASE project, which aims to enhance the conservation and utilization of genetic resources of food legumes. It involves the development of Intelligent Collections comprising nested core collections of single-seed descent-purified accessions supported by exhaustive genotypic and phenotypic characterization that allows for efficient management of genetic resources (Bellucci et al. 2021). Similarly, high-throughput phenotyping technologies are being integrated into genebanks to assess the traits of legume accessions more efficiently (Cortés and Barnaby 2023). This technology allows for the rapid collection of data on plant characteristics, which can be crucial for identifying desirable traits for breeding programs.

6 Conclusions

To sum up, minor legume species are an untapped source of genetic diversity that could be very important for the development of future forage crop breeding. Their adaptability to many conditions and advantageous traits including pest resistance, drought tolerance, and nitrogen fixation make them invaluable resources for sustainable farming methods. By incorporating these species into breeding programs, superior forage varieties that are more productive and better adapted to the unique requirements of livestock systems and climate change can be developed. Furthermore, maintaining biodiversity and guaranteeing food security depend on the management and maintenance of these genetic resources. Harnessing the genetic potential of minor legumes will be essential to developing robust and productive forage systems that can support future livestock production needs as the agricultural landscape changes. We can effectively utilize minor legume species by providing research and development in this field paramount importance. This will ensure the incorporation of these species into common forage approaches and add to the overall sustainability of agricultural systems.

Author Contributions Conception and development of the idea was done by AC and VCSK. Literature survey and drafting were done by SSB. Illustrations and pictorial representations were completed by NRT and PRR. Proofing was done by AC and VCSK. The final manuscript was edited by PRR, NRT, and VAU.

References

Ajlouni MA, El-Oqlah A, Al-Ghzawi AL, Al-Tawaha AA, Amri A (2010) Diversity and in-situ conservation of legume species in the Mediterranean ecosystem. Jordan J Agric Sci. https://archives.ju.edu.jo/index.php/jjas/article/view/1817

Allen DE, Singh BP, Dalal RC (2011) Soil health indicators under climate change: a review of current knowledge. In: Soil biology. Springer, Berlin, Heidelberg, pp 25–45. https://doi.org/10.1007/978-3-642-20256-8_2

Aubry S (2023) Genebanking plant genetic resources in the postgenomic era. Agric Hum Values 40(3):961–971. https://doi.org/10.1007/s10460-023-10417-7

Badhan S, Ball AS, Mantri N (2021) First report of CRISPR/CAS9 mediated DNA-free editing of 4CL and RVE7 genes in chickpea protoplasts. Int J Mol Sci 22(1):396. https://doi.org/10.3390/ijms22010396

Bellucci E, Aguilar OM, Alseekh S, Bett K, Brezeanu C, Cook D, De La Rosa L, Delledonne M, Dostatny D, Ferreira J, Geffroy V, Ghitarrini S, Kroc M, Agrawal SK, Logozzo G, Marino M, Mary-huard T, Mcclean P, Meglič V et al (2021) The INCREASE project: intelligent collections of food-legume genetic resources for European agrofood systems. Plant J 108(3):646–660. https://doi.org/10.1111/tpj.15472

Bhandari K, Sharma KD, Rao BH, Siddique KHM, Gaur P, Agrawal SK, Nair RM, Nayyar H (2017) Temperature sensitivity of food legumes: a physiological insight. Acta Physiol Plant 39(3). https://doi.org/10.1007/s11738-017-2361-5

Boddey RM, Casagrande DR, Homem BGC, Alves BJR (2020) Forage legumes in grass pastures in tropical Brazil and likely impacts on greenhouse gas emissions: a review. Grass Forage Sci 75(4):357–371. https://doi.org/10.1111/gfs.12498

Cole LJ, Baddeley JA, Robertson D, Topp CF, Walker RL, Watson CA (2022) Supporting wild pollinators in agricultural landscapes through targeted legume mixtures. Agric Ecosyst Environ 323:107648. https://doi.org/10.1016/j.agee.2021.107648

Cortés AJ, Barnaby JY (2023) Editorial: harnessing genebanks: high-throughput phenotyping and genotyping of crop wild relatives and landraces. Front Plant Sci 14. https://doi.org/10.3389/fpls.2023.1149469

Crews T, Peoples M (2004) Legume versus fertilizer sources of nitrogen: ecological tradeoffs and human needs. Agric Ecosyst Environ 102(3):279–297. https://doi.org/10.1016/j.agee.2003.09.018

De Koning R, Daryanavard H, Garmyn J, Kiekens R, Toili MEM, Angenon G (2023) Fine-tuning CRISPR/Cas9 gene editing in common bean (Phaseolus vulgaris L.) using a hairy root transformation system and in silico prediction models. Front Plant Sci 14. https://doi.org/10.3389/fpls.2023.1233418

Duncan AJ, Öborn I, Nziguheba G, Temesgen T, Muoni T, Okeyo I et al (2019) Supporting smallholder farmers' decisions on legume use in East Africa – the LegumeCHOICE approach. Asp Appl Biol 138:85–92

Frenda AS, Ruisi P, Saia S, Frangipane B, Di Miceli G, Amato G, Giambalvo D (2013) The critical period of weed control in faba bean and chickpea in Mediterranean areas. Weed Sci 61(3):452–459. https://doi.org/10.1614/ws-d-12-00137.1

Galluzzi G, Seyoum A, Halewood M, Noriega IL, Welch EW (2020) The role of genetic resources in breeding for climate change: the case of public breeding programmes in eighteen developing countries. Plants 9(9):1129. https://doi.org/10.3390/plants9091129

Gayacharan N, Parida SK, Mondal N, Yadav R, Vishwakarma H, Rana JC (2023) Mining legume germplasm for genetic gains: an Indian perspective. Front Genet 14. https://doi.org/10.3389/fgene.2023.996828

Gore PG, Singh N, Gupta V (2022) Grain legumes diversity of Indian National Genebank: a potential resource for food and nutritional security. Indian J Plant Genet Resour 35(3):191–193

Hwang YJ, Hwang H, Go H, Park N, Hwang K (2023) Sword bean (Canavalia gladiata) pods induce differentiation in MC3T3-E1 osteoblast cells by activating the BMP2/SMAD/RUNX2 pathway. Nutrients 15(20):4372. https://doi.org/10.3390/nu15204372

Iannucci A, Terribile MR, Martiniello P (2008) Effects of temperature and photoperiod on flowering time of forage legumes in a Mediterranean environment. Field Crop Res 106(2):156–162. https://doi.org/10.1016/j.fcr.2007.11.005

Jabbar A, Ibrahim H, Mahmood A, Javaid MM, Wasaya A, Nadeem MA, Ismael M, Awais M, Khan SR (2023) Modern agronomic measurement for climate-resilient agriculture. In: Climate-resilient agriculture. Springer, Cham, pp 81–105. https://doi.org/10.1007/978-3-031-37428-9_5

Kägi C, Petitpierre B, Meyer P, Lötscher Y, Eggenberg S, Aubry S (2023) Fostering in situ conservation of wild relatives of forage crops. Front Plant Sci 14. https://doi.org/10.3389/fpls.2023.1287430

Kanishka R, Gayacharan, Basavaraja T, Chandora R, Rana JC (2023) Moth bean (Vigna aconitifolia): a minor legume with major potential to address global agricultural challenges. Front Plant Sci 14:1179547. https://doi.org/10.3389/fpls.2023.1179547

Krishna K, Chavan S, Jagadeesh K, Niranjan T, Bruno A, Reddy YVK et al (2024) Genomic selection-driven wheat breeding for superior genetic gains: status quo and future steps. In: Advances in wheat breeding: towards climate resilience and nutrient security. Springer Nature Singapore, Singapore, pp 721–765

Mapiye C, Foti R, Chikumba N, Poshiwa X, Mwale M, Chivuraise C, Mupangwa JF (2006) Constraints to adoption of forage and browse legumes by smallholder dairy farmers in Zimbabwe. Livest Res Rural Dev 18(12):2006

Marshall AH, Collins RP, Humphreys MW, Scullion J (2016) A new emphasis on root traits for perennial grass and legume varieties with environmental and ecological benefits. Food Energy Secur 5(1):26–39. https://doi.org/10.1002/fes3.78

Menesch J, Godde C, Venables W, Renard D, Richardson A, Cobelli O, Waha K (2023) Agricultural diversification for crop yield stability: a smallholder adaptation strategy to climate variability in Ethiopia. Reg Environ Chang 23(1). https://doi.org/10.1007/s10113-022-02021-y

Nayak SP, Lone RA, Fakhrah S, Chauhan A, Sarvendra K, Mohanty CS (2022) Mainstreaming underutilized legumes for providing nutritional security. In: Future foods: global trends, opportunities, and sustainability challenges. Elsevier, pp 151–163. https://doi.org/10.1016/b978-0-323-91001-9.00023-2

Papastylianou P, Vlachostergios DN, Dordas C, Tigka E, Papakaloudis P, Kargiotidou A, Pratsinakis E, Koskosidis A, Pankou C, Kousta A, Mylonas I, Tani E, Abraham EM, Karatassiou M, Kostoula S (2021) Genotype X environment interaction analysis of faba bean (Vicia faba L.) for biomass and seed yield across different environments. Sustainability 13(5):2586. https://doi.org/10.3390/su13052586

Prakash D, Niranjan A, Tewari SK, Pushpangadan P (2001) Underutilised legumes: potential sources for low-cost protein. Int J Food Sci Nutr 52(4):337–341. https://doi.org/10.1080/09637480120057521. PMID: 11474898

Rehling F, Sandner TM, Matthies D (2021) Biomass partitioning in response to intraspecific competition depends on nutrients and species characteristics: a study of 43 plant species. J Ecol 109(5):2219–2233. https://doi.org/10.1111/1365-2745.13635

Robertson LD, Bounejmate M (1999) Genetic erosion of pasture and forage legumes in the Mediterranean basin. In: Genetic resources of Mediterranean pasture and forage legumes, Current plant science and biotechnology in agriculture. Springer, Dordrecht, pp 67–77. https://doi.org/10.1007/978-94-011-4776-7_6

Rubiales D, Annicchiarico P, Patto MCV, Julier B (2021) Legume breeding for the agroecological transition of global agri-food systems: a European perspective. Front Plant Sci 12. https://doi.org/10.3389/fpls.2021.782574

Salgotra RK, Chauhan BS (2023) Genetic diversity, conservation, and utilization of plant genetic resources. Genes 14(1):174. https://doi.org/10.3390/genes14010174

Samal I, Bhoi TK, Raj MN, Majhi PK, Murmu S, Pradhan AK, Kumar D, Paschapur AU, Joshi DC, Guru PN (2023) Underutilized legumes: nutrient status and advanced breeding approaches for qualitative and quantitative enhancement. Front Nutr 10. https://doi.org/10.3389/fnut.2023.1110750

Sarıkaya MF, Tatar M (2023) Global climate change and crop production. Uluslararası Sivas Bilim ve Teknoloji Üniversitesi Dergisi 2(2):65–78

Sharma S, Upadhyaya HD, Varshney RK, Gowda CLL (2013) Pre-breeding for diversification of primary gene pool and genetic enhancement of grain legumes. Front Plant Sci 4. https://doi.org/10.3389/fpls.2013.00309

Singh G, Gudi S, Amandeep N, Upadhyay P, Shekhawat PK, Nayak G, Goyal L, Kumar D, Kumar P, Kamboj A, Thada A, Shekhar S, Koli GK, Meghana D, Halladakeri P, Kaur R, Kumar S, Saini P, Singh I, Ayoubi H (2022a) Unlocking the hidden variation from wild repository for accelerating genetic gain in legumes. Front Plant Sci 13. https://doi.org/10.3389/fpls.2022.1035878

Singh RK, Singh C, Ambika N, Chandana BS, Mahto RK, Patial R, Gupta A, Gahlaut V, Gayacharan N, Hamwieh A, Upadhyaya HD, Kumar R (2022b) Exploring chickpea germplasm diversity for broadening the genetic base utilizing genomic resourses. Front Genet 13. https://doi.org/10.3389/fgene.2022.905771

Teso MLR, Lara-Romero C, Rubiales D, Parra-Quijano M, Iriondo JM (2022) Searching for abiotic tolerant and biotic stress resistant wild lentils for introgression breeding through predictive characterization. Front Plant Sci 13. https://doi.org/10.3389/fpls.2022.817849

Thakur NR, Ingle KP, Sargar PR, Baraskar SS, Kasanaboina K, Awio B, Pranati J, Abdi G (2024) Sustainable utilization of wild germplasm resources. In: Sustainable utilization and conservation of plant genetic diversity, Sustainable development and biodiversity. Springer, Singapore, pp 551–590. https://doi.org/10.1007/978-981-99-5245-8_16

Tripathi K, Gore PG, Singh M, Pamarthi RK, Mehra R, Gayacharan C (2020) Legume genetic resources: status and opportunities for sustainability. In: Legume crops – prospects, production and uses. IntechOpen. https://doi.org/10.5772/intechopen.91777

Washaya S, Masunda B, Ngongoni N (2024) Impact and adoption of feed technologies at Nharira-Lancashire Dairy Scheme. J Appl Anim Nutr:1–8. https://doi.org/10.1163/2049257x-20231002

Yao D, Kouassi K, Erba D, Scazzina F, Pellegrini N, Casiraghi M (2015) Nutritive evaluation of the Bambara groundnut Ci12 landrace [Vigna subterranea (L.) Verdc. (Fabaceae)] produced in Côte d'Ivoire. Int J Mol Sci 16(9):21428–21441. https://doi.org/10.3390/ijms160921428

Recent Advancements in Breeding and Genomics of Minor Grass Species

Sadia Zafar, Javed Iqbal, Banzeer Ahsan Abbasi, Rizwan Rasheed, Muhammad Arslan Ashraf, Hadiqa Afzal, Shoaib Khan, Shumaila Ijaz, Zakir Ullah, Ghulam Murtaza, Rashid Iqbal, Hassan Sher, Zahid Ullah, and Ahmad Ali

Abstract

As the global population continues to grow and become wealthier, the demand for dairy products, such as milk and meat, is expected to rise significantly. Meeting this demand requires not only the expansion of production but also a shift toward more sustainable agricultural practices. Minor grass species, with their potential to serve as diverse and resilient forage options, are increasingly recognized as critical to this effort. However, traditional breeding methods have been insufficient in optimizing these species for large-scale production due to their inherent complexities. In tropical regions, where pasture lands play a vital role in sustaining high levels of animal output, there is an urgent need to improve the breeding and genomic strategies applied to these minor grass species. Over

S. Zafar · H. Afzal
Department of Botany, Division of Science and Technology, University of Education, Lahore, Punjab, Pakistan
e-mail: sadia.zafar@ue.edu.pk

J. Iqbal (✉)
Department of Botany, Bacha Khan University, Charsadda, Khyber Pakhtunkhwa, Pakistan

B. A. Abbasi
Department of Botany, Rawalpindi Women University, Rawalpindi, Pakistan

R. Rasheed · M. A. Ashraf
Department of Botany, Government College University, Faisalabad, Pakistan
e-mail: drrizwan@gcuf.edu.pk; marslanashraf@gcuf.edu.pk

S. Khan
Department of Chemistry, Abbottabad University of Science and Technology (AUST), Abbottabad, Pakistan

S. Ijaz
Department of Burn and Plastic Surgery, Shenzhen Institute of Translational Medicine, The First Affiliated Hospital of Shenzhen University, School of Biomedical Engineering, Shenzhen University Medical School, Shenzhen University, Shenzhen, China

F. S. Baloch et al. (eds.), *Forage Crops in the Omics Era*,
https://doi.org/10.1007/978-981-95-9249-4_12

the last few decades, conventional breeding programs have released well-developed cultivars, but the genetic gains achieved have not kept pace with the growing global demand. The challenges in breeding these species such as poor seed set, apomixis, photoperiod sensitivity, and thermal sensitivity make the design and implementation of effective breeding programs particularly difficult. This chapter investigates the recent advancements in the breeding and genomics of minor grass species, emphasizing the necessity of these innovations. It explores the latest developments in germplasm accumulation and utilization, as well as the significant achievements made possible through genetic and biotechnological research. These advancements hold the potential to overcome the limitations of conventional breeding, paving the way for the development of more resilient and productive forage species that can thrive in tropical environments.

Keywords

Minor grass · Traditional breeding · Genomics · Germplasm · Biotechnological research

1 Introduction

Grasses are a broad group of genera that include more than 600 species that are currently utilized as animal feed and for grazing (Ravagnani et al. 2012). Most warm season genera of grasses that grow well in tropical climates are Andropogoneae, Paniceae, Eragrosteae, and Chlorideae (Davis et al. 2009). Tropical grass species genera began from Africa, while there are two significant genera that are indigenous to tropical and sub-tropical region which are Paspalum and Axonopus. With a diverse spectrum of eco climate ranging from humid tropical to semi-arid, temperate to Alpine, the Indian subcontinent is one of the primary sites of crop origin and plant diversity in the world. Grass diversity in Indian subcontinent includes 1256 species and 245 genera (Karthikeyan 2005).

Z. Ullah
Department of Plant Sciences, Faculty of Biological Sciences, Quaid-i-Azam University, Islamabad, Pakistan
e-mail: zakirullah@bs.qau.edu.pk

G. Murtaza
School of Agriculture, Yunnan University, Kunming, Yunnan, China

R. Iqbal
Department of Agronomy, Faculty of Agriculture and Environment, The Islamia University of Bahawalpur, Bahawalpur, Pakistan

Department of Life Sciences, Western Caspian University, Baku, Azerbaijan
e-mail: rashid.iqbal@iub.edu.pk

H. Sher · Z. Ullah · A. Ali
Center for Plant Sciences and Biodiversity, University of Swat, Charbagh Swat, Pakistan

Almost all the grasses including lasiurus, panicum, cynodon, bothriochloa, Pennisetum, and dichanthium are known for their fodder value. Along with micro centers associated with particular species, the main center of genetic diversity is North-Eastern Region for subtropical types and the Peninsular India for tropical region. As compared to cultivated cereals, tropical perennial forage grasses have a very different breeding strategy. All three forms of reproduction, that is, sexual, asexual, and clonal are present in temperate grasses. Approximately 60% of tropical grasses are apomictic, meaning they reproduce exactly like their mother plant. The benefit of this approach is that with a thorough understanding of apomixes, hybrid vigor can be restored. Vigorous plant selection and intensity selection have significantly related to advancement in genetics. Through phenotypic breeding, we can improve feed productivity, quality, and adaptability. Cultivar development brought about by direct selection of plants from population has increased digestibility by 26% and produced leaf yield that are 80–130% more productive.

Maximizing the yield of metabolizable energy is the main priority even though there may be a negative correlation between the larger dry matter yield and nutritious quality. There are several grass types that are commonly used for agriculture and non-agriculture process belonging to cool season minor grass species that enhance the variety of crop. Minor grasses is a catchall term that defines the apparent level of regard in which these species are held by researchers, plant developers, germplasm preservationists, and industry. The species discussed in this chapter are also characterized as "secondary" in their agricultural significance (Spedding and Diekmahns 1972). Minor grass species are suited to a variety of environmental condition and can be found in a variety of natural, semi-natural, and planted grassland types, including everlasting grazing area and lawns and meadows. Some examples of these species are *Trisetion flavescentis*, *Cynosution cristati*, and *Arrhenatherion elatioris*. These are frequently dominant or diagnostic species of vegetative unit.

The emergence of several genetic markers as a result of molecular biology advancements has transformed our understanding of the structure and evolution of crop genomes (Altaf et al. 2025). Finding genetic variety in crops provides a chance to comprehend the molecular basis of several biological occurrences (Adhikari et al. 2017). The use of molecular markers in phylogenetic analysis of various organisms has become more significant in recent years (Yalinkiliç et al. 2024). Fast DNA sequencing methods' accessibility and the advancement of reliable statistical analysis techniques gave this discipline a new start. Traditional classification systems of organisms based on morphology have major drawbacks despite still being frequently used. On the other hand, it seems that the classic morphology-based method for phylogenetic investigations can be supplemented via the use of molecular markers, despite recent popularity and the fact that they are not without flaws. Nuclear genome sequencing (NGS) and EST (expressed sequence tag) programs should enhance the number of genes used for phylogenetic analysis of crops, animals, and microorganisms. As they exhibit significant commonality across a wide range of organisms, the power of genes involved in an organism's physiology, such as salt tolerance genes, the cell division (cdc) genes, heat shock genes, receptor genes, homeotic genes, etc., should also be investigated (Patwardhan et al. 2014).

The development of molecular markers is now moving in new directions owing to public genomic databases, which have also changed the sorts of PCR-based procedures frequently employed in plant science (Sun et al. 2024). In addition to the frequently used DNA markers, different approaches have been proposed. Targeted fingerprinting methods include non-identical methodological improvements such as incorporating gene or promoter components into primers using amplified DNA technology applications. Although these semi-random markers have promising qualities, they can cause collision and heterogeneous issues similar to those discovered with randomly generated fingerprints. Many elements in motion found in plant genomes can also be used to produce fingerprints. By exploiting certain targeted sites, these markers boost genome coverage and generate bands that primarily resemble one another (Poczai et al. 2013).

In its first stage, plant breeding was a naïve process that has evolved over the last decades into a modern, sophisticated system. One of the foundations of contemporary plant breeding is the 1900 revision of Mendel's rules of genetics. These days, it can also count using contemporary methods like automated phenotyping, bioinformatics, and high-throughput sequencing. Keeping in mind the parameters found in the so-called "breeder's equation." These tactics could be used in breeding pipelines to increase rates of genetic gain. By altering the breeder's equation's variables, it will be possible to add more potential for obtaining genetic gain, reducing, as a consequence, time spent in cultivar development (Pereira et al. 2018).

2 Techniques and Objectives for Breeding

In order to create better fodder varieties in the future, breeding is essential. Plant selection has been a highly active process that has significantly advanced genetics (Busey 1989). Globally focused efforts have been made to generate all types of major grasses for a variety of growing environment such reduce and transport pastureland and for grassland purpose (Fig. 1).

2.1 Selection

In order to ensure 80–130% greater leaf production large germplasm bases direct selection in phenotypic traits and the identification of superior accession number are essential (Burton 1989; Burton et al. 1993). The majority of commercially marketed tropical grass cultivars are natural ecotypes that were chosen from natural diversity (Hacker and Jank 1998). Many widely adaptable and farmer-preferred grass types have emerged in India as a result of the use of the natural variability found in the germplasm collection. The most basic method of breeding for the enhancement of fodder species is bulking, harvesting, and mass selection of seed without evaluating the progeny. In addition, a number of breeding philosophies have been used to

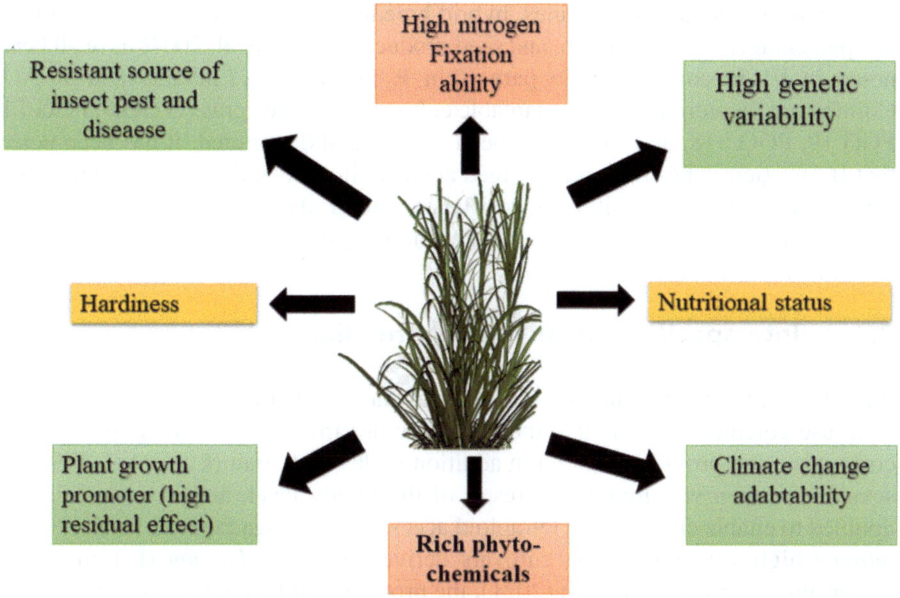

Fig. 1 The different key attributes of a hybrid grasses

Table 1 Applications of grass biodiversity in varietal development programs

Tropical grass species	Varietal development
Pennisetum pedicellatum	Bundel-2, Jawahar Pennisetum-12, Bundel-1, Pusadeenanath grass
Panicum maximum	Bundel Guinea-1, Bundel Guinea-2, Harithasree, Punjab Guinea grass-1
Cenchrus ciliaris	Bundel Anjan-1, Marwar Anjan 1, CO-1, Bundel Anjan-3
Cenchrus setigerus	Marwar Dhaman, CAZRI-76
Chrysopogon fulvus	Bundel Dhawalu Ghas-1

create superior cultivars and variations. Additive genes, dominant genes, or varying degrees of dominance provide the foundation for resistance to numerous diseases and insect pests. Applications of Grass Biodiversity in Varietal Development Programs details are given (Table 1).

2.2 Hybridization

Apomictic reproductive behavior is the largest obstacle in recombinant breeding in tropical grasses; therefore, identifying and producing sexual lines in apomictic grasses may open up a new field of study. By utilizing apomixis, the advantages of hybrid vigor can be fixed provided suitable sexual lines are available. In certain species, such as Bracharia, buffel grass, and guinea grass, when there are viable facultative or obligate sexual lines, hybrid breeding has been used. Maximum foliage percentage, faster regrowth, and seed production (Jank et al. 2004) have all been noted in F1 in comparison to parents in P. From PAU, Ludiana, and TNAU, Coimbatore, numerous widely adaptable cultivars of guinea grass (PGG 9, PGG 14, PGG 19, PGG 518, and CO_2) have been created and distributed. It has been noted that B. ruziziensis has high-quality feed and a well-defined flowering cycle, B. brizantha has a yield and spittlebug resistance, and B. decumbens has vigor and an adaptation to acidic, infertile soils (Peters and Lascano 2003).

2.3 Interspecific and Polyploidy Breeding

It has been attempted to directly cross sexual and apomictic plant of *B. ruzuzuensis* or to use colchicnine to double the chromosomes in order to make grasses cross-compatible and provide variation in addition to desirable trait (Lutts et al. 1994). A sexual accession was found as a result of the ploidy level, and it was artificially doubled to enable crosses with tetraploid accessions (Hojsgaard et al. 2008). *P. purpureum* high seed production cultivars derived from *P. glaucum* (L.) crossings. According to Pereira and Lédo (2021), the progenies of this hybrid are triploid and sterile at first, but after being doubled by colchicine, they become hexaploid and fertile. ICAR-IGFRI in Jhansi, India, the ploidy series of grasses was constructed in order to decipher the genomic complexity in the native stage of the majority of grass species. In order to study the ploidy-regulated phenomenon at the genetic level, ICARIGFRI has created the largest ploidy series of Panicum maximum in the world. Through interspecific hybridization, pearl millet can produce superior fodder plant varieties. P. Orientale × P. glaucoma at IGFRI, Jhansi; *P. glaucoma* x *P. squamulatum*. In comparison to other tropical grasses, the resulting plant has demonstrated increased biomass, better adaptation, better tolerance for moisture deficit, and improved forage qualities (IGFRI, Annual Report 2014–15).

3 Genetic Resources and Utilization

In European gene banks and organizations, there are no sizable minor agricultural species, from germplasm collection which also included grasses. In ex situ plant collections across Europe, 1615 accessions of the genus Agrostis the best-represented among minor grasses have been preserved, compared to 11,228 accessions of the genus Lolium (EURISCO 2009). According to GRIN 2009, the genus Bromus has the greatest number of accessions among minor grasses in the USA germplasm collection. The abundance of accessions in the Australian and germplasm collections in New Zeeland further suggests the widespread interest in Bromus (Williams 1996; Margot Forde Germ. Cent. 2009). Majority of germplasm accessions has a wild sample status. Minor grass landraces rarely found and, if they are, have long since vanished. The oldest cultivars' germplasm has frequently been lost as well, either because new cultivars have supplanted them or because the grass species was removed from past crop production (*P. arundinacea* in Germany, for example). Nonetheless, wild populations can still be found in large numbers in their native habitats for the majority of minor grass species, meaning that they can be gathered to protect their genetic variety for use in the present and the future.

4 Forage Crop Genomics Study

Biotechnology-based new methods are becoming more widely available; one such method is functional genomics, which enables marker-assisted introgression and selection. In contrast to other crops, tropical grasses have a higher genetic diversity because of their polyploidy, apomictic character, high potential for interspecies hybridization, lack of domestication-related genetic bottlenecks, and high potential for genetic diversity. Allele mining can benefit greatly from the biodiversity conservation across a variety of geographic and biological niches, as it will enable a more robust response to upcoming problems such shifting climatic conditions and altered land use. Advancing genetic techniques based on traits, including mapping and expression profiling, have proven effective in identifying novel genes or genomic areas that impact the quality of fodder. It is easy to communicate relevant information between related forage species. Functional genomics' increased comprehension of the biochemical significance of these genes which will lead to the rapid discovery of novel candidate genes for various phenotypes. In theory, any fodder crop's genes of interest can be found by utilizing data derived from conserved map positions or sequence homology.

5 Studies on the Association Between Genotype and Trait

Developments in DNA resequencing, SNPs in crops have been found more quickly. This has made it easier to identify the genes or loci linked to a range of significant features, improving our knowledge of crops and speeding up their breeding (Huang and Han 2014). Genome-wide association studies (GWASs) have discovered the areas linked to particular phenotypes. In addition to loci linked to climatic adaptation, a number of genomic areas have been linked to field variables, including NUE, components of yield, and metabolite content. After analyzing 38 crops, 8332 loci corresponding to 367 phenotypes were found, of which 6573 loci were linked to 257 field attributes. The study focused on four main phenotypic categories: metabolite diversity, resistance to biotic and abiotic stimuli, plant design, and yield.

6 Improvements in Yield

Many studies have been conducted on yield-related features since they are very important economically. Using GWAS, numerous significant loci associated with yield have been found. For instance, the qSW locus has been mapped to a QTL region linked to weight of soybean seed. Grain length of rice is regulated and yield is increased by genes expressing APETALA2/ethylene-responsive 125 element binding protein and OsSPL13, a plant-specific transcription factor discovered by Si et al. (2016) and Yu et al. (2017). Ma et al. (2019) discovered OsSNB, a different gene that regulates the grain size of rice which can result in an increase in grain weight, width, and length. It has been discovered that transcription factors are significant in crop yield characteristics (Hu et al. 2021). Rice grain weight was related to transcription factor called MYB (Huang et al. 2012). Furthermore, it was discovered that the genes for ribosome recycling factor and ARF18 were connected to the length of silique and oil seed weight (Hu et al. 2022). In tartary buckwheat (*Fagopyrum tataricum*), an AP2 transcription factors and a tryptophan amino transferases-related protein that are encoded by genes were shown to be potential factors that influence the weight of seed and size of seed correspondingly (Zhang et al. 2021). An AP2/ETHYLENE response factor-encoding gene was found to confer an increase in lint production, a drop in seed index, and a decrease in the number of bolls per plant. Similarly, a transcription factor AP2 in cotton has been linked to yield attributes (Fang et al. 2017). Castor bean seed size was discovered to be correlated with the WRKY transcription factors (Cai et al. 2021).

7 Plant Architecture

Numerous complicated factors, such as height of the plant, structure of branches, angle of tillers, and panicle shape, influence the architecture of plants (Cheng 2022). The discovery of numerous genes controlling plant architecture has been accelerated by the use of plant genome resequencing. Using GWAS, loci associated with tiller angle in rice were found, and TAC3 was identified as a potential causative gene (Dong et al. 2016) for increased tiller angle. Likewise, it was discovered that the rice leaf angle was correlated with three basic helix loop (BHLH) genes family: OsbHLH153, OsbHLH173, and OsbHLH174 (Dong et al. 2018). There have also been reports of genes having pleiotropic effects, such as OsNAL1, which is linked to number of panicle and width of leaf blade (Yano et al. 2016). A GWAS made it easier to identify the rice architecture-controlling an encoding gene N-acetyl glucosamine transferase (OsSPY) (Yano et al. 2019). It was discovered that certain areas linked to leaf angle and tiller angle in several investigations overlapped (Dong et al. 2018). Figure 2 represent population genomic approaches in plant breeding to cope up with changing climatic conditions.

8 Effects of Various Abiotic and Biotic Components on the Production of Feed Crops

Due to the effects of global warming and potential climatic anomalies, forage crops frequently encounter an increasing number of biotic and abiotic obstacles that severely hinder their growth and output. It has been demonstrated that when abiotic

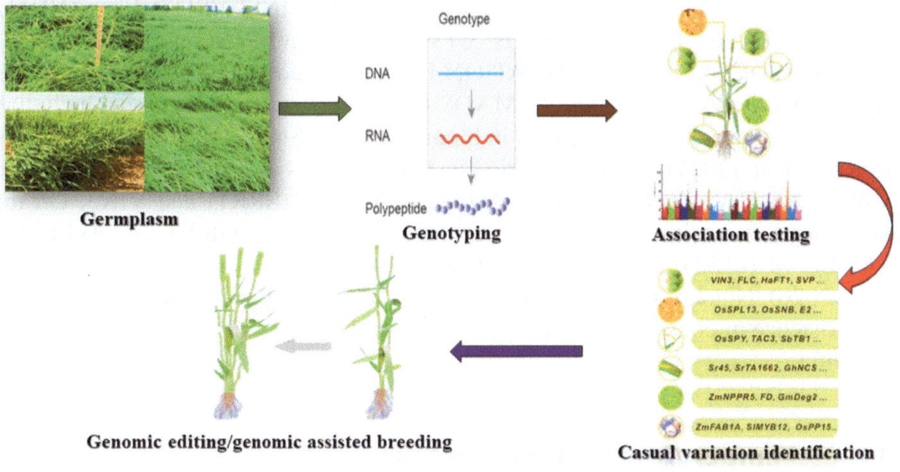

Fig. 2 Diagram of population genomic approaches in plant breeding to cope up with changing climate

challenges like drought, temperature, soil moisture, and other variables occur together with biotic pressures at different stages of crop growth, the crop yield is more severely harmed. In simple terms, biotic factors living or biological forces that have an impact on agricultural output or practice affect agricultural productivity. Weeds, insects, and illnesses are known to be impacted by salt, high and low temperatures, and drought (biotic variables).

9 Reactions of Genomics to Environmental Changes

Change breeding crops that are suited to new and more varied climates is desperately needed, as the effects of climate change on agriculture become more pronounced (Sarıkaya and Tatar 2023; Abberton et al. 2016; Anderson et al. 2020). Identification of appropriate adaptation alleles is improved by an understanding of species-wide variety (Marsh et al. 2021; Petereit et al. 2022). Data sets with a high range of habitat resequencing gives a valuable insight into how genome changes in response to changing environment.

10 Flowering Time

One significant trait linked to plant adaptation is flowering time. Significant relationships between Arabidopsis flowering timing and genes involved in flowering regulation and chilling treatment were found by the 1001 Genomes Project. A few noteworthy instances are VERNALIZATION INSENSITIVE3 (VIN3), VERNALIZATION INSENSITIVE 3 (VIN3), FLOWERING LOCUS T (FT), FLC, SVP, DELAY OF GERMINATION 1 (DOG1), and FLOWERING LOCUS C (FLC). It was discovered that 28 MADS-box genes in *Brassica napus*, including FLC, were connected to variations in blooming time (Song et al. 2020). According to Kang et al. (2021), flowering period in *B. juncea* was also found to be correlated with SENSITIVITY TO RED LIGHT REDUCED 1 (SRR1) and VIN3. On the other hand, flowering was linked to genomic areas containing FT, FLOWERING LOCUS M (FLM), and EARLY FLOWERING7 (ELF7) in sunflower (Todesco et al. 2020). Numerous of these genes are shared by many plant species, indicating the existence of conserved processes that control flowering time and environmental adaption.

11 Temperature and Precipitation

Traits related to abiotic stress resistance are important targets for crop improvement and a major research focus. For instance, the gene HEAT-INTOLERANT1 (HIT1) responsible for thermos tolerance was discovered to be related to number of climatic variables that are linked to temperature in the study of sunflower by Todesco et al. (2020). Also, 141 loci were identified in *Setaria viridis*, located in the genes or their surroundings related to the maintenance of organellar genomes, breakdown of proteins, homeostasis of gibberellic acid, and metabolism of fucose, which in some manner are linked to their ability to adapt to severe weather conditions (Mamidi et al. 2020). In the fonio millet study, genes linked to Harmon biosynthesis, the process of metabolism, and the chromosomal region that is linked to heat stress were discovered to maintain the shoot apical meristem and to develop root growth; a locus linked to drought sensitivity had a homolog of Arabidopsis WSD1 which is known to enhance the drought tolerance through wax buildup. Such genes and all other factors had been identified from the genomic loci in ear fonio and homologs of these genes are known to be involved in chilling stress tolerance in Arabidopsis (Zheng et al. 2020; Ji et al. 2021).

12 Resistance to Abiotic Stress

Dealing with different types of abiotic stress, such as high temperature, salinity, and water stress, has been considered a key factor in the growth and adaptation processes of plants. GWAS has been useful in finding candidate loci and genes that are linked to tolerance against such stresses. For instance, GWAS has successfully identified three loci in cotton with G-type lectin receptor kinase family genes that are associated to high-temperature stress (Ma et al. 2021). Furthermore, a GWAS identified OsSAP16 to be an important gene for regulating germination at low temperatures in rice (Wang et al. 2018). Lastly, GWAS was used to identify a gene in Arabidopsis, NIP1;1, that confers sensitivity to toxic H2O2 (Sadhukhan et al.

2017). Several transporters have been identified by GWAS to be related to salt tolerance in various plants. For example, OsHKT1;1, OSHKT1;4, OsHKT1;5, OsHKT2;1, OsMOLYBDATE TRANSPORTER1;1, OsHAK5, OsHAK6, and OsZmHAK4 were discovered by some studies related to salt tolerance (Meyer et al. 2016; Campbell et al. 2017; Zhang et al. 2019; Hartley et al. 2020). These studies highlight how non-destructive phenotyping can be efficiently done and for the discovery of genes and loci linked to crop tolerance to abiotic stress in conjunction with GWAS.

13 Challenges

The gene banks that hold genetic resources for forage crops and grasses have a lot of climate-resilient traits in them. Most target genes associated with such allelic variants in most forage crops and grasses are not yet disclosed (Hanley et al. 2021). This fact presented several unknown and specific challenges to the identification of useful allelic variants and their efficient exploitation in forage crops and grasses. The selection of germplasm to be used in allele mining is the first significant obstacle. Screening crops whole germplasm increase the likelihood of finding uncommon alleles although it is thought to be laborious and wasteful process. The development of accurate phenotype methods, core and mini-core collection, and effective computational tools can increase the likelihood of mining potential allelic variations (Kumar et al. 2015). Only few grass species have been shown to possess candidate genes that provide resistance to abiotic stress. Furthermore, the majority of grasses lack genomic resources in the form of whole genome sequence data. Another challenge in the case of forage crops and grasses is the handling of genomic resources for exploiting allele mining. The delineation of core promoter regions from the regulatory elements is also a difficult task in mining promoters. This makes it very difficult to achieve a fully sequenced genome in the characterization of cis-acting regulatory elements, since this causes transcriptional differences due to various trans-acting factors and environmental conditions. In addition to the significant time and efforts involved in allele mining, another crucial factor is the high cost of sequencing platforms.

14 Genomics-Assisted Breeding for Trait Improvement

Identification of genomic markers is the first step in genomics-assisted breeding that is linked to genes or QTLs associated with the desired agronomic characteristic, followed by their implementation in the breeding platform (Singh et al. 2017; Crossa et al. 2017). With NGS-based high genotyping technologies, molecular markers help at the early stage of growth in the breeding cycle by providing a variety of desirable progeny. Genomic selection, marker-assisted recurrent selection and marker assisted backcrossing are GAB techniques that have been used for the improvement of crop. To speed up the breeding process, speed breeding has recently been introduced to the list. In this situation, expanding breeding plans and updating methods are necessary to swiftly tackle with the issue that affects the agriculture.

15 Backcrossing with Marker Assistance and Recurrent Selection

Marker-assisted back-crossing (MABC) refers to the process of introducing a specific genomic region such as QTL, Locus, or gee that carries a desired trait from a donor genotype into breeding line or elite cultivar. This is achieved through multiple generations of back-crossing. This process avoids linkage drag. The entire genomes of an elite parent, along with the gene(s) or QTL(s) from the donor parent that contribute to the desired phenotype, are present in the final MABC product (Gupta et al. 2010). The effectiveness of Marker-Assisted Back-crossing (MABC) depends on factors such as the number of molecular markers utilized, the strength of marker association with desired phenotype, the presence of undesirable linkage drag, and the population size used in each back-crossing generation. This technique has been extensively employed to develop superior cultivars with enhanced tolerance to biotic and abiotic stresses. For instance, root QTLs from the drought-tolerant japonica upland cultivar CT9993 were incorporated into drought susceptible lowland indica cultivars IR20 to produce drought-tolerant near-isogenic lines (NILs) (Suji et al. 2012). Yang et al. (2019) combined genomics-based background selection with MABC to introduce the blast resistance gene Pi2 into highly developed strains. By using the fast and advanced generation approach, Kang et al. (2019) created near isogenic lines (NIL) from parents resistant to the rice stripe virus and for those which are susceptible to it more quickly. This was done by using MABC to create a virus-resistant line quickly. Details related to quantitative trait locus mapping linked to intricate agronomic traits are given in (Fig. 3).

16 Speed Breeding and Genomics Selection

Large-scale DNA markers scattered across the genome are used in genomic selection (GS) and genomic-wide selection to create superior germplasm lines. Due to this, the genomi selection method may be able to identify several widely dispersed QTLs or genes with slightly additive effect. In a breeding population, intensive phenotyping is not required, and genotype predictions: a combination of genomic and pedigree data spanning multiple generations of the breeding cycle are the main focus of subsequent offspring selection (Nakaya and Isobe 2012). Recurrent selection is based on genomic estimated breeding vale (GEBV), which is the product of the information index and the combined influence of genome-wide molecular marker (Singh et al. 2017). Precise GEBV and consequently GWS require high-density molecular markers with each QTL in linkage with at least one single genomic marker (Habier et al. 2007). When it comes to breeding, time plays a significant role in determine when genotypes are released to farmers. In conventional

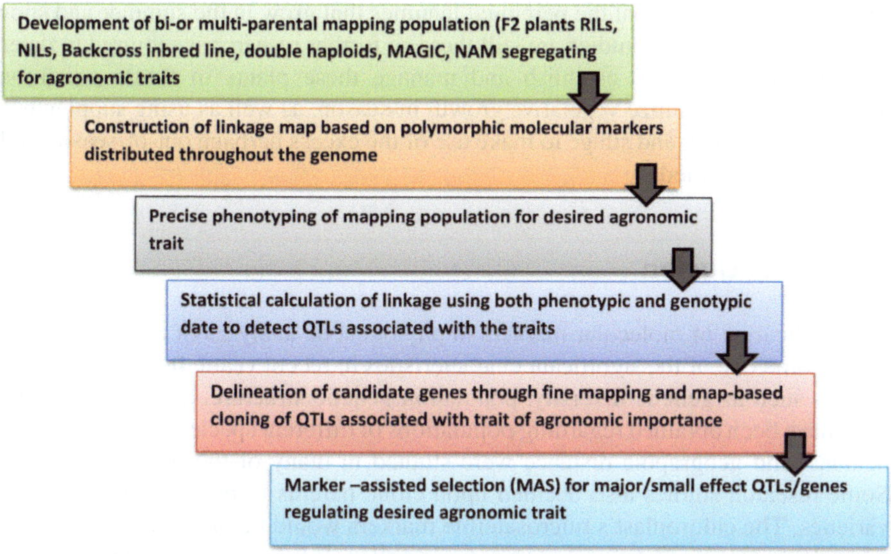

Fig. 3 Quantitative trait locus mapping linked to intricate agronomic traits

breeding, crossing studies takes 3–7 years, yield, disease, and quality testing takes 4–5 years, and variety release takes further 3–5 years. It takes a lot of time and effort to improve a single genotype or variation. In light of this, the strategy known as speed breeding which involves varying day length and light level to speed up the life cycle (Watson et al. 2018; Ghosh et al. 2018).

17 Future Research

With climate already showing increased variability in wet and dry periods, heat stress, and increased levels of carbon dioxide in the atmosphere, the changed circumstances of future environments are going to increase the tropical grass meadows function. The establishment and growth of tropical species are probably going to benefit from these altered rainfall patterns as well as the anticipated temperature increases. However, such potential for tropical pastures remains largely untapped. Red meat demand is expected to high, with both domestic and export markets requiring greater quantities. In order to meet the increased demand, we need to enhance both the quality and quantity of pastures to provide adequate feed and also livestock numbers have to rise. Tropical pastures contribute importantly to an improved supply of forage in the summer-dominant rainfall areas throughout the year (Boschma et al. 2010).

The possibility of developing tropical grass pastures has attracted some interest in the past few years, but it needs to be documented and verified as we need to improve their productivity and persistence, even in challenging environments.

These results suggest that the perennial legumes that grow in the summer and show promise in our current studies are still there so further research is required to determine the best ways to establish and manage these plants in future condition. Techniques to maximize vegetative growth in-season, as well as using supplements and conserving hay and silage to make use of the excess herbage out-of-season, still remains to be studied.

18 Conclusion

The application of molecular markers in phylogenetic analyses of many organisms has become one of the significant characteristics in recent years. In studies, various aspects such as genetic diversity, population genetic structure, and recombination rates have been obtained regarding populations of different species in detail. Genetic distance and geographic distance were studied in many of the types of research. Some research studies also decided upon clone parents to produce new synthetic varieties. The chloroplast's microsatellite markers would be of great help in identifying those species that received transgenic pollen. It is important to identify germplasms accurately and determine their rates of variability for the development of effective conservation and breeding programs. The adoption of molecular genetics methods by the use of microsatellite markers in the initial assessment of germplasms helps to identify the species and assess the likelihood of successful hybridization. Improvement in tropical grasses was achieved mainly by the exploitation of the available variability through conventional methods of plant improvement. The strategies for germplasm exploration, characterization, and utilization need to be reinforced, with greater emphasis laid on enhancing the collection in gene banks.

Author Contributions S.Z., J.I., B.A.A., R.R., M.A.A., H.A., designed and conceived the chapter idea. S.Z., J.I., M.A.A., R.I., wrote the original draft. S.K., S.I., Z.U., G.M., R.I., H.S., Z.U., and A.A., reviewed and edited the manuscript and provided valuable Insights. All authors have made valuable revisions and edited the manuscript and approved the last version.

References

Abberton M, Batley J, Bentley A, Bryant J, Cai H, Cockram J, Costa de Oliveira A, Cseke LJ, Dempewolf H, De Pace C et al (2016) Global agricultural intensification during climate change: a role for genomics. Plant Biotechnol J 14:1095–1098

Adhikari S, Saha S, Biswas A, Rana TS, Bandyopadhyay TK, Ghosh P (2017) Application of molecular markers in plant genome analysis: a review. Nucleus 60(3):283–297

Altaf MT, Nadeem MA, Ali A, Liaqat W, Bedir M, Baran N et al (2025) Applicability of Start Codon Targeted (SCoT) markers for the assessment of genetic diversity in bread wheat germplasm. Genet Resour Crop Evol 72(1):1205–1218

Anderson R, Bayer PE, Edwards D (2020) Climate change and the need for agricultural adaptation. Curr Opin Plant Biol 56:197–202

Boschma SP, Lodge GM, Harden S (2010) Seedling competition of lucerne in mixtures with temperate and tropical pasture species. Crop Pasture Sci 61:411–419

Burton GW (1989) Registration of 'Tifton 9' Pensacola bahiagrass. Crop Sci 29:1326
Burton GW, Gates RN, Hill GM (1993) Registration of 'Tifton 85' bermudagrass. Crop Sci 33:644
Busey P (1989) Cultural management of turfgrass salinity. Crop Science, 29(4), 1003–1007
Cai X, Sun X, Xu C, Sun H, Wang X, Ge C, Zhang Z, Wang Q, Fei Z, Jiao C, Wang Q (2021) Genomic analyses provide insights into spinach domestication and the genetic basis of agronomic traits. Nat Commun 12:7246
Campbell MT, Bandillo N, Al Shiblawi FRA et al (2017) Allelic variants of OsHKT1; 1 underlie the divergence between indica and japonica subspecies of rice (Oryza sativa) for root sodium content. PLoS Genet 13:e1006823
Cheng S (2022) Gregor Mendel: the father of genetics who opened a biological world full of wonders. Mol Plant 15:1641–1645
Crossa J, Pérez-Rodríguez P, Cuevas J et al (2017) Genomic selection in plant breeding: methods, models, and perspectives. Trends Plant Sci 22:961–975
Davis AP, Govaerts R, Bridson DM, Ruhsam M, Moat J, Brummitt NA (2009) A global assessment of distribution, diversity, endemism, and taxonomic effort in the Rubiaceae. Ann Mo Bot Gard 96(1):68–78
Dong H, Zhao H, Xie W, Han Z, Li G, Yao W, Bai X, Hu Y, Guo Z, Lu K et al (2016) A novel tiller angle gene, TAC3, together with TAC1 and D2 largely determine the natural variation of tiller angle in rice cultivars. PLoS Genet 12:e1006412
Dong H, Zhao H, Li S, Han Z, Hu G, Liu C, Yang G, Wang G, Xie W, Xing Y (2018) Genome-wide association studies reveal that members of bHLH subfamily 16 share a conserved function in regulating flag leaf angle in rice (Oryza sativa). PLoS Genet 14:e1007323
EURISCO Catalogue (2009). http://eurisco.ecpgr.org. Accessed 30 Dec 2008
Fang C, Ma Y, Wu S, Liu Z, Wang Z, Yang R, Hu G, Zhou Z, Yu H, Zhang M et al (2017) Genome-wide association studies dissect the genetic networks underlying agronomical traits in soybean. Genome Biol 18:161
Ghosh S, Watson A, Gonzalez-Navarro OE, Ramirez-Gonzalez RH et al (2018) Speed breeding in growth chambers and glasshouses for crop breeding and model plant research. Nat Protoc 13:2944–2963
Gupta PK, Kumar J, Mir RR, Kumar A (2010) Marker assisted selection as a component of conventional plant breeding. Plant Breed Rev 33:145–217
Habier D, Fernando RL, Dekkers JC (2007) The impact of genetic relationship information on genome-assisted breeding values. Genetics 177:2389–2397
Hacker JB, Jank L (1998) Breeding tropical and subtropical forage plants. In: Cherney JH, Cherney DJR (eds) Grass for dairy cattle. CABI, Wallingford, pp 49–71
Hanley SJ, Pellny TK, de Vega JJ, Castiblanco V, Arango J, Eastmond PJ et al (2021) Allele mining in diverse accessions of tropical grasses to improve forage quality and reduce environmental impact. Ann Bot 128(5):627–637
Hartley TN, Thomas AS, Maathuis FJ (2020) A role for the OsHKT 2; 1 sodium transporter in potassium use efficiency in rice. J Exp Bot 71:699–706
Hojsgaard D, Schegg E, Valls JF, Martínez EJ, Quarin CL (2008) Sexuality, apomixis, ploidy levels, and genomic relationships among four paspalum species of the subgenus Anachyris (Poaceae). Flora Morphol Distrib Funct Ecol Plants 203(7):535–547
Hu W, Ji C, Liang Z, Ye J, Ou W, Ding Z, Zhou G, Tie W, Yan Y, Yang J et al (2021) Resequencing of 388 cassava accessions identifies valuable loci and selection for variation in heterozygosity. Genome Biol 22:316
Hu J, Chen B, Zhao J, Zhang F, Xie T, Xu K, Gao G, Yan G, Li H, Li L et al (2022) Genomic selection and genetic architecture of agronomic traits during modern rapeseed breeding. Nat Genet 54:694–704
Huang X, Han B (2014) Natural variations and genome-wide association studies in crop plants. Annu Rev Plant Biol 65:531–551
Huang X, Zhao Y, Wei X, Li C, Wang A, Zhao Q, Li W, Guo Y, Deng L, Zhu C et al (2012) Genome-wide association study of f lowering time and grain yield traits in a worldwide collection of rice germplasm. Nat Genet 44:32–39

Jank L, Resende RMS, Calixto S, Gontijo Neto MM, Laura VA, Macedo MCM, Do Valle CB (2004) Preliminary performance of Panicum maximum accessions and hybrids in Brazil. In: Proceedings of the XX International Grassland Congress, Dublin

Ji F, Ma Q, Zhang W, Liu J, Feng Y, Zhao P, Song X, Chen J, Zhang J, Wei X et al (2021) A genome variation map provides insights into the genetics of walnut adaptation and agronomic traits. Genome Biol 22:300

Kang JW, Shin D, Cho JH et al (2019) Accelerated development of rice stripe virus-resistant, near-isogenic rice lines through marker-assisted backcrossing. PLoS One 14:e0225974

Kang L, Qiǎn L, Zheng M, Chen L, Chen H, Yang L, You L, Yang B, Yan M, Gu Y et al (2021) Genomic insights into the origin, domestication and diversification of Brassica juncea. Nat Genet 53:1392–1402

Karthikeyan S (2005) Common tropical and sub tropical sedges and grasses: an illustrated account: review. Rheedea 15(2):141–142

Kumar V, Singh A, Mithra SVA, Krishnamurthy SL, Parida SK, Jain S, ... Mohapatra T (2015) Genome-wide association mapping of salinity tolerance in rice (Oryza sativa). DNA Research, 22(2), 133–145

Lutts S, Kinet JM, & Bouharmont J (1994) NaCl-induced senescence in leaves of rice (Oryza sativa L.) cultivars differing in salinity resistance. Annals of Botany, 78(3), 389–398

Ma X, Feng F, Zhang Y et al (2019) A novel rice grain size gene OsSNB was identified by genome-wide association study in natural population. PLoS Genet 15:e1008191

Ma Y, Min L, Wang J, Li Y, Wu Y, Hu Q, Ding Y, Wang M, Liang Y, Gong Z et al (2021) A combination of genome-wide and transcriptome-wide association studies reveals genetic elements leading to male sterility during high temperature stress in cotton. New Phytol 231(1):165–181. https://doi.org/10.1111/nph.17325

Mamidi S, Healey A, Huang P, Grimwood J, Jenkins J, Barry K, Sreedasyam A, Shu S, Lovell JT, Feldman M et al (2020) A genome resource for green millet Setaria viridis enables discovery of agronomically valuable loci. Nat Biotechnol 38:1203–1210

Margot Forde Germplasm Centre (2009). http://www.agresearch.co.nz/seeds/(S(dgbwvanh45qqcnzvw dgkfi45))/frmquery.aspx

Marsh JI, Hu H, Gill M, Batley J, Edwards D (2021) Crop breeding for a changing climate: integrating phenomics and genomics with bioinformatics. Theor Appl Genet 134:1677–1690

Meyer RS, Choi JY, Sanches M, Plessis A, Flowers JM, Amas J, Dorph K, Barretto A, Gross B, Fuller DQ et al (2016) Domestication history and geographical adaptation inferred from a SNP mapof African rice. Nat Genet 48:1083–1088

Nakaya A, Isobe SN (2012) Will genomic selection be a practical method for plant breeding? Ann Bot 110:1303–1316

Patwardhan A, Ray S, Roy A (2014) Molecular markers in phylogenetic studies- a review. J Phylogen Evol Biol 2:131

Pereira WVS, & Lédo FJS (2021) Advances in forage crop breeding and genetic resources. Grass and Forage Science, 76(3), 321–335

Pereira WVS, et al. (2018) Genetic diversity and breeding perspectives in forage crops. Crop Breeding and Applied Biotechnology, 18(4), 395–402

Petereit J, Bayer PE, Thomas WJW, Tay Fernandez CG, Amas J, Zhang Y, Batley J, Edwards D (2022) Pangenomics and crop genome adaptation in a changing climate. Plants 11:1949

Peters M, Lascano CE (2003) Forage technology adoption: linking on-station research with participatory methods. Trop Grassl 37:197–203

Poczai P, Varga I, Laos M, Cseh A, Bell N, Valkonen J, Hyvönen J (2013) Advances in plant gene-targeted and functional markers: a review. Plant Methods 9(1):1–32

Ravagnani A, Michael T, Abberton SL (2012) Development of genomic resources in the species of Trifolium L. and its application in forage legume breeding. Agronomy 2:116–131

Sadhukhan A, Kobayashi Y, Nakano Y, Iuchi S, Kobayashi M, Sahoo L, Koyama H (2017) Genome-wide association study reveals that the aquaporin NIP1; 1 contributes to variation in hydrogen peroxide sensitivity in arabidopsis thaliana. Mol Plant 10:1082–1094

Sarıkaya MF, Tatar M (2023) Global climate change and crop production. Uluslararası Sivas Bilim ve Teknoloji Üniversitesi Dergisi 2(2):65–78

Si L, Chen J, Huang X et al (2016) OsSPL13 controls grain size in cultivated rice. Nat Genet 48:447–456

Singh RK, Sahu PP, Muthamilarasan M, Dhaka A, Prasad M (2017) Genomics-assisted breeding for improving stress tolerance of graminaceous crops to biotic and abiotic stresses: progress and prospects. In: Senthil-Kumar M (ed) Plant tolerance to individual and concurrent stresses, pp 59–81. https://doi.org/10.1007/978-81-322-3706-8_5

Song JM, Guan Z, Hu J, Guo C, Yang Z, Wang S, Liu D, Wang B, Lu S, Zhou R et al (2020) Eight high-quality genomes reveal pan-genome architecture and ecotype differentiation of Brassica napus. Nat Plants 6:34–45

Spedding CRW, & Diekmahns EC (1972) Grassland ecology. Oxford, UK: Oxford University Press

Suji KK, Prince KSJ, Mankhar PS, Kanagaraj P, Poornima R, Amutha K, Kavitha S, Biji KR, Gomez M, Babu RC (2012) Evaluation of rice (*Oryza sativa* L.) near isogenic lines with root QTLs for plant production and root traits in rainfed target populations of environment. Field Crop Res 137:89–96

Sun L, Lai M, Ghouri F, Nawaz MA, Ali F, Baloch FS, Nadeem MA, Aasim M, Shahid MQ (2024) Modern plant breeding techniques in crop improvement and genetic diversity: from molecular markers and gene editing to artificial intelligence—a critical review. Plants 13(19):2676

Todesco M, Owens GL, Bercovich N, Le'garé JS, Soudi S, Burge DO, Huang K, Ostevik KL, Drummond EBM, Imerovski I et al (2020) Massive haplotypes underlie ecotypic differentiation in sunflowers. Nature 584:602–607

Wang X, Zou B, Shao Q, Cui Y, Lu S, Zhang Y, Huang Q (2018) Natural variation reveals that OsSAP16 controls low temperature germination in rice. J Exp Bot 69:413–421

Watson A, Ghosh S, Williams MJ, Cuddy WS, Simmonds J et al (2018) Speed breeding is a powerful tool to accelerate crop research and breeding. Nat Plants 4:23–29

Williams WM (1996) Genetic resources of temperate native and low-input grasses in New Zealand and Australian collections. NZ J Agric Res 39:513–526

Yalinkiliç NA, Başbağ S, Altaf MT, Ali A, Nadeem MA, Baloch FS (2024) Applicability of SCoT markers in unraveling genetic variation and population structure among sugar beet (Beta vulgaris L.) germplasm. Mol Biol Rep 51(1):584

Yang D, Tang J, Yang D, Chen Y, Ali J, Mou T (2019) Improving rice blast resistance of Feng39S through molecular marker-assisted backcrossing. Rice (N Y) 12:70

Yano K, Morinaka Y, Wang F et al (2019) GWAS with principal component analysis identifies a gene comprehensively controlling rice architecture. Proc Natl Acad Sci USA 116:21262–21267

Yano K, Yamamoto E, Aya K, Takeuchi H, Lo PC, Hu L, … Matsuoka M (2016) Genome-wide association study using whole-genome sequencing identifies candidate genes for agronomic traits in rice. Nature Genetics, 48(8), 927–934

Yu J, Xiong H, Zhu X et al (2017) OsLG3 contributing to rice grain length and yield was mined by Ho-LAMap. BMC Biol 15:1–18

Zhang M, Liang X, Wang L, Cao Y, Song W, Shi J, Lai J, Jiang C (2019) Variation of salt tolerance in maize. Nat Plants 5:1297–1308

Zhang K, He M, Fan Y, Zhao H, Gao B, Yang K, Li F, Tang Y, Gao Q, Lin T et al (2021) Resequencing of global Tartary buckwheat accessions reveals multiple domestication events and key loci associated with agronomic traits. Genome Biol 22:23

Zheng X, Chen L, Li M, Lou Q, Xia H, Wang P, … Luo L (2020) Transcriptional responses to salinity stress in rice: A genome-wide association and transcriptomic analysis. Plant Physiology, 182(2), 699–715

Recent Advancements in Breeding and Genomics of Bluegrass

Priya Yadav, Noopur Srivastava, Pragati, Shashi Ranjan, and Sayanti Mandal

Abstract

Cool-season turfgrass known as Kentucky bluegrass (*Poa pratensis* L.) is often found in parks, urban green areas, and athletic fields around the globe. It is a significant temperate perennial grass species that is facultatively apomictic and utilized for both farmed turf and fodder. However, it is vulnerable to adverse abiotic stress which reduces turf quality in bluegrass by affecting the texture, shoot density, uniformity, colour, growth habit, and restoration capacity. In this regard, addressing the abiotic stress challenge involves sophisticated applications of molecular genetics and genomic techniques. Moreover, research with the availability of reference genome sequences and the development of new breeding techniques (genome editing) it possible to study the genes involved in stress tolerance and yield components, which has created new opportunities for bluegrass improvement. Gene editing tools such as CRISPR/Cas can assist in the

P. Yadav
Department of Chemistry, Amity Institute of Applied Sciences, Amity University, Noida, Uttar Pradesh, India

N. Srivastava · S. Mandal (✉)
Department of Chemistry & Biochemistry, Sharda School of Engineering & Science, Sharda University, Greater Noida, Uttar Pradesh, India

Pragati
Institute of Plant and Microbial Biology, Academia Sinica, Taipei, Taiwan

Molecular and Biological Agricultural Sciences, The Taiwan International Graduate Program, Academia Sinica, Taipei, Taiwan

Graduate Institute of Biotechnology, National Chung-Hsing University, Taichung, Taiwan

S. Ranjan
Amrita Centre for Nanosciences and Molecular Medicine, Amrita Vishwa Vidyapeetham, Kochi, Kerala, India

F. S. Baloch et al. (eds.), *Forage Crops in the Omics Era*,
https://doi.org/10.1007/978-981-95-9249-4_13

production of transgene-free bluegrass varieties. Its versatility has made it the most popular crop enhancement tool. It will expedite turfgrass breeding with precise gene editing. CRISPR/Cas-mediated genome editing has been used to elucidate gene function, establish stress-resistant cultivars, and aid in the advancement of resilient bluegrass varieties.

Keywords

Poaceae (Gramineae) · Kentucky bluegrass · Turfgrass · QTL · Genomics · CRISPR · Climate resilient

1 Introduction

Kentucky bluegrass (*Poa pratensis L.*) (KBG) is an amazing C-3 cool-season turfgrass native to Europe and Asia and is widely utilized in urban green areas, parks, sports fields, and residential lawns worldwide. It is a rhizomatous perennial grass distinguished by its fine texture, thick growth, and striking blue-green hue (Phillips et al. 2023). It is highly appreciated for its capacity to produce a dense, lush turf that is both resilient and visually beautiful. This turfgrass species is well-known for its adaptation to cool temperate areas and ability to survive low mowing heights, making it ideal for lawns and sports fields (Braun et al. 2022). It is noted for its high-quality appearance and cool-season grasses, making it a popular choice in locations with moderate temperatures and appropriate precipitation. Overall, Kentucky bluegrass is a versatile and visually beautiful turfgrass species valued for its longevity, density, and visual appeal in various landscape and recreational environments (Cui et al. 2021a, b). A huge landmass, approximately 40 million acres, equivalent to the area of Florida, is currently maintained as turf in the United States, which may serve as a significant carbon sink. Its beneficial role is now outweighed by the considerable amounts of water and fertilizer needed (Milesi et al. 2005; Phillips et al. 2023).

Ongoing breeding initiatives aim to improve plants' ability to tolerate environmental stress, combat diseases, and fight insects, increase the quality and quantity of yield and ensure consistency and stability of desirable characteristics (Bonos and Huff 2013). Despite having a high commercial value, *P. pratensis* is a very invasive plant that has rapidly colonized the Northern Great Plains of North America over the last 30 years, changing the dynamics of nutrients and pollinators while also lowering plant diversity and ecosystem function (Kral-O'Brien et al. 2019; Hendrickson et al. 2021; Phillips et al. 2023). Further investigation is required to explore the genetic variation of wild *P. pratensis* to comprehend the genes and mechanisms responsible for stress resilience. Analysing *P. pratensis* may also facilitate the discovery of disease-resistant or environmentally resilient ecotypes that may be used to breed turfgrass.

Poa pratensis presents difficulties for genetic investigation and improvement because of apomixis and multiple chromosome sets, that is, polyploidy (Phillips et al. 2023). *P. pratensis* can reproduce either sexually or asexually through aposporous apomixis (asexual development of seed). Additionally, it is a polyploid organism that often exhibits aneuploidy (Brown 1939). While apomixis is advantageous for seed production, its frequent occurrence hinders the recombination necessary for genetic analysis and the integration of favourable features in a single cultivar (Bonos and Huff 2013). The copy number variation of regions of interest and non-Mendelian inheritance that results from double reduction further exacerbates these difficulties. Polyploidy and aneuploidy are also contributing factors. However, there have been advancements in the management of apomixis, such as the identification of its genetic basis. Furthermore, additional genomic and molecular tools are required in *P. pratensis* to advance genetic analysis and breeding endeavours conquering its complicated biology (Albertini et al. 2004; Marconi et al. 2020). Conventional breeding methods have significantly contributed to the genetic enhancement of the current agricultural crop cultivars. These conventional breeding techniques include broad crossings, introgression from wild-crop via hybrid breeding, mutation breeding, etc. Using these cutting-edge methods, researchers can speed crop development efforts while addressing the problems posed by changing environmental conditions and rising food demand (Kopecký and Studer 2014). Several molecular genetics tools offer benefits to traditional turfgrass breeding production (Sun et al. 2025). They help in allowing for the identification and characterization of specific genes or genomic regions associated with desirable traits in the turfgrass region. Utilizing molecular markers, such as SSR markers, can expedite the breeders to select their desired traits in turfgrass cultivars MAS enables the identification of genotypes that carry favourable alleles without the need for time-consuming field, which helps the breeding process. SSR markers also help in the facilitation of the molecular genetic characterization of turfgrass germplasm allowing the breeders to assess genetic diversity, relatedness among cultivators and population structure. Several traits of interest in turfgrass, such as disease resistance or abiotic stress tolerance, were controlled by multiple genes. For instance, Arabidopsis *DREB1A* gene was overexpressed in Kentucky bluegrass, the transgenic grass exhibited greater drought tolerance than the control plants (Xu et al. 2017). The upregulation of the cytokinin synthesising gene isopentenyl transferase in creeping bentgrass led to the control of the expression of certain transcription factor genes such as bHLH, MYB, and WRKY, thereby improving drought tolerance (Xu et al. 2017). Moreover, the genetic dissection of complex traits through techniques like genetics mapping and quantitative trait loci (QTL) analysis. Classical field breeding techniques can enhance the efficiency and effectiveness of turfgrass breeding programmes. Traditional methods rely on phenotypic evaluations and molecular tools that provide a deeper understanding of the genetic basis of traits allowing, which allows for more informed decision-making in cultivar development. There have been several specific molecular genetics techniques used in turfgrass breeding production. Microsatellite markers (SSR; Simple Sequence Repeat) used for genotyping Kentucky bluegrass cultivars, collections, experimental selections, and hybrids. It's been valuable for assessing genetic relationships, population structure, and genetic diversity in turfgrass. AFLP Marker is used for genetic mapping in creeping bentgrass. It provides high-resolution genetic maps and aids in QTL analysis. CSIP

(Chromosome Specific Intron Polymorphism) and Intron Length polymorphism markers utilized in constructing a genetic linkage map of creeping bentgrass help to assess syntonies relationships between turfgrass species and model cereal crops. RAPD Markers have been used earlier to characterize Kentucky bluegrass cultivars These techniques enhance the understanding of genetic diversity, relatedness among cultivars, and the importance of traits in turfgrass species, ultimately contributing to improving turfgrass breeding production (Joshua Andrew Honig 2015).

Molecular breeding methods, including marker-assisted selection, gene selection, and state-of-the-art biotechnologies like CRISPR/Cas gene editing, hold the potential to overcome the constraints of traditional breeding. These tools allow for precise and targeted genetic improvements. The primary hindrance in the breeding process is the extended generation period. Conventional breeding often takes more than 10 years to generate and distribute new cultivars for the majority of agricultural plants (Jähne et al. 2020). It usually takes four or six generations to attain a fixed homozygous condition after crossing chosen parent plants, at this point, a better genotype may be identified. Moreover, multi-year testing in repeated field trials at several sites is part of this approach to find candidate genotypes in various situations (Voss-Fels et al. 2019). Genome editing may enable genome modification by precisely altering specific genomic areas. Furthermore, the reagents required for this procedure can be introduced into the cell without introducing DNA into the genome (Zhang et al. 2016; Bandyopadhyay et al. 2020; Ma et al. 2020). The changes arising by genome editing resemble those that naturally occur, possibly facilitating their regulation, in contrast to conventional genetically modified organisms (GMOs). Another fantastic characteristic of this technique is that it generates inheritable mutations in the genome with low risk of producing off-targets. DNA alterations caused by genome editing include large fragment replacement, single nucleotide substitution (SNPs), insertions, and deletions. Site-specific nucleases (SSNs) that cause nucleotide excision include CRISPR-associated protein (Cas), transcription activator-like effector nucleases (TALENs), zinc-finger nucleases (Kim et al. 1996; Christian et al. 2010; Jinek et al. 2012), and engineered homing endonucleases, also known as mega-nucleases (MNs) (Smith et al. 2006). This chapter presents a comprehensive summary of bluegrass response to the impacts of abiotic stress caused by climate change. Furthermore, we provide a concise overview of the current status of genome-editing applications in bluegrass. This chapter also presents a proposal on the use of genome editing to develop bluegrass with enhanced resistance to environmental stressors.

2 Abiotic Stress and Bluegrass Production

Abiotic stressors including drought, salinity, heat, and cold are important limiting variables impacting Kentucky bluegrass productivity both qualitatively and quantitatively (Mondal et al. 2024). These environmental stresses are anticipated to become significantly more serious due to the expanding human population and climate change (Sarıkaya and Tatar 2023). Abiotic stressors substantially impact turfgrass growth, high shoot density, leaf quality, and general health. It significantly lowers downs its morphological and agronomic traits and hence decreases the productivity (Huang et al. 2014).

2.1 Drought Stress

Drought stress causes reduction in growth rates, which results in shortness in development. Due to low water availability, the grass may experience diminished leaf elongation and overall biomass accumulation (Charagh et al. 2024; Raza et al. 2024a). During drought conditions, Kentucky bluegrass undergoes physiological modifications to preserve water. This involves stomatal closure to limit transpiration, which may result in lower photosynthesis and energy output. The activation state of Rubisco (the enzyme responsible for carbon fixation in photosynthesis) may be altered, resulting in poorer carboxylation efficiency and photosynthetic rates (Fan et al. 2020). Drought stress in Kentucky bluegrass cultivars may have higher water use efficiency, allowing plants to maintain growth and quality under low-water situations. Even during a drought, these cultivars can improve water loss and photosynthetic efficiency. Kentucky bluegrass has the potential to go quiescent during lengthy droughts. In this state, the grass lowers metabolic activity and saves energy until water becomes accessible again. Drought-tolerant cultivars can recover more successfully after re-watering, allowing them to resume growth and photosynthesis. Drought stress can cause apparent indicators such as leaf wilting, browning, and a general decrease in lawn quality. This has an impact on the aesthetic appeal of lawns and sports fields, which is especially significant for recreation and ornamentation. Drought-stressed in Kentucky bluegrass may become more sensitive to diseases and pests when physiological circumstances deteriorate. LEA genes (including *RAB15* and *HVA1*), HSP 70, and SRK2A genes which are linked to ABA-induced stomatal closure have all recently been discovered to be elevated in drought-tolerant accessions of Kentucky bluegrass (Bushman et al. 2021). In a similar vein, under drought stress, Kentucky bluegrass, creeping bentgrass, and perennial ryegrass all shared certain genes linked to ABA-responsive LEA homologs (Bushman et al. 2022). Stress can weaken a plant's defenses, leaving it prone to diseases and pest infestations. During drought stress, Kentucky bluegrass may store carbohydrates in response to restricted photosynthesis. These stored carbohydrates can be used as energy reserves during recuperation following re-watering, which aids in the regrowth process. During drought stress, Kentucky bluegrass may store carbohydrates in response to restricted photosynthesis. These stored carbohydrates can act as energy reserves during recuperation following re-watering, assisting with the regrowth process (Fan et al. 2020).

2.2 Salinity Stress

Salinity stress can have a substantial negative influence on Kentucky bluegrass, affecting growth, physiological processes, and overall health. Salinity stress often reduces shoot and root growth in Kentucky bluegrass (Liu et al. 2022; Zuo et al. 2023). High salt concentrations can limit cell division and elongation, leading to stunted growth and decreased accumulation of biomass. Salinity alters the plant's water relations, resulting in osmotic stress. Because of the high osmotic potential of salty soils, this can reduce water intake, limiting the plant's ability to obtain water. The plant may also undergo ion toxicity, mainly from sodium (Na^+) and chloride (Cl^-) ions, which can disrupt cellular activities and cause physiological imbalances (Liu et al. 2022). Salinity stress can reduce chlorophyll concentration and photosynthesis efficiency. This leads to lower carbon fixation and energy generation. Kentucky bluegrass may collect carbohydrates in response to salt stress, which can be used as energy reserves when photosynthesis is limited. However, high salinity can still reduce overall carbohydrate synthesis. Leaf fire, or browning and drying of the leaf tips, is one of the apparent indicators of salt stress in Kentucky bluegrass (Mastalerczuk et al. 2019). This not only has an impact on the turf's aesthetics, but it may also reflect underlying physiological stress. In terms of saline tolerance, Kentucky bluegrass cultivars exhibit great genetic diversity. Some cultivars have been discovered as more tolerant to saline conditions, with improved growth and reduced leaf firing under salt stress. This variability opens up possibilities for breeding initiatives aiming at producing more salt-tolerant grass species (Zuo et al. 2023). Salinity stress can cause changes in root morphology, such as increased root length and root-to-shoot biomass ratios in some tolerant genotypes. This modification may allow the plant to acquire water and nutrients more effectively under saline circumstances. The ability of Kentucky bluegrass to recover from salt stress varies by cultivar. Drought-tolerant and salinity-tolerant cultivars may demonstrate faster recovery rates when re-watered following exposure to saline conditions, restarting growth and photosynthesis more effectively than sensitive kinds (Zuo et al. 2023).

2.3 Heat Stress

Heat stress has a substantial impact on Kentucky bluegrass in a variety of ways, including physiological, biochemical, and growth responses. High temperatures increase transpiration rates, resulting in increased water loss (Ye et al. 2016). If water is not delivered enough, the grass may become dehydrated and wilt. To conserve water, Kentucky bluegrass may close its stomata, reducing carbon dioxide intake and, as a result, photosynthesis, resulting in decreased growth and vigour. Prolonged exposure to high temperatures can inhibit growth, resulting in thinner turf and lower overall quality. Symptoms of the grass may include leaf bending and browning, particularly at the tips. High temperatures can cause senescence in leaves, resulting in premature aging and death of plant components (Feng et al. 2021). This is frequently accompanied with a disturbance of water and nutrient uptake. Heat stress can cause a considerable drop in total protein content in leaves. For example, research has demonstrated that the major subunit of Rubisco, a crucial enzyme in photosynthesis, degrades under heat stress conditions. The grass may produce more antioxidant enzymes to combat oxidative stress produced by high temperatures. This reaction is critical for safeguarding cellular structures and sustaining metabolic activities. Kentucky bluegrass may go dormant in excessive heat, causing it to turn brown and stop growing. This dormancy permits the plant to live until better conditions return. Kentucky bluegrass cultivars possess variable tolerance to heat. Some cultivars have increased protein content and endure less severe protein degradation during heat stress, which increases their resilience. Heat-stressed may become more susceptible to diseases and pests as a result of weaker physiological conditions, thus compromising turf quality. High temperatures can also have an impact on soil moisture and microbial activity, both of which are important for nutrient availability and plant health (Noor et al. 2022). Heat stress is a major concern for Kentucky bluegrass, affecting growth, physiological functioning, and overall health. Effective management measures, such as proper irrigation and the use of heat-tolerant cultivars, are critical for sustaining healthy turf in high-temperature environments (Mondal et al. 2024).

2.4 Low Temperature Stress/Cold Stress

Low temperature stress can have a substantial influence on Kentucky bluegrass (*Poa pratensis*), affecting growth, physiological functions, and general health. Kentucky bluegrass goes through cold adaptation, which involves modifying its metabolic pathways to prepare for low temperatures (Cui et al. 2021a, b). Changes in membrane fluidity, as well as the buildup of protective proteins and solutes, enable the plant endure freezing temperatures. If temperatures change, particularly in late winter or early spring, the grass may become deacclimated, leaving it more sensitive to future cold spells. This can raise the risk of injury. Low temperatures can cause winter damage, which is characterized by browning and dieback of grass blades. This injury is sometimes aggravated by conditions such as desiccation and ice cover, which inhibit the grass from absorbing moisture. Kentucky bluegrass has been subjected to considerable cold stress; it may green up later and recover from dormancy later in the spring. This has the potential to impact the turf's overall aesthetic quality. Low temperatures can impede metabolic processes and limit development rates (Liu et al. 2008). This can lead to weaker turf and diminished overall vigour during the growth season. Cold stress can have a negative impact on root growth, which is necessary for nutrient and water acquisition. Poor root development might lead to additional stress during the warmer months. Low temperatures can disrupt cell membranes, resulting in leakage of cellular contents and loss of functionality. This is especially problematic during freezing temperatures, when ice crystals can form within and between cells. Chilling stress can cause oxidative stress due to metabolic abnormalities, resulting in the generation of reactive oxygen species (ROS), which can harm cellular components. Kentucky bluegrass is generally hardy and can recover from low temperature stress if the damage is minimal (Sachdev et al. 2021). As temperatures rise, photosynthetic activity and root growth resume. Kentucky bluegrass cultivars vary in their cold tolerance. Some cultivars are more adapted to cold temperatures and recover faster from winter injury (Cui et al. 2021a, b).

3 Conventional and Modern Molecular Breeding Methods for Bluegrass Improvement

Agriculture emerged when early humans domesticated wild plants. Farmers chose the most attractive plants and seeds for cultivation in later seasons, providing the groundwork for selective breeding. As agriculture advanced, farmers found that they could artificially mate or cross-pollinate different plants to combine desired features. This practice facilitated the production of new kinds with improved traits. Plant breeders improved their grasp of inheritance and the genetic basis of traits as modern genetics emerged in the twentieth century (Sun et al. 2023; Abdul Aziz and Masmoudi 2024). This understanding enabled them to better choose excellent plants and develop new and improved types. The creation of hybrid varieties was a key milestone in plant breeding. By crossing pure lines, breeders were able to create F1 hybrids with increased vigour, greater yields, and superior disease resistance than open-pollinated (OP) types. The extensive adoption of hybrid seeds, particularly in crops like maize, has resulted in much higher agricultural production (Curry 2023). For instance, hybridization and improved agricultural practices have boosted corn grain yields in the United States to more than triple over the past 50 years (Garcia et al. 2022). While conventional plant breeding has greatly increased agricultural production, it does have limitations. Breeding can only take place between plants that can sexually reproduce, which limits the introduction of new traits. Furthermore, crossing plants can unintentionally impart unwanted features alongside the desired ones. Conventional breeding is choosing from current populations plants with desired features. Breeders analyse these plants' efficiency in terms of yield, quality, disease resistance, and adaptability to environmental circumstances. This selecting technique improves crop performance across generations. Breeders utilize regulated mating strategies to cross plants that have complementing characteristics. This process combines beneficial qualities from many parent lines, resulting in the generation of new varieties with superior traits such as increased yield or stress tolerance. Conventional breeding contributes to preserving and utilizing genetic diversity within plant species. Breeders can improve the genetic foundation of cultivated crops by adding features from landraces or wild cousins, making them more resistant to environmental stresses and disease tolerance (Anand et al. 2022). This technique is frequently used for transferring a specific trait from a donor plant into an elite variety while preserving the elite variety's desired qualities. Backcrossing helps to recover the elite parent's high yield and quality while adding new features like insect resistance. This strategy uses many rounds of selection and recombination within a population to raise the frequency of beneficial alleles. It is especially successful at improving qualities influenced by numerous genes, such as yield and quality traits. Conventional breeding enables the creation of crop varieties that are carefully tailored to local environmental variables such as soil type, climate, and pest threats. This targeted strategy has the potential to boost agricultural output and sustainability (Ahmar et al. 2020a, b). Conventional breeding is a time-tested strategy that has resulted in major advances in crop types throughout the years. It helped contribute to the Green Revolution and remains a significant strategy for improving food security and agricultural resilience. Molecular techniques, like as molecular markers, enable breeders to discover and select for specific traits at the genetic level. This precision allows for the identification of desirable qualities such as disease resistance, drought tolerance, and increased yield without depending exclusively on phenotypic data. Using molecular markers connected to desirable features, breeders can efficiently choose plants species with these traits early in the breeding process. This speeds up the production of novel kinds by lowering the time required for traditional selection methods, which can involve multiple generations to evaluate features (Wanga et al. 2021). Molecular approaches enable the transfer of genetic variation from wild relatives or other species into cultivated crops. This increases the genetic diversity of breeding programs, making crops more resistant to environmental challenges and disease. This innovative method employs genome-wide markers to predict the success of breeding lines based on their genetic composition. It enables breeders to make informed judgements about which lines to progress in the breeding process, thereby increasing the efficiency of creating high-performance cultivars. Molecular breeding sheds light on how certain genes function and how they influence plant development and stress responses. This knowledge can help guide the selection of features that improve crop production under a variety of situations (Alemu et al. 2024). Molecular techniques allow for the construction of genetically modified organisms (GMOs) by adding certain genes that give desirable features, such as insect resistance or increased nutritional content. This can result in the rapid development of new crop species that address specific agricultural concerns. Molecular tools play an important part in modern plant breeding approaches because they allow for exact tracking and modulation of genetic characteristics (Sinha et al. 2023). They enable breeders to track the dynamics of genomic recombination and implement a gene-by-gene breeding strategy. Molecular markers are used to generate genetic maps that aid in the identification of genes associated with specific phenotypes. This method involves genotyping molecular markers in plant populations and using recombination frequencies to determine their order and distance on chromosomes, resulting in the creation of quantitative trait loci (QTL) maps. Once QTLs have been identified, breeders can employ the closest molecular markers associated with desirable traits for selection. This makes breeding more efficient by focusing on specific genetic differences that contribute to desirable qualities like yield or disease resistance (Budhlakoti et al. 2022). Molecular techniques allow breeders to select for features more accurately, lowering the time and resources required to generate new cultivars. This precision is especially critical when dealing with issues of environmental sustainability and food security.

Molecular tools can permit the introduction of exotic genetic material into elite germplasm, thus broadening the genetic basis of cultivated crops and improving their resilience to biotic and abiotic pressures (Breseghello and Coelho 2013). Over the past few decades, traditional breeding has significantly improved agriculture; however, because of its low efficiency, labour intensity, dependence on environmental factors, and time consumption, scientists are looking at new options for crop improvement in sustainable agriculture (Ajeesh Krishna et al. 2023). Breeding efficiency is increased by the improvements in breeding technology, especially molecular marker-assisted selection (MAS) (Anders et al. 2021). Moreover, it is possible to achieve elite bluegrass genotypes with desired attributes by genome editing, which might support resilient and sustainable turfgrass systems.

4 The Potential for Bluegrass Improvement Via Genomic-Assisted Breeding

Through advanced breeding techniques, genomics offers tools to improve yield quantity, quality, and stability of production. The application of DNA markers in crossbreeding has proven beneficial in assisting MAS for crop enhancement. Plant genomics advancements have provided new approaches to enhance our understanding of crop diversity at both the species and gene levels. These developments have also led to the identification of DNA markers, which have notably accelerated genetic improvement efforts in crops (Muthamilarasan et al. 2013, 2014). To develop novel cultivars that are 'climate change ready', a genomics-led breeding method begins by determining the stress(es) that would likely impact agricultural yield and productivity under various climate change scenarios (Varshney et al. 2005; Varshney et al. 2021; Charagh et al. 2024; Raza et al. 2024b). By analysing data collected, using the data gathered from testing in several extreme conditions, one can predict the 'stress-impacts' of environments on specific populations of crops and targeted demographics. Plant breeders and gene bank curators will explore for physiological and morphological features in accessible germplasm that might promote agricultural resilience in response to such climatic unpredictability (Singh et al. 2021). In this respect, crop physiology may be able to assist in defining the ideotypes that should be sought for the purpose of improving such adaptability. In addition, accessions from stressful environments can be identified by using passport data and geographic information systems, while DNA fingerprinting and evaluation data along with mapping of desired genes or quantitative trait loci (QTL) can be used to select promising accessions for further screening against specific stressors (Ahmar et al. 2020a, b) (Fig. 1). Similar to this, precise phenotypic evaluations and appropriate biometric examination will assist in identifying the distinct reactions of a group of genotypes at a particular physiological phase that is impacted by shifting weather patterns. Genomics-assisted breeding techniques, such as genome-wide selection of potential germplasm for use in crop breeding in the future to enhance the population and release cultivars, will make further use of this knowledge (Samantara et al. 2022). For the first time, agricultural traits have been rapidly

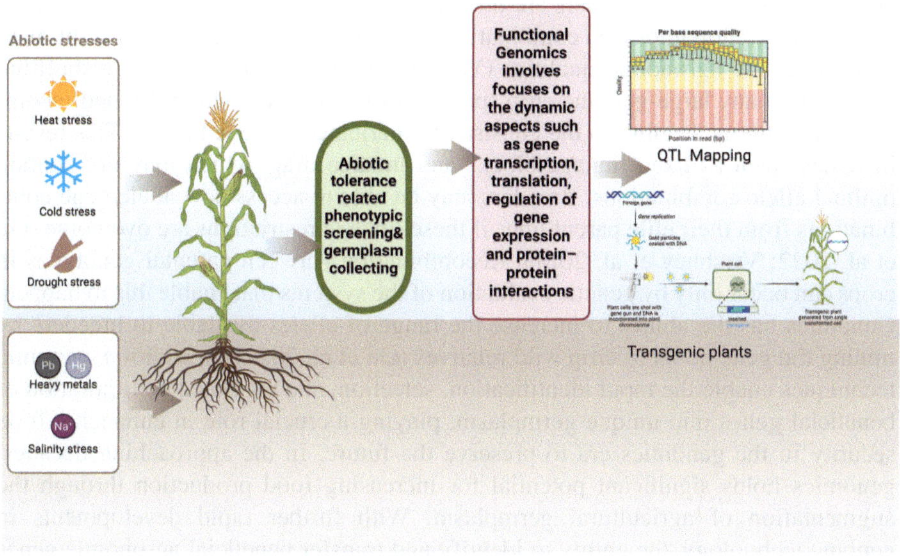

Fig. 1 Application of genomic based approaches to improve bluegrass for abiotic stress tolerance

dissected through QTL and genetic mapping analysis using bi-parental or association-mapping (AM) populations. This has made it possible to directly calculate and select high-value genotypes for breeding programs. (Kushanov et al. 2021). Until recently, AM and GS were hindered by the need for exceptionally high marker density coverage of the genome. The advancement of next-generation sequencing (NGS) technology has aided the production of a significant number of genetic markers, like single nucleotide polymorphisms (SNPs), insertion-deletions (InDels), and other genetic markers, even in crop species that have received little attention (Ahmar et al. 2020a, b; Samantara et al. 2022). Then, genotyping-by-sequencing (GBS) techniques might then be used to discover new genes/alleles for any given characteristic or traits of interest. In the same way, genome-wide association studies (GWAS) could be used to find the genomic regions that control traits of interest by looking for statistical links between DNA polymorphisms and trait variations in a large number of germplasms that have been phenotyped and genotyped for traits of interest. Moreover, NGS in conjunction with GWAS improves the mapping resolution for pinpointing genes/alleles/QTL (Ma et al. 2015; Liu et al. 2013; Varshney et al. 2014; Varshney et al. 2021; Samantara et al. 2022). In the process of evolution, new genes have been created by nature, and then these genes have been rearranged and chosen in a broad variety of situations to generate the genetic diversity that is seen in wild species. However, the selection and domestication of crops by humans is a comparatively recent phenomenon, having taken place about 10,000 years ago. Major crops' genetic diversity has been drastically reduced as they've been bred for higher yields in highly controlled agricultural contexts as part of the domestication and breeding process. Breeders are now rearranging a very small number of genes in order to develop improved combination traits that give greater yield in addition to other desirable agronomic qualities (Varshney et al. 2021). Even this reshuffling process in many large-genome crop species is often constrained by limited recombination patterns, leading to the consistent inheritance of gene blocks. This results in issues, such as fixed linkage blocks and linkage drag, which may not contain optimal allele combinations. Breeders may be able to access new allele/gene combinations from their elite parent lines if these linkage restrictions are overcome (Gu et al. 2022; Varshney et al. 2021). Recombination between parental genotypes in crops can occur only by genetic evaluation of the systems that enable this to happen. Genomics has the ability to increase the range of alleles available to breeders by mining the gene pools of crop wild relatives (Gu et al. 2022). In addition, genomic techniques enable the rapid identification, selection, and regulated incorporation of beneficial genes into unique germplasm, playing a crucial role in enhancing food security in the genomics era to preserve the future. In the approaching decades, genomics holds significant potential for increasing food production through the augmentation of agricultural germplasm. With further rapid developments in genome technology, the ability to identify and transfer beneficial agronomic genes from allied gene pools and crop cousins to top crops will accelerate, contributing to global food security efforts.

5 Genomics Tools for Bluegrass Trait Enhancement

The quantitative response of plants to abiotic stressors is regulated by a collection of genes known as QTLs that are dispersed across the genome. The identification of QTLs is crucial for improving stress tolerance characteristics via marker-assisted selection (MAS) (Fig. 1). Several articles on QTL mapping for a seemingly limitless range of morpho-physiological features have been extensively evaluated (Sarkar et al. 2023). Because there is a lot of uncertainty, especially with low heritability, traditional QTL mapping studies with a small population size and few markers (poor resolution) are not very helpful for breeding. Genes in large QTL areas should be studied since they may include numerous potential candidates for further study. Genotyping by sequencing (GBS) and fixed-content high-density Axiom arrays have recently been employed for high-throughput evaluation of DNA sequence markers, like single nucleotide polymorphisms (SNPs) (Li et al. 2023). It is now feasible to generate genetic maps with high resolution and exact mapping of QTLs, as well as the identification of genes inside the QTL, utilizing such dense and widely dispersed DNA markers (Si et al. 2022). Genomic selection is now being used in breeding efforts for a variety of crops in order to speed up the process. During genomic selection, a test population (the training set) is genotyped and phenotyped to represent the genetic diversity of a larger breeding population (the reference set). Moreover, the genomic estimated breeding value is calculated using marker–trait association analysis, which forecasts phenotypic performance based on marker frequencies (Sinha et al. 2023; Alemu et al. 2024). The reference set is then genotyped, and the breeding value is then applied to the reference set to forecast the phenotypes of the breeding population's lines. Recently, 'Speed breeding', which combines genomic selection, rapid generation progression of segregating plants, and reduced phenotyping time (which is often challenging), has shortened crop breeding cycles, enhancing genetic gain in a shorter period (Varshney et al. 2014).

To address the disadvantages of traditional QTL mapping techniques, next-generation sequencing (NGS)-based trait mapping methodologies have recently been applied in several crops (Khan et al. 2021). One of these methods is QTL-Seq, which is better than bulked segregant analysis and can quickly find plant QTLs by resequencing the whole genomes of 20–50 individuals from each of two populations whose members have traits that are very different for a given phenotype (Zhao et al. 2019). In addition to SNP ratio mapping, SHOREmap, and next-generation mapping are examples of novel trait mapping methodologies (Schneeberger et al. 2009; Li et al. 2022). Whole-genome resequencing (WGRS) or low-coverage sequencing of the whole mapping population or bulks is now feasible since reference genomes for a number of crop species are now available. In a mapping population, the GBS or WGRS of each individual provides large-scale genome-wide SNPs for high-resolution mapping of the desired features (Susmitha et al. 2023). These genome-wide SNPs or CNVs have substantially facilitated genome-wide association mapping (GWAS), allowing the exploration of particular haplotypes in varied germplasm sets by limiting the potential areas (Yasir et al. 2022). Three experimental elements are necessary for GWAS to be successful: plants/samples, phenotyping, and genotyping. Plants indicate the number of genotypes/accessions examined as well as the population structure. Phenotyping involves genetic complexity and heritability of recorded characteristics, whereas genotyping refers to the density of markers or quality of sequence reads (Nybom and Lācis 2021). Phenotyping comprises the genetic complexity and heritability of the reported attributes as well as the density and quality of the sequence reads that make up the genotyping component (Alseekh et al. 2021; Xiao et al. 2021). The magnitude and frequency of a phenotypic impact of a variable in natural diversity panels are critical for GWAS success. Success in GWAS depends on the extent and frequency of a trait's phenotypic influence in natural diversity panels. Nevertheless, many of the abiotic stress tolerance features are governed by several genes containing either a high number of common variations with relatively little effects or a small number of uncommon variants with comparatively significant impacts (Xiao et al. 2021).

6 Crop Improvement Using Gene Editing Tools

6.1 TALEN (Transcription Activator-Like Effector Nucleases)

TALEN technology is important for crop development due to its ability to do precise genetic changes. TALENs enable precise editing of specific genes related with desired features including disease resistance, drought tolerance, and increased nutritional content (Ahmar et al. 2020a, b). By wiping out or changing specific genes, researchers can improve crop performance. By targeting and changing genes that make crops susceptible to infections, TALENs can be utilized to create disease-resistant cultivars. This minimizes the need for chemical pesticides, potentially leading to more sustainable agriculture methods. TALENs can be used to improve features that have a direct impact on crop output and quality, such as fruit or grain size, flavour, and nutritional profiles. Modifications, for example, can be made to boost the levels of vital vitamins and minerals in basic crops (Malzahn et al. 2017). TALEN technology can help create crops that are more resistant to environmental challenges like drought, salinity, and severe temperatures (Sprink et al. 2015). Crops can be made more resilient to adverse situations by altering genes involved in stress response pathways, ensuring food security in changing climates. Unlike classic genetic modification methods, TALENs can make modifications without leaving marker genes in the plant genome. This is especially significant for regulatory approval and consumer acceptance, as marker-free crops are frequently perceived more positively. In comparison to conventional breeding procedures, TALENs allow for the quick generation of new crop varieties. This quickness is critical in responding to developing agricultural concerns, such as new pests or changing weather conditions (Wei et al. 2013).

6.2 ZFNs (Zinc Finger Nucleases)

Zinc finger nucleases (ZFNs) are designed restriction enzymes with two primary components: a sequence-specific DNA binding domain composed of zinc finger protein domains and a nonspecific cleavage domain that causes double-strand breaks (DSBs) in the DNA (Iqbal et al. 2020). Each zinc finger domain recognizes and binds to a specific DNA sequence of about three base pairs, allowing ZFNs to be designed to target specific genomic regions. An array of zinc finger domains recognizes nearby DNA sequences and binds them together (Bhuyan et al. 2023). Once coupled to the target DNA, the ZFN's nuclease domain causes a double-strand break in the specific spot. This DSB is a key event that activates the cell's DNA repair machinery. After this, the cell tries to repair the DSB via two main pathways first Non-Homologous End Joining (NHEJ) which provides an error-prone repair method causes insertions or deletions (indels) at the break site, resulting in gene-specific alterations. Homology-Directed Repair (HDR) in which if a donor DNA template is provided that is homologous to the sequences bordering the DSB, the cell can use it to properly repair the break, allowing for precise changes such as gene substitution or the introduction of new genetic material (Petolino 2015) The use of zinc finger nucleases (ZFNs) for genetic alteration in plants provides various benefits, it enable precision targeting of certain genomic regions, allowing for the introduction of specific mutations, deletions, and insertions. This precision is essential for investigating gene activity and creating organisms with desired characteristics. They can be programmed to identify practically any DNA sequence, making them extremely useful genome-editing tools. This versatility enables researchers to target different genes or regulatory regions during a single experiment. ZFNs can support both non-homologous end joining (NHEJ) and homology-directed repair (HDR), allowing for a variety of genetic alterations. This flexibility can increase the efficiency of making the needed genetic modifications (Paschon et al. 2019) (Table 1).

Table 1 Comparison between TALEN, CRISPR, and ZNFs in genome modification

Function	ZNFs	TALEN	CRISPR	Reference
Mechanism of action	ZFNs are zinc finger proteins that can be designed to bind to specific DNA sequences. Each zinc finger identifies a triplet of DNA nucleotides. ZFNs, like TALENs, are fused to a FokI nuclease, which causes a double-strand break when they attach to their target sites.	TALENs use a pair of engineered transcription activator-like effectors that attach to specific sequences of DNA. Each has a repeat domain that detects a particular base pair in the target DNA. When the TALEs connect to their target sites, they activate a FokI nuclease, causing a double-strand break (DSB) in the DNA.	The guide RNA (gRNA) used by CRISPR/Cas9 complements the target DNA sequence. When the Cas9 nuclease reaches a particular location in the genome, guided by the gRNA, it produces a DSB. Cas9 recognizes the target DNA in the presence of a protospacer adjacent motif (PAM) sequence.	Sauer et al. (2016), Kumar and Jain (2015), Bhardwaj and Nain (2021)
Design complexity	ZFNs are more challenging to design than TALENs and CRISPR/Cas9 because they involve the engineering of numerous zinc finger domains to attain target sequence specificity. This level of intricacy can make it difficult to achieve great specificity.	Designing TALENs is easier since the TALE repeat domains may be built to target specific sequences. However, developing the required constructs can be time-consuming.	CRISPR/Cas9 is often regarded as easier and faster to design. Researchers simply need to generate a small RNA sequence (gRNA) to target the desired DNA, making it a more accessible genome- editing technique.	Sauer et al. (2016), Li et al. (2013), Shamshirgaran et al. (2022)
Targeting specificity	ZFNs are highly specialized, their design complexity can result in performance unpredictability. Off-target effects do occur, but they are less common than with CRISPR/Cas9.	TALENs have great specificity because to the protein–DNA interaction, off-target effects can occur, particularly if the target location is identical to other sequences in the genome.	CRISPR/Cas9 has great specificity, it is also known to cause off-target effects, especially if the gRNA has partial complementarity with adjacent genomic areas. Advances in gRNA design and Cas9 variations aim to increase specificity.	Khatodia et al. (2016), Safari et al. (2019)

(continued)

Table 1 (continued)

Function	ZNFs	TALEN	CRISPR	Reference
Time/cost	ZFNs take longer and cost more to produce	It takes more time and resources to design and build.	It is typically faster and less expensive to design and deploy, making it more accessible to researchers.	Ndudzo et al. (2024), Carroll (2008)
Efficiency of editing	Although ZFNs are efficient, their performance is frequently uneven. The necessity for several zinc finger domains complicates design and reduces overall efficiency when compared to TALENs and CRISPR/Cas9.	Effective in creating targeted alterations, however the effectiveness varies greatly depending on the specific TALEN pair and target locus. Some TALENs may fail to elicit changes entirely due to sequence or chromatin environment.	Known for its remarkable efficiency in creating double-strand breaks and subsequent gene changes. It has high editing rates in numerous cell types and organisms, making it a popular choice for a variety of applications	Castro et al. (2021), Shamshirgaran et al. (2022), Gaj et al. (2013)
Delivery methods	ZFNs can be delivered via plasmids or viral vectors. However, the larger size of ZFN constructs can limit their delivery efficiency.	Commonly delivered via plasmid transfection, viral vectors, or ribonucleoprotein (RNP) complexes. The greater size of TALEN structures presents delivery problems, particularly in vivo.	CRISPR/Cas9 can be delivered as plasmids, RNPs, or through viral vectors. The smaller size of the gRNA and Cas9 protein allows for more flexible delivery options, including lipid nanoparticles and electroporation.	Shamshirgaran et al. (2022), Gaj et al. (2013)
Off-target effects	Typically have lesser off-target effects than CRISPR/Cas, however the intricacy of developing ZFNs can result in variability in specificity.	Off-target effects are generally less common than with CRISPR/Cas9, although they can still occur, especially if the target site is related to other genomic areas.	CRISPR/Cas9 is known for having off-target consequences due to its reliance on RNA-DNA hybridization. Advances in gRNA design, Cas9 variations, and computational methods are intended to mitigate these consequences.	Shamshirgaran et al. (2022), Cui et al. (2021a, b)

6.3 CRISPR/Cas (Clustered Regularly Interspaced Short-Palindromic Repeats)

CRISPR/Cas, which stands for Clustered Regularly Interspaced Short-Palindromic Repeats, along with the CRISPR-associated protein (Cas), forms a prokaryotic adaptive immune mechanism. This system is used to cleave invading RNA or DNA and is found across bacteria and archaea. The CRISPR-Cas systems are categorized based on their structural components and mechanisms. For instance, Class I systems have multi-protein effector complexes, while Class II systems have a single effector protein. The specificity of CRISPR/Cas9 relies on nucleic acid base pairing rather than protein–DNA recognition. This system uses DNA–RNA recognition and binding to achieve sequence-specific nucleic acid cleavage on the hybrid strand, creating a DSB at the target site with minimal cost (Saini et al. 2023).

The type IIA CRISPR/Cas system, particularly CRISPR/Cas9 from *Streptococcus pyogenes*, is the most extensively studied. In this system, Cas9 is the sole protein needed for the interference step. The CRISPR/Cas9 complex consists of a single guide RNA (sgRNA), which is a fusion of tracrRNA and crRNA, and the Cas9 protein, a 160 kDa DNA endonuclease enzyme that precisely cleaves each strand of double-stranded DNA. RNaseIII assists in transforming pre-crRNA into mature sgRNA, whereas Csn2, Cas1, and Cas2 are necessary for the DNA acquisition phase (Bhattacharya and Satpati 2023).

By introducing the DNA target protospacer sequence into crRNAs or sgRNAs, CRISPR/Cas systems may be constructed. The discovery of various PAM (protospacer adjacent motif) specific Cas orthologs and polymorphisms has enhanced the editing capabilities of these tools. This technique allows for the specific interference with foreign nucleic acids based on the sequence of sgRNAs. Genome alteration at the target locus is achieved via CRISPR/Cas9, which induces DSBs through site-specific nucleases. These DSBs can lead to gene deletion or fusion through non-homologous end joining (NHEJ), which occurs without donor DNA. Alternatively, homology-directed repair (HDR) uses homologous sequences to precisely adjust gene sequences in response to even minor changes in either DNA strand (Saini et al. 2023).

Genome editing has enabled the creation of plants with heritable genome changes, including (1) point mutations, (2) small, random insertions or deletions (indels), (3) DNA fragment insertions, (4) DNA fragment deletions, and (5) targeted chromosomal rearrangements. Before modifying the genome with CRISPR/Cas, a single guide RNA (sgRNA) must be designed to target the desired gene. The core component of the ribonucleoprotein (RNP) complex is the CRISPR/Cas complex, which consists of the guide RNA (gRNA) and the Cas nuclease. The 3′ ends of DNA targets require 5'-NGG-3′ sequences containing the protospacer adjacent motif (PAM). The targeting sequence (crRNA), located 20 nucleotides upstream of the PAM sequence, is cleaved by the Cas9 nuclease into approximately three bases

(Fig. 2). Only when the gRNA contains a certain PAM can it attach to the genomic DNA. The DNA is subsequently split into two strands by the Cas9 nuclease. A specific sgRNA including a Cas9 nuclease-recruiting domain and a targeting sequence (crRNA sequence) connected to the tracrRNA is necessary for the CRISPR/Cas9 system (Saini et al. 2023).

CRISPR/Cas is a potent gene-editing tool that enables researchers to precisely modify certain DNA regions. The system consists of two key components: (1) Cas9 Protein: It is an enzyme which acts as a molecular scissors, which cuts DNA at predetermined locations. It contains two endonuclease domains, RuvC and HNH, which work together to create double-stranded breaks (DSBs) in the target DNA. (2) Single Guide RNA (sgRNA): This RNA molecule serves as a guide, which directs the Cas9 protein to the desired target DNA sequence. It consists of a scaffold sequence that binds to Cas9 and a spacer sequence that is complementary to the target DNA. Once the Cas9 protein is guided to the target site by the sgRNA, it cuts the DNA, resulting in creation of a DSB. Cas9, a nuclease enzyme, binds to the gRNA and the target DNA, forming a complex that leads to the cleavage of the target DNA. This cleavage induces DNA repair mechanisms, resulting in a targeted mutations or genome editing. The process involves various interactions between Cas9, gRNA, and target DNA, as well as the presence of Mg^{2+} for catalytic activity. Cell attempts to repair their breakage using any one of two pathways either through NHEJ. This pathway is error-prone, often leading to random insertions or deletions of nucleotides (indels) at the cut site. This can disrupt the gene's function, making it a useful tool for gene knockout experiments or through HDR. This pathway is more precise, using a donor DNA template to repair the DSB without errors. By providing a donor template with the desired edit, scientists can introduce specific changes into the genome. There are several uses for CRISPR/Cas9 in a variety of field, including as agriculture, health, and research. It offers the potential to treat genetic diseases, develop new therapies, and improve crop yields (Bhattacharya and Satpati 2023; Uddin et al. 2020).

6.3.1 Limitations of CRISPR

CRISPR/Cas9 systems, while powerful tools for crop genetic improvement, face certain limitations. One significant constraint is the requirement for a specific Protospacer Adjacent Motif (PAM) sequence, typically NGG, immediately adjacent to the target DNA sequence. This limits the range of genes that can be effectively edited. To overcome this limitation, researchers have been working to develop Cas9-like systems that can recognize a broader range of PAM sequences. For instance, xCas9 is a modified version of SpCas9 that can identify PAM sequences with GAT, NG, and GAA, expanding its applicability in mammalian cells (Wang et al. 2022).

In plants, several efforts have been made to combine xCas9 with Cas9-NG, a variant that recognizes NGN PAMs. This hybrid system enhances the scope of CRISPR/Cas9-mediated gene editing in plants. Recently, a near-PAMless SpCas9

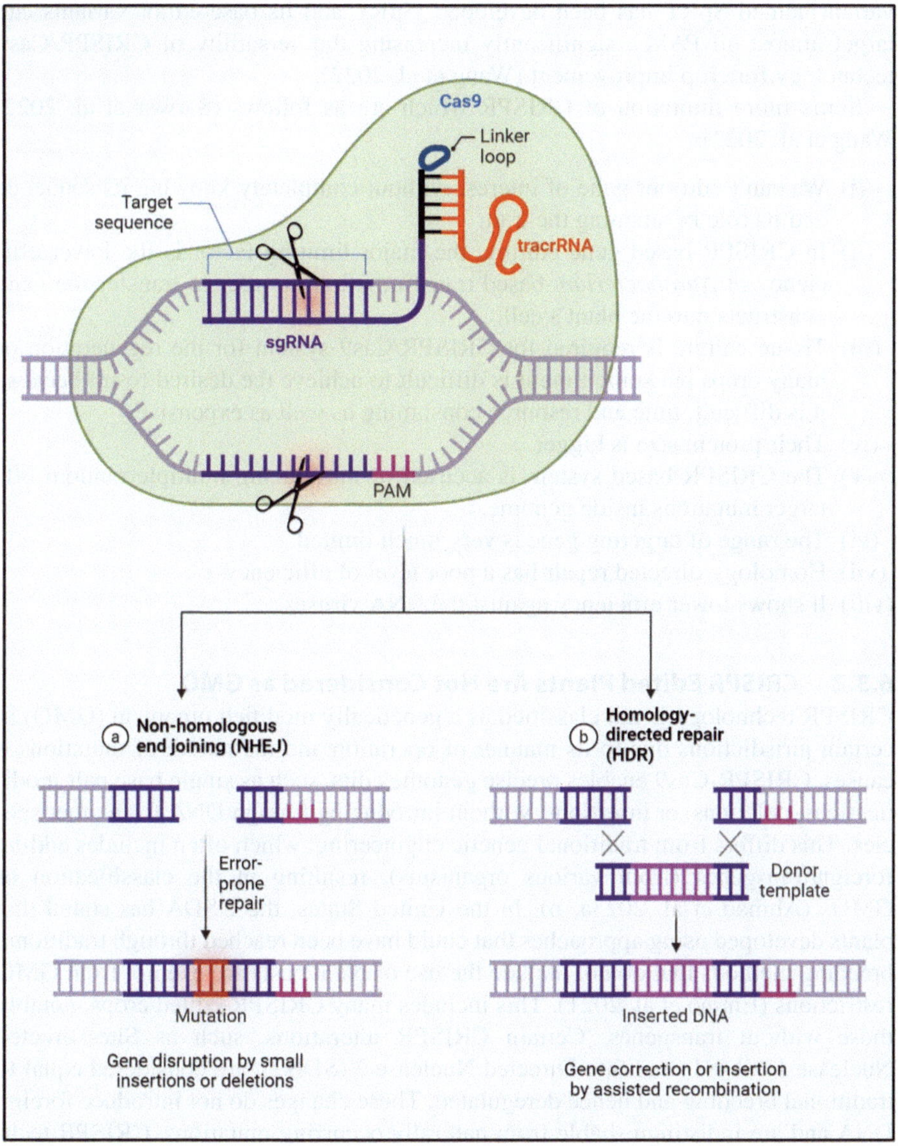

Fig. 2 A schematic representation of CRISPR/Cas9 genome editing and its working mechanism. When coupled with sgRNA, the single Cas9 endonuclease hooks to the target DNA and generates triple-base upstream of protospacer adjacent motif (PAM) double-stranded breaks (DSBs). DSBs are corrected in mammalian cells either homologously directed repair (HDR) or non-homologous end joining (NHEJ). Following CRISPR/Cas9 intervention, NHEJ creates knock out genes and HDR creates knock in genes. (Images created with BioRender)

variant named SpRY has been developed. SpRY and its base-editor variants can target almost all PAMs, significantly increasing the versatility of CRISPR/Cas9 technology for crop improvement (Wang et al. 2022).

Some more limitation of CRISPR which are as follows (Sarwar et al. 2022; Wang et al. 2022):

(i) We can't edit our gene of interest without completely knowing its sequence and its role in retaining the trait.
(ii) In CRISPR-based gene editing, the major limiting factor is the lower efficiency of *Agrobacterium*-based transformation in order to transfer the gene constructs into the plant's cell.
(iii) Tissue culture is required for CRISPR/Cas9 system for the regeneration of many crops but some time it is difficult to achieve the desired result because it is difficult, time and resource consuming as well as expensive.
(iv) Their protein size is bigger.
(v) The CRISPR-based system is accused of introducing multiple random off-target mutations inside genome.
(vi) The range of targeting gene is very much limited.
(vii) Homology-directed repair has a poor level of efficiency.
(viii) It shows lower efficiency against the RNA viruses.

6.3.2 CRISPR Edited Plants Are Not Considered as GMO

CRISPR technology is not classified as a genetically modified organism (GMO) in certain jurisdictions due to its manner of operation and the nature of mutations it causes. CRISPR-Cas9 enables precise genome edits, such as single base pair modifications, deletions, or insertions, without introducing foreign DNA from other species. This differs from traditional genetic engineering, which often includes adding foreign/transgenes (from various organisms), resulting in the classification of GMOs (Ahmad et al. 2023a, b). In the United States, the USDA has stated that plants developed using approaches that could have been reached through traditional breeding methods that do not include the use of plant pests are exempt from GMO restrictions (Entine et al. 2021). This includes many CRISPR-edited crops, notably those without transgenes. Certain CRISPR alterations, such as Site-Directed Nuclease 1 (SDN1) and Site-Directed Nuclease 2 (SDN2), are considered equal to traditional breeding and hence deregulated. These changes do not introduce foreign DNA and are indistinguishable from naturally occurring mutations. CRISPR technology is frequently perceived as having a lower risk profile than classic GMOs, especially when no foreign DNA is inserted. This utilitarian stance lends weight to the claim that CRISPR-edited creatures pose less risk to human health and the environment, making them more regulatory-friendly (Kawall et al. 2020). Different countries have different definitions and rules for GMOs. For example, while the EU views all gene-edited plants as GMOs, the United States and certain other nations do not categorize specific CRISPR-edited crops as GMOs, creating a confusing regulatory landscape.

7 Targeting Disease Resistance, Yield, and Stability in Bluegrasses Via Genome Editing

Genome editing is a potential method for improving disease resistance, yield, and stability in bluegrass species, especially *Poa annua*. CRISPR/Cas9 and other genome-editing technologies can be used to change specific genes associated with disease resistance in bluegrass. Researchers can create more resilient bluegrass types by finding and targeting genes that confer resistance to infections like as anthracnose, pink snow mould, and dollar spot. Several factors can influence turf-grass yield, including growth rate, tiller density, and overall plant vigour (Riaz et al. 2022). Gene editing can be utilized to modify features that lead to increased biomass production and better turf quality. For example, genes involved with growth regulation and nutrient uptake can be targeted to improve these characteristics, resulting in stronger bluegrass cultivars.

In breeding bluegrass, particularly greens-type *Poa annua*, desirable features are unstable between generations. Genome editing can help stabilize certain features by accurately altering the genomic pathways that control trait expression. For example, targeting genes associated with blooming period and growth habit may help preserve the greens-type phenotype under various environmental conditions (May et al. 2023). The improvement of agronomic qualities of bluegrass, particularly *Poa annua*, has been a research priority due to its importance in turf management, particularly for golf courses. *Poa annua* has a wide variety of genetic diversity, including disease resistance, salt tolerance, and specific flowering requirements. Understanding this genetic variability is critical for breeding efforts designed to improve specific agronomic features. The greens-type variety of *Poa annua* has been shown to have desirable characteristics such as fine leaf texture, high tiller density, and the capacity to thrive at low mowing heights (Bao et al. 2022). However, the stability of these features has been a concern in breeding attempts. Populations might revert to less desirable annual-type characteristics when left unmoved. Historically, attempts to produce stable greens-type cultivars have been problematic, owing principally to the volatility of the greens-type morphology. The study suggests that a single recessive gene may control the greens-type phenotype. However, advanced generations have revealed that inheritance patterns do not follow classic genetic models, implying a more sophisticated genetic process at work. The greens-type phenotype's volatility may be modified by non-Mendelian genetic pathways, complicating breeding attempts (Yoosefzadeh Najafabadi et al. 2022). This instability has been found in progeny that exhibit somatic reversions to annual-type traits, demonstrating that environmental conditions and genetic interactions play substantial roles in trait expression (Robbins et al. 2022). Identifying and choosing genotypes with desirable agronomic properties, such as increased seed yield, better drought resistance, and superior growth characteristics, is critical. The study used seed storage protein profiles to aid in genotype discrimination, which can help identify superior cultivars for breeding programs. Combining traditional breeding procedures with new technology, such as molecular markers (e.g., SDS-PAGE for protein profiling), can improve breeding programme efficiency. This enables the selection of features that would otherwise be difficult to detect via morphological assessments. Understanding the genetic variety of current genotypes is critical for breeding. The study discovered limited genetic variation among the evaluated genotypes, implying that introducing new genetic material or using wild relatives could enhance diversity and improve agronomic qualities. Breeding for features that improve adaptation to changing environmental conditions, such as drought tolerance and disease resistance, is critical. The study underlined the impact of global warming on water supplies, emphasizing the importance of genotypes that can live under such conditions. Combining biochemical tests (such as seed storage protein profiling) with conventional agronomic assessments can provide a more complete picture of genotype performance (Salgotra and Chauhan 2022). This integrated approach can help to make better-informed cultivar selection and breeding decisions. Extensive field trials over several years enable the study of agronomic properties under different environmental circumstances. This long-term evaluation can assist in discovering stable and high-performance genotypes (Sandhya 2020).

8 Improving Bluegrass Grain Quality with Gene Editing

Gene editing enables precise alterations to specific genes related to grain quality parameters such as protein content, starch composition, and disease resistance. Breeders can increase grain quality by knocking out or altering these genes. Gene editing can improve the nutritional content of bluegrass grains by increasing the quantities of key amino acids, vitamins, and minerals, making them more suitable for animal feed or human consumption. Breeders can create bluegrass cultivars with consistent production and quality by altering genes associated to abiotic stress tolerance (e.g., drought, salt). Gene editing can considerably reduce the time necessary to generate new cultivars with superior grain quality attributes when compared to traditional breeding methods, allowing desired traits to be introduced into bluegrass populations more rapidly (Ku and Ha 2020). Gene editing can also target genes associated for anti-nutritional components in bluegrass grains, thereby enhancing their overall quality and palatability to animals. Gene editing is an effective method for improving grain quality in bluegrass species, notably in fodder and turf grasses. While bluegrass is mostly known for its use in lawns and sports fields, advances in genetic technologies, such as gene editing, can boost its overall quality and adaptability (Nagamine and Ezura 2022). Gene editing tools, such as CRISPR/Cas9, enable precise alterations to specific genes linked with grain quality attributes. This can involve improving features like nutrient content, digestibility, and resistance to illnesses or pests. Researchers can increase the grain quality of bluegrass types by targeting genes that affect these properties. Genome editing can be used to improve the nutritional value of bluegrass grain. Modifications can be made to increase the levels of critical amino acids, vitamins, and minerals in grains, making them more beneficial for cattle and potentially for human consumption in niche markets.

Bluegrass forages are commonly utilized as animal feed. Gene editing can improve the digestibility of bluegrass grains, making them more appealing and nutritious to animals (Kumar et al. 2022). This can result in improved feed conversion rates and overall animal health. Bluegrass species can be prone to a variety of diseases that degrade grain quality. Gene editing can be used to establish or strengthen resistance to specific infections, lowering the need for chemical treatments while enhancing crop health overall. This can result in larger yields and improved grain quality. Climate change and environmental stressors can have a negative impact on grain quality. Gene editing can help create bluegrass types that are more resistant to abiotic challenges including drought, salinity, and severe temperatures (Manzoor et al. 2024).

9 Challenges Associated with Bluegrass Genome Editing

Bluegrass frequently have complex and polyploid genomes, complicating the editing procedure. Multiple copies of a gene can cause unintended consequences or make it harder to achieve consistent modifications across all genome copies. Bluegrasses have fewer molecular genetic resources than main crops, including genomic sequences, genetic maps, and functional annotations. This scarcity of resources can make it difficult to select target genes for editing and comprehend the potential ramifications of changes. The regulatory environment for genetically modified organisms (GMOs) can be strict and complicated (May et al. 2023). Even if genome-editing techniques like CRISPR/Cas9 do not entail the introduction of foreign DNA, they may undergo regulatory scrutiny, which can delay research and commercialization efforts. Achieving effective transformation in bluegrass species can be complex. Many turfgrass species, especially bluegrasses, have slow regeneration rates and transformation efficiencies, which can limit the number of successfully modified plants generated (Benson et al. 2023). The phenotypic manifestation of altered traits can be modified by environmental circumstances and genetic background, resulting in diversity in outcomes. This diversity can make it difficult to measure the efficacy of genome-editing initiatives. There may be widespread hostility to genetically engineered organisms, particularly those created by genome editing. Addressing concerns about safety, environmental effects, and ethical considerations is critical for garnering acceptability from consumers and stakeholders (Phillips et al. 2023). Many bluegrass species are polyploid, which means they have multiple sets of chromosomes. This intricacy might hamper genome- editing attempts since targeting individual genes may necessitate detailed knowledge of the various alleles found on each chromosome pair. The ability to regrow entire plants from modified cells is essential for successful genome editing. Bluegrasses frequently have limited regeneration rates after tissue culture, making it difficult to generate stable, altered plants. Improving tissue culture procedures is critical to increasing regeneration efficiency (Schaart et al. 2021; Ahmad et al. 2023a, b).

10 Conclusion

Significant advances and challenges in Kentucky bluegrass response to drought stress. It underlines the significance of comprehending the morphological, physiological, and molecular mechanisms that contribute to the drought tolerance of this commonly used turfgrass species. As urbanization grows and water scarcity becomes a significant issue, new cultivars that can survive dry conditions are critical. But this can be overcome by the integration of genomics with diverse fields of crop breeding has been made possible because of the advances in NGS technology. In breeding, large-scale genetic markers and high-throughput genotyping have sped up the quantity and cost of cultivar development. Similarly, functional and comparative genomics have served as a foundation for gene identification and functional characterization.

Bluegrass germplasm screening is a cost-effective and feasible method of discovering crop genotypes that are resistant to this combination. Further identification of adaptive characteristics that contribute to plant adaptation to diverse stress may be possible via the exploration of diverse gene pools. Approaches such as traditional QTL mapping, genome-wide association studies, and whole-genome linkage scans (WGRS) might throw light on the QTLs, genetic determinants, and haplotypes that underlie stress tolerance throughout the whole genome. Genetic-assisted breeding (GAB) might be used to transfer these genomic areas into superior agricultural cultivars.

Genome editing is largely regarded as the most effective and adaptable approach for improving agricultural attributes. The technologies presented, particularly CRISPR/Cas9, have considerable advantages in terms of efficiency, specificity, and the ability to do multiplex edits, which can speed up the breeding of superior crop types. Advances in genome editing are paving the way for novel approaches to plant breeding, which are critical for solving the global food security issues faced by a growing population and changing climate. Furthermore, regulatory roadblocks stand in the way of broad adoption of CRISPR-edited crops. Notable progress has been made in the development of revolutionary alternative genetic engineering approaches such as HI-Edit, which completely eliminates the need for the introduction of foreign CRISPR machinery and so sidesteps GMO regulation. To ensure global food security in the face of the rising incidence of de novo abiotic stresses, a comprehensive strategy combining plant breeding, molecular biology, physiology, and multiple 'omics' is required. The field investigations of crop stress responses by plant 'omics' scientists are more valuable than laboratory research in improving our knowledge of the complex stress tolerance mechanisms and to accelerate the development of stress-tolerant cultivars.

11 Future Perspective

There is a considerable need for continued breeding programmes to generate Kentucky bluegrass cultivars with improved drought resistance. This includes finding and selecting certain morphological and physiological features that improve water-use efficiency and resistance during droughts. As water shortage becomes a more pressing concern, future research should prioritize the development of Kentucky bluegrass variants that use less water while maintaining aesthetic and functional qualities. This is consistent with broader sustainability goals in turf management and environmental conservation. As CRISPR technology advances, we should expect greater precision and efficiency. Newer approaches, such as base editing and prime editing, may allow for even more precise alterations without creating double-strand breaks, lowering the chance of off-target effects and unwanted mutations.

References

Abdul Aziz M, Masmoudi K (2024) Molecular breakthroughs in modern plant breeding techniques. Hortic Plant J. https://doi.org/10.1016/j.hpj.2024.01.004

Ahmad N, Fatima S, Mehmood MA, Zaman QU, Atif RM, Zhou W, Gill RA (2023a) Targeted genome editing in polyploids: lessons from brassica. Front Plant Sci 14. https://doi.org/10.3389/fpls.2023.1152468

Ahmad A, Jamil A, Munawar N (2023b) GMOs or non-GMOs? The CRISPR conundrum. Front Plant Sci 14:1232938. https://doi.org/10.3389/fpls.2023.1232938

Ahmar S, Saeed S, Khan MH, Ullah Khan S, Kamran M, Faheem A, Maqsood A, Rauf M, Saleem S, Hong W, Jung K (2020a) A revolution toward gene-editing technology and its application to crop improvement. Int J Mol Sci 21(16):5665. https://doi.org/10.3390/ijms21165665

Ahmar S, Gill RA, Jung H, Faheem A, Qasim MU, Mubeen M, Zhou W (2020b) Conventional and molecular techniques from simple breeding to speed breeding in crop plants: recent advances and future outlook. Int J Mol Sci 21(7). https://doi.org/10.3390/ijms21072590

Ajeesh Krishna TP, Veeramuthu D, Maharajan T, Soosaimanickam M (2023) The era of plant breeding: conventional breeding to genomics-assisted breeding for crop improvement. Curr Genomics 24(1):24–35. https://doi.org/10.2174/1389202924666230517115912

Albertini E, Marconi G, Barcaccia G et al (2004) Isolation of candidate genes for apomixis in *Poa pratensis* L. Plant Mol Biol 56:879–894. https://doi.org/10.1007/s11103-004-5211-y

Alemu A, Åstrand J, Montesinos-López OA, Isidro y Sánchez J, Fernández-Gónzalez J, Tadesse W, Vetukuri RR, Carlsson AS, Ceplitis A, Crossa J, Ortiz R, Chawade A (2024) Genomic selection in plant breeding: key factors shaping two decades of progress. Mol Plant 17(4):552–578. https://doi.org/10.1016/j.molp.2024.03.007

Alseekh S, Kostova D, Bulut M, Fernie AR (2021) Genome-wide association studies: assessing trait characteristics in model and crop plants. Cell Mol Life Sci 78(15):5743–5754. https://doi.org/10.1007/s00018-021-03868-w

Anand A, Subramanian M, Kar D (2022) Breeding techniques to dispense higher genetic gains. Front Plant Sci 13. https://doi.org/10.3389/fpls.2022.1076094

Anders S, Cowling W, Pareek A, Gupta KJ, Singla-Pareek SL, Foyer CH (2021) Gaining acceptance of novel plant breeding technologies. Trends Plant Sci 26(6):575–587. https://doi.org/10.1016/j.tplants.2021.03.004

Bandyopadhyay A, Kancharla N, Javalkote VS, Dasgupta S, Brutnell TP (2020) CRISPR-Cas12a (Cpf1): a versatile tool in the plant genome editing tool box for agricultural advancement. Front Plant Sci 11:584151

Bao Q, Wolabu TW, Zhang Q, Zhang T, Liu Z, Sun J, Wang Y (2022) Application of CRISPR/Cas9 technology in forages. Grassl Res 1(4):244–251. https://doi.org/10.1002/glr2.12036

Benson CW, Sheltra MR, Maughan PJ et al (2023) Homoeologous evolution of the allotetraploid genome of *Poa annua* L. BMC Genomics 24:350. https://doi.org/10.1186/s12864-023-09456-5

Bhardwaj A, Nain V (2021) TALENs—an indispensable tool in the era of CRISPR: a mini review. J Genet Eng Biotechnol 19(1):1–10. https://doi.org/10.1186/s43141-021-00225-z

Bhattacharya S, Satpati P (2023) Insights into the mechanism of CRISPR/Cas9-based genome editing from molecular dynamics simulations. ACS Omega 8(2):1817–1837. https://doi.org/10.1021/ACSOMEGA.2C05583

Bhuyan SJ, Kumar M, Ramrao Devde P, Rai AC, Mishra AK, Singh PK, Siddique KH (2023) Progress in gene editing tools, implications and success in plants: a review. Front Genome Ed 5:1272678. https://doi.org/10.3389/fgeed.2023.1272678

Bonos SA, Huff DR (2013) Cool-season grasses: biology and breeding. In: Turfgrass: biology, use, and management, vol 56, pp 591–660. https://doi.org/10.2134/agronmonogr56.c17

Braun RC, Bremer DJ, Ebdon JS, Fry JD, Patton AJ (2022) Review of cool-season turfgrass water use and requirements: II. Responses to drought stress. Crop Sci 62(5):1685–1701. https://doi.org/10.1002/csc2.20790

Breseghello F, Coelho AS (2013) Traditional and modern plant breeding methods with examples in rice (*Oryza sativa* L.). J Agric Food Chem 61(35):8277–8286. https://doi.org/10.1021/jf305531j

Brown WL (1939) Chromosome complements of five species of Poa with an analysis of variation in Poa pratensis. Am J Bot 26:717–723

Budhlakoti N, Kushwaha AK, Rai A, Chaturvedi KK, Kumar A, Pradhan AK, Kumar U, Kumar RR, Juliana P, Mishra DC, Kumar S (2022) Genomic selection: a tool for accelerating the efficiency of molecular breeding for development of climate-resilient crops. Front Genet 13:832153. https://doi.org/10.3389/fgene.2022.832153

Bushman BS, Robbins MD, Thorsted K, Robins JG, Warnke SE, Martin R, Harris-Shultz K (2021) Transcript responses to drought in Kentucky bluegrass (*Poa pratensis* L.) germplasm varying in their tolerance to drought stress. Environ Exp Bot 190:104571. https://doi.org/10.1016/j.envexpbot.2021.104571

Bushman BS, Robbins MD, Warnke SE, Martin R, Harris-Shultz K, Amundsen KE (2022) Gene expression differences for drought stress response in three cool-season turfgrasses. Int Turfgrass Soc Res J 14(1):694–700. https://doi.org/10.1002/its2.25

Carroll D (2008) Zinc-finger nucleases as gene therapy agents. Gene Ther 15(22):1463. https://doi.org/10.1038/gt.2008.145

Castro NG, Bjelic J, Malhotra G, Huang C, Alsaffar SH (2021) Comparison of the feasibility, efficiency, and safety of genome editing technologies. Int J Mol Sci 22(19). https://doi.org/10.3390/ijms221910355

Charagh S, Wang H, Wang J et al (2024) Leveraging multi-omics tools to comprehend responses and tolerance mechanisms of heavy metals in crop plants. Funct Integr Genomics 24:194. https://doi.org/10.1007/s10142-024-01481-1

Christian M, Cermak T, Doyle EL, Schmidt C, Zhang F, Hummel A, Bogdanove AJ, Voytas DF (2010) Targeting DNA double-Strand breaks with TAL effector nucleases. Genetics 186(2):757–761. https://doi.org/10.1534/genetics.110.120717

Cui Z, Liu H, Zhang H, Huang Z, Tian R, Li L, Fan W, Chen Y, Chen L, Zhang S, Das BC, Severinov K, Hitzeroth II, Debata PR, Jin Z, Liu J, Huang Z, Xie W, Xie H et al (2021a) The comparison of ZFNs, TALENs, and SpCas9 by GUIDE-seq in HPV-targeted gene therapy. Mol Ther Nucl Acids 26:1466–1478. https://doi.org/10.1016/j.omtn.2021.08.008

Cui J, Shah S, Fahad S, Chen Y (2021b) A review on Kentucky bluegrass responses and tolerance to drought stress. IntechOpen. https://doi.org/10.5772/intechopen.93812

Curry HA (2023) Breeding confusion: hybrid seeds and histories of agriculture. J Peasant Stud 50(3):1037–1055. https://doi.org/10.1080/03066150.2023.2180357

Entine J, Felipe MSS, Groenewald H, Kershen DL, Lema M, McHughen A, Nepomuceno AL, Ohsawa R, Ordonio RL, Parrott WA, Quemada H, Ramage C, Slamet-Loedin I, Smyth SJ, Wray-Cahen D (2021) Regulatory approaches for genome edited agricultural plants in select countries and jurisdictions around the world. Transgenic Res 30(4):551–584. https://doi.org/10.1007/s11248-021-00257-8

Fan J, Zhang W, Amombo E, Hu L, Kjorven JO, Chen L (2020) Mechanisms of environmental stress tolerance in turfgrass. Agronomy 10(4):522. https://doi.org/10.3390/agronomy10040522

Feng G, Xiao P, Wang X, Huang L, Nie G, Li Z, Peng Y, Li D, Zhang X (2021) Comprehensive transcriptome analysis uncovers distinct expression patterns associated with early salinity stress in annual ryegrass (*Lolium Multiflorum* L.). Int J Mol Sci 23(6):3279. https://doi.org/10.3390/ijms23063279

Gaj T, Gersbach CA, Barbas CF (2013) ZFN, TALEN, and CRISPR/Cas-based methods for genome engineering. Trends Biotechnol 31(7):397–405. https://doi.org/10.1016/j.tibtech.2013.04.004

Garcia A, Gaju O, Bowerman AF, Buck SA, Evans JR, Furbank RT, Gilliham M, Millar AH, Pogson BJ, Reynolds MP, Ruan L, Taylor NL, Tyerman SD, Atkin OK (2022) Enhancing crop yields through improvements in the efficiency of photosynthesis and respiration. New Phytol 237(1):60–77. https://doi.org/10.1111/nph.18545

Gu H, Liang S, Zhao J (2022) Novel sequencing and genomic technologies revolutionized rice genomic study and breeding. Agronomy 12(1):218. https://doi.org/10.3390/agronomy12010218

Hendrickson J, Liebig M, Printz J, Toledo D, Halvorson J, Christensen R, Kronberg S (2021) Kentucky bluegrass impacts diversity and carbon and nitrogen dynamics in a northern great plains rangeland. Rangel Ecol Manag 79:36–42. https://doi.org/10.1016/j.rama.2021.07.005

Huang B, DaCosta M, Jiang Y (2014) Research advances in mechanisms of turfgrass tolerance to abiotic stresses: from physiology to molecular biology. Crit Rev Plant Sci 33(2–3):141–189. https://doi.org/10.1080/07352689.2014.870411

Honig A, Marton I, Rosenthal M, Smith JJ, Nicholson M G, Jantz D, Zuker A, Vainstein A (2015) Transient Expression of Virally Delivered Meganuclease In Planta Generates Inherited Genomic Deletions. Molecular plant, 8(8):1292–1294. https://doi.org/10.1016/j.molp.2015.04.001

Iqbal Z, Iqbal MS, Ahmad A, Memon AG, Ansari MI (2020) New prospects on the horizon: genome editing to engineer plants for desirable traits. Curr Plant Biol 24:100171. https://doi.org/10.1016/j.cpb.2020.100171

Jinek M, Chylinski K, Fonfara I, Hauer M, Doudna JA, Charpentier E (2012) A programmable dual RNA-guided DNA endonuclease in adaptive bacterial immunity. Science (New York, N.Y.) 337(6096):816. https://doi.org/10.1126/science.1225829

Jähne F, Hahn V, Würschum T (2020) Speed breeding short-day crops by LED-controlled light schemes. Theor Appl Genet 133:2335–2342. https://doi.org/10.1007/s00122-020-03601-4

Kawall K, Cotter J, Then C (2020) Broadening the GMO risk assessment in the EU for genome editing technologies in agriculture. Environ Sci Eur 32(1):1–24. https://doi.org/10.1186/s12302-020-00361-2

Khan SU, Saeed S, Ullah Khan MH, Fan C, Ahmar S, Arriagada O, Shahzad R, Branca F, Mora-Poblete F (2021) Advances and challenges for QTL analysis and GWAS in the plant-breeding of high-yielding: a focus on rapeseed. Biomolecules 11(10). https://doi.org/10.3390/biom11101516

Khatodia S, Bhatotia K, Passricha N, Khurana SMP, Tuteja N (2016) The CRISPR/Cas genome-editing tool: application in improvement of crops. Front Plant Sci 7:506

Kim YG, Cha J, Chandrasegaran S (1996) Hybrid restriction enzymes: zinc finger fusions to Fok I cleavage domain. Proc Natl Acad Sci USA 93(3):1156–1160. https://doi.org/10.1073/pnas.93.3.1156

Kopecký D, Studer B (2014) Emerging technologies advancing forage and turf grass genomics. Biotechnol Adv 32(1):190–199. https://doi.org/10.1016/j.biotechadv.2013.11.010

Kral-O'Brien KC, Limb RF, Hovick TJ, Harmon JP (2019) Compositional shifts in forb and butterfly communities associated with Kentucky bluegrass invasions. Rangel Ecol Manag 72(2):301–309. https://doi.org/10.1016/j.rama.2018.10.003

Ku K, Ha H (2020) Improving nutritional and functional quality by genome editing of crops: status and perspectives. Front Plant Sci 11. https://doi.org/10.3389/fpls.2020.577313

Kumar V, Jain M (2015) The CRISPR-Cas system for plant genome editing: advances and opportunities. J Exp Bot 66:47–57

Kumar D, Yadav A, Ahmad R, Dwivedi UN, Yadav K (2022) CRISPR-based genome editing for nutrient enrichment in crops: a promising approach toward global food security. Front Genet 13:932859. https://doi.org/10.3389/fgene.2022.932859

Kushanov FN, Turaev OS, Ernazarova DK, Gapparov BM, Oripova BB, Kudratova MK, Rafieva FU, Khalikov KK, Erjigitov DS, Khidirov MT, Kholova MD, Khusenov NN, Amanboyeva RS, Saha S, Yu JZ, Abdurakhmonov IY (2021) Genetic diversity, QTL mapping, and marker-assisted selection technology in cotton (*Gossypium* spp.). Front Plant Sci 12:779386. https://doi.org/10.3389/fpls.2021.779386

Li JF, Norville JE, Aach J, McCormack M, Zhang D, Bush J, Church GM, Sheen J (2013) Multiplex and homologous recombination-mediated genome editing in Arabidopsis and *Nicotiana benthamiana* using guide RNAandCas9. Nat Biotechnol 31:688–691

Li P, Li G, Zhang Y, Zuo J, Liu J, Zhang Y (2022) A combinatorial strategy to identify various types of QTLs for quantitative traits using extreme phenotype individuals in an F_2 population. Plant Commun 3(3):100319. https://doi.org/10.1016/j.xplc.2022.100319

Li X, Liu X, Fan Y, Li S, Yu M, Qian M, Chen Y, Chen H, Li X, Liu B, Xu X, Qu C, Li J, Lu K (2023) Development of a target capture sequencing SNP genotyping platform for genetic analysis and genomic breeding in rapeseed. Crop J 11(2):499–510. https://doi.org/10.1016/j.cj.2022.08.008

Liu J, Xie X, Du J, Sun J, Bai X (2008) Effects of simultaneous drought and heat stress on Kentucky bluegrass. Sci Hortic 115(2):190–195. https://doi.org/10.1016/j.scienta.2007.08.003

Liu S, Wang X, Wang H, Xin H, Yang X, Yan J et al (2013) Genome-wide analysis of *ZmDREB* genes and their association with natural variation in drought tolerance at seedling stage of *Zea mays* L. PLoS Genet 9:e1003790. https://doi.org/10.1371/journal.pgen.1003790

Liu H, Todd JL, Luo H (2022) Turfgrass salinity stress and tolerance—A review. Plants 12(4):925. https://doi.org/10.3390/plants12040925

Ma X, Zhang X, Liu H, Li Z (2020) Highly efficient DNA-free plant genome editing using virally delivered CRISPR-Cas9. Nat Plants 6(7):773–779. https://doi.org/10.1038/s41477-020-0704-5

Malzahn A, Lowder L, Qi Y (2017) Plant genome editing with TALEN and CRISPR. Cell Biosci 7:21. https://doi.org/10.1186/s13578-017-0148-4

Manzoor S, Nabi SU, Rather TR, Gani G, Mir ZA, Wani AW, Ali S, Tyagi A, Manzar N (2024) Advancing crop disease resistance through genome editing: a promising approach for enhancing agricultural production. Front Genome Ed 6:1399051. https://doi.org/10.3389/fgeed.2024.1399051

Marconi G, Aiello D, Kindiger B, Storchi L, Marrone A, Reale L, Terzaroli N, Albertini E (2020) The role of APOSTART in switching between sexuality and apomixis in Poa pratensis. Genes 11(8). https://doi.org/10.3390/genes11080941

Mastalerczuk G, Borawska-Jarmułowicz B, Kalaji HM (2019) How Kentucky bluegrass tolerate stress caused by sodium chloride used for road de-icing? Environ Sci Pollut Res Int 26(1):913–922. https://doi.org/10.1007/s11356-018-3640-4

May D, Sanchez S, Gilby J, Altpeter F (2023) Multi-allelic gene editing in an apomictic, tetraploid turf and forage grass (*Paspalum notatum* Flüggé) using CRISPR/Cas9. Front Plant Sci 14. https://doi.org/10.3389/fpls.2023.1225775

Ma X, Zhang Q, Zhu Q, Liu W, Chen Y, Qiu R, Wang B., Yang, Z, Li H, Lin Y, Xie Y, Shen R, Chen S, Wang Z, Chen Y, Guo J, Chen L, Zhao X, Dong Z, Liu Y (2015) A Robust CRISPR/Cas9 System for Convenient, High-Efficiency Multiplex Genome Editing in Monocot and Dicot Plants. Molecular Plant, 8(8):1274–1284. https://doi.org/10.1016/j.molp.2015.04.007

Milesi C, Running SW, Elvidge CD et al (2005) Mapping and Modeling the biogeochemical cycling of turf grasses in the United States. Environ Manag 36:426–438. https://doi.org/10.1007/s00267-004-0316-2

Mondal S, Seelam RT, Mondal B, Jespersen D (2024) Understanding abiotic stress tolerance mechanisms in non-food grass species through omics approaches. Plant Stress 13:100516. https://doi.org/10.1016/j.stress.2024.100516

Muthamilarasan M, Theriappan P, Prasad M (2013) Recent advances in crop genomics for ensuring food security. Curr Sci 105:155–158

Muthamilarasan M, Venkata Suresh B, Pandey G, Kumari K, Parida SK, Prasad M (2014) Development of 5123 intron-length polymorphic markers for large-scale genotyping applications in foxtail millet. DNA Res 21:41–52. https://doi.org/10.1093/dnares/dst039

Nagamine A, Ezura H (2022) Genome editing for improving crop nutrition. Front Genome Ed 4. https://doi.org/10.3389/fgeed.2022.850104

Ndudzo A, Sibanda Makuvise A, Moyo S, Bobo ED (2024) CRISPR-Cas9 genome editing in crop breeding for climate change resilience: implications for smallholder farmers in Africa. J Agric Food Res 16:101132. https://doi.org/10.1016/j.jafr.2024.101132

Noor M, Fan J, Zhang J, Zhang C, Sun S, Gan L, Yan X (2022) Bermudagrass responses and tolerance to salt stress by the physiological, molecular mechanisms and proteomic perspectives of salinity adaptation. Agronomy 13(1):174. https://doi.org/10.3390/agronomy13010174

Nybom H, Lācis G (2021) Recent large-scale genotyping and phenotyping of plant genetic resources of vegetatively propagated crops. Plants 10(2). https://doi.org/10.3390/plants10020415

Paschon DE, Lussier S, Wangzor T, Xia DF, Li PW, Hinkley SJ, Scarlott NA, Lam SC, Waite AJ, Truong LN, Gandhi N, Kadam BN, Patil DP, Shivak DA, Lee GK, Holmes MC, Zhang L, Miller JC, Rebar EJ (2019) Diversifying the structure of zinc finger nucleases for high-precision genome editing. Nat Commun 10(1):1–12. https://doi.org/10.1038/s41467-019-08867-x

Petolino JF (2015) Genome editing in plants via designed zinc finger nucleases. In Vitro Cell Dev Biol Plant 51(1):1–8. https://doi.org/10.1007/s11627-015-9663-3. Epub 2015 Jan 29

Phillips AR, Seetharam AS, Albert PS, AuBuchon-Elder T, Birchler JA, Buckler ES, Gillespie LJ, Hufford MB, Llaca V, Romay MC, Soreng RJ, Kellogg EA, Ross-Ibarra J (2023) A happy accident: a novel turfgrass reference genome. G3 13(6). https://doi.org/10.1093/g3journal/jkad073

Raza A, Bhardwaj S, Rahman MA, García-Caparrós P, Copeland RGR, Charagh S, Rivero RM, Gopalakrishnan S, Corpas FJ, Siddique KHM, Hu Z (2024a) Fighting to thrive via plant growth regulators: green chemical strategies for drought stress tolerance. Physiol Plant 176(6):e14605. https://doi.org/10.1111/ppl.14605

Raza A, Gangurde SS, Singh Sandhu K, Lv Y (2024b) Omics-assisted crop improvement under abiotic stress conditions. Plant Stress 100626. https://doi.org/10.1016/j.stress.2024.100626

Riaz A, Kanwal F, Ahmad I, Ahmad S, Farooq A, Madsen CK, Brinch-Pedersen H, Bekalu ZE, Dai F, Zhang G, Alqudah AM (2022) New hope for genome editing in cultivated grasses: CRISPR variants and application. Front Genet 13. https://doi.org/10.3389/fgene.2022.866121

Robbins MD, Bushman BS, Huff DR, Benson CW, Warnke SE, Maughan CA, Jellen EN, Johnson PG, Maughan PJ (2022) Chromosome-scale genome assembly and annotation of allotetraploid annual bluegrass (*Poa annua* L.). Genome Biol Evol 15(1). https://doi.org/10.1093/gbe/evac180

Sachdev S, Ansari SA, Ansari MI, Fujita M, Hasanuzzaman M (2021) Abiotic stress and reactive oxygen species: generation, signaling, and defense mechanisms. Antioxidants 10(2). https://doi.org/10.3390/antiox10020277

Safari F, Zare K, Negahdaripour M, Barekati-Mowahed M, Ghasemi Y (2019) CRISPR Cpf1 proteins: structure, function and implications for genome editing. Cell Biosci 9:36

Saini H, Thakur R, Gill R, Tyagi K, Goswami M (2023) CRISPR/Cas9-gene editing approaches in plant breeding. GM Crops Food 14(1):1–17. https://doi.org/10.1080/21645698.2023.2256930

Salgotra RK, Chauhan BS (2022) Genetic diversity, conservation, and utilization of plant genetic resources. Genes 14(1). https://doi.org/10.3390/genes14010174

Samantara K, Bohra A, Mohapatra SR, Prihatini R, Asibe F, Singh L, Reyes VP, Tiwari A, Maurya AK, Croser JS, Wani SH, Siddique KHM, Varshney RK (2022) Breeding more crops in less

time: a perspective on speed breeding. Biology (Basel) 11(2):275. https://doi.org/10.3390/biology11020275
Sarıkaya MF, Tatar M (2023) Global climate change and crop production. Uluslararası Sivas Bilim ve Teknoloji Üniversitesi Dergisi 2(2):65–78
Sarkar B, Varalaxmi Y, Vanaja M, RaviKumar N, Prabhakar M, Yadav SK, Maheswari M, Singh VK (2023) Mapping of QTLs for morphophysiological and yield traits under water-deficit stress and well-watered conditions in maize. Front Plant Sci 14:1124619. https://doi.org/10.3389/fpls.2023.1124619
Sarwar N, Atique-ur-Rehman, Ahmad S, Hasanuzzaman M (2022) Modern techniques of rice crop production. In: Modern techniques of rice crop production. Springer, Singapore. https://doi.org/10.1007/978-981-16-4955-4
Sauer NJ, Mozoruk J, Miller RB, Warburg ZJ, Walker KA, Beetham PR, Schöpke CR, Gocal GFW (2016) Oligonucleotide-directed mutagenesis for precision gene editing. Plant Biotechnol J 14:496–502
Sandhya D, Jogam P, Allini V R, Abbagani S, Alok A (2020) The present and potential future methods for delivering CRISPR/Cas9 components in plants. Journal of Genetic Engineering & Biotechnology, 18:25. https://doi.org/10.1186/s43141-020-00036-8
Sun F, Wang C, Lin Y (2025)Sweet potato gene editing and breeding technologies: Research progress and future prospects. Advances in Resources Research, 5(2):852–871. https://doi.org/10.50908/ARR.5.2_852
Schaart JG, Smulders M, M. J. (2021) Genome editing of polyploid crops: prospects, achievements and bottlenecks. Transgenic Res 30(4):337–351. https://doi.org/10.1007/s11248-021-00251-0
Schneeberger K, Ossowski S, Lanz C, Juul T, Petersen AH, Nielsen KL, Jørgensen J, Weigel D, Andersen SU (2009) SHOREmap: simultaneous mapping and mutation identification by deep sequencing. Nat Methods 6(8):550–551. https://doi.org/10.1038/nmeth0809-550
Shamshirgaran Y, Liu J, Sumer H, Verma PJ, Taheri-Ghahfarokhi A (2022) Tools for efficient genome editing; ZFN, TALEN, and CRISPR. In: Verma PJ, Sumer H, Liu J (eds) Applications of genome modulation and editing. Methods in molecular biology, vol 2495. Humana, New York. https://doi.org/10.1007/978-1-0716-2301-5_2
Si Z, Jin S, Chen J et al (2022) Construction of a high-density genetic map and identification of QTLs related to agronomic and physiological traits in an interspecific (*Gossypium hirsutum* × *Gossypium barbadense*) F_2 population. BMC Genomics 23:307. https://doi.org/10.1186/s12864-022-08528-2
Singh RK, Muthamilarasan M, Prasad M (2021) Biotechnological approaches to dissect climate-resilient traits in millets and their application in crop improvement. J Biotechnol 327:64–73. https://doi.org/10.1016/j.jbiotec.2021.01.002. Epub 2021 Jan 7. PMID: 33422569
Sinha D, Maurya AK, Abdi G, Majeed M, Agarwal R, Mukherjee R, Ganguly S, Aziz R, Bhatia M, Majgaonkar A, Seal S, Das M, Banerjee S, Chowdhury S, Adeyemi SB, Chen T (2023) Integrated genomic selection for accelerating breeding programs of climate-smart cereals. Genes 14(7). https://doi.org/10.3390/genes14071484
Smith J, Grizot S, Arnould S, Duclert A, Epinat C, Chames P, Prieto J, Redondo P, Blanco FJ, Bravo J, Montoya G, Pâques F, Duchateau P (2006) A combinatorial approach to create artificial homing endonucleases cleaving chosen sequences. Nucleic Acids Res 34(22):e149. https://doi.org/10.1093/nar/gkl720
Sprink T, Metje J, Hartung F (2015) Plant genome editing by novel tools: TALEN and other sequence specific nucleases. Curr Opin Biotechnol 32:47–53. https://doi.org/10.1016/j.copbio.2014.11.010
Sun L, Lai M, Ghouri F, Nawaz MA, Ali F, Baloch FS, Nadeem MA, Aasim M, Shahid MQ (2023) Modern plant breeding techniques in crop improvement and genetic diversity: from molecular markers and gene editing to artificial intelligence—a critical review. Plants 13(19):2676. https://doi.org/10.3390/plants13192676
Susmitha P, Kumar P, Yadav P, Sahoo S, Kaur G, Pandey MK, Singh V, Tseng TM, Gangurde SS (2023) Genome-wide association study as a powerful tool for dissecting competitive traits in legumes. Front Plant Sci 14:1123631. https://doi.org/10.3389/fpls.2023.1123631

Uddin F, Rudin CM, Sen T (2020) CRISPR gene therapy: applications, limitations, and implications for the future. Front Oncol 10. https://doi.org/10.3389/fonc.2020.01387

Varshney RK, Graner A, Sorrells ME (2005) Genomics-assisted breeding for crop improvement. Trends Plant Sci 10:621–630. https://doi.org/10.1016/j.tplants.2005.10.004

Varshney RK, Terauchi R, McCouch SR (2014) Harvesting the promising fruits of genomics: applying genome sequencing technologies to crop breeding. PLoS Biol 12:e1001883. https://doi.org/10.1371/journal.pbio.1001883

Varshney RK, Bohra A, Yu J, Graner A, Zhang Q, Sorrells ME (2021) Designing future crops: genomics-assisted breeding comes of age. Trends Plant Sci 26(6):631–649. https://doi.org/10.1016/j.tplants.2021.03.010

Voss-Fels KP, Stahl A, Wittkop B, Lichthardt C, Nagler S, Rose T, Chen TW, Zetzsche H, Seddig S, Majid Baig M, Ballvora A, Frisch M, Ross E, Hayes BJ, Haydenx MJ, Ordon F, Leon J, Kage, H, Friedt W, Stützel H, … Snowdon RJ (2019) Breeding improves wheat productivity under contrasting agrochemical input levels. Nature plants, 5(7):706–714. https://doi.org/10.1038/s41477-019-0445-5

Wang Y, Zafar N, Ali Q, Manghwar H, Wang G, Yu L, Ding X, Ding F, Hong N, Wang G, Jin S (2022) CRISPR/Cas genome editing technologies for plant limitations, and future perspectives. Cells 11:3928

Wanga MA, Shimelis H, Mashilo J, Laing MD (2021) Opportunities and challenges of speed breeding: a review. Plant Breed 140(2):185–194. https://doi.org/10.1111/pbr.12909

Wei C, Liu J, Yu Z, Zhang B, Gao G, Jiao R (2013) TALEN or Cas9—rapid, efficient and specific choices for genome modifications. J Genet Genomics 40(6):281–289. https://doi.org/10.1016/j.jgg.2013.03.013

Xiao Q, Bai X, Zhang C, He Y (2021) Advanced high-throughput plant phenotyping techniques for genome-wide association studies: a review. J Adv Res 35:215–230. https://doi.org/10.1016/j.jare.2021.05.002

Xu L, Li F, Han L, Song G, Zhang X (2017) Overexpression of Arabidopsis *DREB1A* gene in transgenic Poa pratensis: impacts on osmotic adjustment and hormone metabolism under drought. Int Turfgrass Soc Res J 13:527–536

Yasir M, Kanwal HH, Hussain Q, Riaz MW, Sajjad M, Rong J, Jiang Y (2022) Status and prospects of genome-wide association studies in cotton. Front Plant Sci 13:1019347. https://doi.org/10.3389/fpls.2022.1019347

Ye T, Shi H, Wang Y, Yang F, Chan Z (2016) Contrasting proteomic and metabolomic responses of bermudagrass to drought and salt stresses. Front Plant Sci 7:227879. https://doi.org/10.3389/fpls.2016.01694

Yoosefzadeh Najafabadi M, Hesami M, Rajcan I (2022) Unveiling the mysteries of non-mendelian heredity in plant breeding. Plants 12(10):1956. https://doi.org/10.3390/plants12101956

Zhang Y, Liang Z, Zong Y, Wang Y, Liu J, Chen K, Qiu J, Gao C (2016) Efficient and transgene-free genome editing in wheat through transient expression of CRISPR/Cas9 DNA or RNA. Nat Commun 7(1):1–8. https://doi.org/10.1038/ncomms12617

Zhao Y, Ma J, Li M, Deng L, Li G, Xia H, Zhao S, Hou L, Li P, Ma C, Yuan M, Ren L, Gu J, Guo B, Zhao C, Wang X (2019) Whole-genome resequencing-based QTL-seq identified AhTc1 gene encoding a R2R3-MYB transcription factor controlling peanut purple testa colour. Plant Biotechnol J 18(1):96–105. https://doi.org/10.1111/pbi.13175

Zuo Z, Li Y, Mi X, Li Y, Zhai C, Yang G, Wang Z, Zhang K (2023) Physiological and lipidomic response of exogenous choline chloride alleviating salt stress injury in Kentucky bluegrass (*Poa pratensis*). Front Plant Sci 14:1269286. https://doi.org/10.3389/fpls.2023.1269286

Recent Advancements in Breeding and Genomics of Wheatgrasses

Meliha Feryal Sarıkaya, Muhammed Tatar, Amjad Ali, Mehmet Bedir, Sarmad Ali Qureshi, Muhammad Tanveer Altaf, Selim Özdemir, Mahmut Kaplan, and Kağan Kökten

Abstract

Grasslands and pastures are vital ecosystems offering multifaceted ecological and economic benefits. Intermediate wheatgrass (*Thinopyrum intermedium*), a significant perennial crop, has garnered attention for its diverse applications in forage, grain, and biofuel production. This chapter explores the ecological roles of grasslands, their vegetation dynamics, and the distribution and traits of intermediate wheatgrass. It details breeding advancements targeting enhanced forage quality, seed yield, and abiotic and biotic stress resilience. Modern breeding tools, including genomic selection, QTL mapping, and genome editing, have accelerated trait improvement and domestication processes. The dual-purpose utility of intermediate wheatgrass as a sustainable grain and biomass crop underscores its potential in addressing global challenges such as food security, environmental degradation, and climate change. This chapter emphasizes the need for interdisciplinary approaches to optimize its productivity and ecological contributions, advocating for sustainable agricultural practices and innovative breeding strategies.

M. F. Sarıkaya (✉) · M. Tatar · A. Ali · M. Bedir · S. A. Qureshi · K. Kökten
Faculty of Agricultural Sciences and Technologies, Sivas University of Science and Technology, Sivas, Turkey

M. T. Altaf
Department of Field Crops, Faculty of Agriculture and Natural Sciences, Recep Tayyip Erdoğan University, Pazar, Rize, Turkey

S. Özdemir
Field Crops Program, Department of Crop and Animal Production, Food, Agriculture and Livestock Vocational School, Bingol University, Bingol, Turkey

M. Kaplan
Department of Field Crops, Faculty of Agriculture, University of Erciyes, Kayseri, Turkey

F. S. Baloch et al. (eds.), *Forage Crops in the Omics Era*,
https://doi.org/10.1007/978-981-95-9249-4_14

Keywords

Intermediate wheatgrass · Forage breeding · Sustainable agriculture · Genomic selection · Dual-purpose crops

1 Introduction

Agricultural areas are defined as lands where crops were cultivated with all sustainable practices. Soil and land are two different concepts. Soil is a mixture containing air, water, organic and inorganic substances (sand, clay, silt) and harbouring many living organisms (Adewara et al. 2024). Land, on the other hand, is the part of the earth that includes soil, its location, water status, vegetation cover, proximity, and distance to main roads and settlements. Different characteristics of land and soil enable the formation of different vegetation covers (Koc et al. 2015). Some lands are not suitable for soil cultivation. These lands, which have one or more of the characteristics, such as erosion damages, drainage problems, short growth period, excessive moisture, stony and rocky, excessive soil salinity and alkalinity, are not used in cultivation, but are used as grassland and/or pastures (Altın et al. 2021). Grassland and pastures are ecologically important natural ecosystems and have many important functions (Sanderson et al. 2007). Vegetation drives the water cycle on Earth. The bare soil surface is usually dry and therefore the amount of water returning to the atmosphere is limited. As long as there is accessible moisture in the root zone, the plant continues to emit water vapour from its leaves to the atmosphere (Gavrilescu 2021). In terms of these qualities, grassland-pasture plants provide water circulation to a significant extent (Aasim et al. 2011). Vegetation covers interrupt the runoff of surface water generated by precipitation, the rate of fall of raindrops and increase the percolation of the soil (Cen et al. 2024). While the fringe-rooted plants in grasslands and pastures increase the water retention capacity in the upper layer of the soil, the taprooted plants allow rainwater to infiltrate into the deep layers of the soil. Thus, rainwater seeps into the soil and reaches groundwater (Altın et al. 2021). However, in a bare land, rainwater passes into surface runoff without being transferred to the soil. Biodiversity is one of the most important natural wealth of a country. Interest in the protection of biodiversity and natural balance is increasing day by day in the world (Raven and Wackernagel 2020). As a matter of fact, the ability of ecosystems to maintain their functions in a healthy way depends first of all on the protection of their biological diversity. Grassland and pasture are very important in terms of living diversity and gene source. Pasture plants have an invaluable value in terms of plant breeding, especially because they contain the ancestors of cultivated plants (Rauf et al. 2016). Again, the fact that it contains many endemic species increases the interest in these areas. On the other hand, the main and most important function of grassland and pasture is animal feeding. There is a positive relationship between the content richness of animal foods and the content of feeds (Koc et al. 2014). The products of animals fed with species-rich feeds are of better quality and balanced. Apart from these, the richness of honeydew and flower powder of grassland-pasture plants enables beekeeping activities and hosts wildlife by providing nutrition, shelter, and breeding areas (Devkota 2020).

Grassland and pasture are essentially natural forage production areas that provide roughage for animals. Although grassland and pastures have similar characteristics in terms of their structure and functions, their utilization patterns are different (Wróbel et al. 2023). Since grasslands are mostly utilized by mowing, the main issue in grassland management is to address the grass produced and the most profitable way of harvesting and preserving it. On the contrary, since pastures are areas that are directly utilized by grazing animals, the vegetation and the animals that utilize it should be considered as a whole, and the relations and balance between them should be well established. Pasture vegetations have been shaped for centuries under climate and soil factors and grazing of wild and domestic animals and have reached a dynamic balance level. Grassland-pasture ecosystems have a dynamic structure and are constantly changing (Kendir et al. 2009; Wezel et al. 2021). The main reason for this change is the changes in the environment and habitat. There is never a stable structure on earth. Differentiation in the environment or habitat makes changes in the vegetation adapted to that area inevitable. Because plant communities compete with each other under the existing ecological factors and form communities with certain structural characteristics (Götzenberger et al. 2012).

Land on Earth covers an area of approximately 13 billion hectares. In these areas, different plant communities are formed depending on changes in climatic factors such as precipitation and temperature. Pastures are very widely distributed grasslands in the world. Pasture vegetation typically occurs in regions with an annual rainfall of 250–1000 mm and an average temperature ranging from 0 to 26 °C (Holechek et al. 1995).

One of the climates where pastures are seen is the Mediterranean climate (Turner 2004). This climate type, which is defined as hot and dry in summers and mild and rainy in winters, is characterized by the southern parts of Turkey, the central parts of Chile in the Southern Hemisphere, and the Republic of South Africa. These areas are suitable for shrubs (maquis, garig) rather than herbaceous plants. During the summer months, when temperatures are high enough for plant growth (at least 10 °C), moisture levels near the soil surface are low. During this period, shrubs with deep roots (more than 1 m) use moisture more efficiently than short and fringe-rooted wheatgrasses. Therefore, shrub pastures are seen in areas with Mediterranean climate.

One of the climates in the world where pastures are seen is steppe (temperate zone continental) climate (Oztas et al. 2003). It is seen in the inner parts of the middle belts. Türkiye (Central Anatolia, Eastern Anatolia, Southeastern Anatolia), Kazakhstan, Pakistan, North Africa (Morocco, Tunisia, Algeria, Libya, Libya, Egypt, Sudan), Brazil (Northeast), the United States (Western inland areas) can be given as examples of areas where this climate is dominant. Because these are areas where pastures are commonly seen. In the steppe climate, convectional precipitation occurs when the sun heats the water masses and the air rises. Due to the low and short-term precipitation, wheatgrass density is observed in these pastures. Because in this rainfall regime, a few centimetres of the upper part of the soil is wetted, therefore, wheatgrasses forming fringe roots in the phyllosphere benefit more from this moisture than legumes and shrubs. These areas are mainly composed of perennial wheatgrasses. Annual species are more common in arid, heavily grazed, and degraded pastures. Apart from the mentioned factors, many factors such as drought, cold and heat resistance, salt resistance, shade resistance, adaptation to soil structure, adaptation to high ground water and flooding, adaptation to land structure affect vegetation. For example, in areas with salinity problems, *Agropyron elongatum*, *Arrhenatherum elatius*, *Bromus inermis*, *Trifolium fragiferum* are observed, while in acidic soils, *Agrostis palustris*, *Lespedeza stipulacea*, *Vigna sinensis*, drought-tolerant *Agropyron cristatum*, *Elymus junceus*, *Agropyron desertorum*, *Agropyron trichophorum*, *Elymus giganteus*, and *Ammophila arenaria* can be seen in sandy areas (Altın et al. 2021).

Grassland flora predominantly comprises numerous perennial species. Perennial plants offer various ecological advantages over annual crop cultivation (Asbjornsen et al. 2013). In perennial crops, biomass distribution to roots is two to four times greater than in annual crops. Extensive root systems enhance nutrient utilization efficiency in plants relative to annuals (Dietzel et al. 2015). This improves the synchronization between plant demand and nutrient supply, which can reduce the need for synthetic fertilizers for perennial crops (Crews et al. 2022). At the same time, water use efficiency is increased due to the root system (Ning et al. 2022). Less tillage and cultivation operations in perennial systems also reduce greenhouse gas emissions by reducing fuel and labour inputs and can result in more net carbon storage due to extensive root systems (Kantar et al. 2016). Apart from these, it provides many benefits to the ecosystem such as protecting the soil from erosion (Ashworth et al. 2022), reducing weed populations (Law et al. 2021), preventing nutrient leaching (Huddell et al. 2022), pest and pathogen control (Jungers et al. 2018), improving soil health (Rakkar et al. 2023).

2 Intermediate Wheatgrass Origin and Distribution

Numerous species within the genus Celand are significant as forage crops in arid and semi-arid regions globally. In North America, the Commonwealth of Independent States and some eastern European countries, the cultivation of celand species is widespread. Many researchers state that there are about 100–150 species of celand on the earth. Two-thirds of these species are of Asian-European origin. About 20–30 species are naturally cultivated in North America. A few species are also found in South America, Africa, and New Zealand. The majority of the species are distributed especially in steppe and semi-desert regions of Asia (Asay 1995). Although there are many species in the genus Agropyron, only a few of them have agricultural importance. Among the most important species of weeds cultivated today are *A. trichophorum*, *A. cristatum*, *A. intermedium*, *A. desertorum*, *A. elongatum*. These species originate from the Old World. They are found naturally in a very wide area extending from Central Europe to the Caucasus, from Asia Minor to

Siberia and Tibet. The intermediate wheatgrass is found in the Black Sea region, the Balkans, and the gently sloping and hillside regions of Anatolia (Dillman 1946).

Intermediate wheatgrass is found in low mountain belts of southern Europe, Middle East, former Soviet Union, and western Pakistan. It is a perennial, allogamous, hexaploid ($2n = 6\times = 42$) species (Dewey 1962). The intermediate wheatgrass complex comprises glabrous and hairy varieties. The hairless species is *A. intermedium*, while the hairy species is *A. trichophorum* (Hitchcock 1951). Dewey (1978) proposed that the hairless and hairy varieties were amalgamated into one species, *Th. intermedium*. Flora North America, the latest classification for intermediate wheatgrass, identifies two subspecies. These include subsp. *intermedium* (glabrous) and subsp. *barbulatum* (hairy) (Barkworth 2007).

It serves as hay and pasture grass in marginal regions where intermediate wheatgrass, *D. glomerata* and *B. inermis* are unsuitable for cultivation (Jensen and Anderson 2014). Intermediate wheatgrass cultivars have been assessed for biological yields (Jensen and Anderson 2014). The study utilized Plant Introduction (PI) 98,568 in Russia as the foundational germplasm, concentrating on seedling establishment, dry matter yield, and seed yield. PI 98568 played a pivotal role in the advancement of cultivars including Ree, Chief, Greenar, Oahe, and Slate. The introduction of the intermediate wheatgrass variety 'Amur' in 1952 resulted from the direct seed increase of PI from China (Alderson and Sharp 1994). The foundational germplasm of the cultivar 'Chief' comprised 'Ree' plants and PI selected for enhanced seed yield from the former Soviet Union.

The cultivar 'Tegmar' was derived from PI 109219, which was collected near Bolu, Türkiye. This cultivar was chosen for its delayed maturation, prolific tillering, robust seedlings, rapid rhizome growth, drought resistance, and compact growth habit (Alderson and Sharp 1994). 'Luna', a hairy variety, was derived from PI 106831 (Turkmenistan, Ashgabad,) and 116,252 (Asay and Jensen 1996). 'Rush' intermediate wheatgrass, chosen for its exceptional seedling emergence and dry matter yield in regions with average annual precipitation exceeding 300 mm was acquired from the German Botanical Garden in Berlin in 1962. The United States introduced 'Reliant', a six-clone synthetic wheatgrass, for its resistance to leaf spot, seed yield, and nutritional quality. 'Manska', 'Haymaker', and 'Beefmaker' were selected for their dry matter and seed yield. 'Manifest', introduced in 2007, was developed by crossing ten collections from the Caucasus region, focusing on dry matter yield, seed yield, and spring recovery.

As previously mentioned, PI has been instrumental in the advancement of wheatgrass cultivars in the United States. The USDA NPGS collection presently contains 269 collections of *Th. intermedium* subsp. *intermedium*, of which 219 are classified as active collections (those with available seeds). All 14 collections of subspecies of *Th. intermedium* are inactive. Descriptive data for the majority of grass wheat collections, encompassing chromosome number, growth form, flowering date, spring recovery, 100-seed weight, leaf count, stem count, flower count, spike date, disease scores, rough protein content, and reliable, were recorded in the NPGS from 1989 to 1998 and are accessible at https://www.ars-grin.gov.

3 Herbal Properties and Uses of Intermediate Wheatgrass

Intermediate wheatgrass is a cool climate wheatgrass forage plant, which forms a strong grass cover with its short and moderately vigorous rhizomes and also forms tangles. It is tall (around 1 m in height). It is adapted to cool and dry climates. However, its resistance to drought is less than grassland weed and more than awnless brome. It is adapted to places with total annual rainfall over 400 mm. They are not selective in terms of soil requirements. Since the seeds are relatively large and seedling development is fast, it is easy to settle. It can be used in grassland-pasture establishment both in the barren and irrigated areas. It forms a good mixture with alfalfa. It rapidly loses its flavour by solidifying rapidly with maturation. It stays green longer in summer as it develops later than grassy bramble and awnless bromine. It is damaged by intensive grazing. It generally becomes sparse after the fourth year (Altın et al. 2021).

Intermediate wheatgrass is a significant perennial crop that has been incorporated into breeding programmes for over 30 years, serving both as a cereal for human consumption and for fodder production. The study aims to improve the economic sustainability of intermediate wheatgrass, a perennial forage grass, by enhancing its grain production attributes (Table 1) and, being one of the nearest perennial relatives of wheat, it was utilized to convey advantageous traits to annual wheat. Intermediate wheatgrass has been extensively utilized as an external genetic resource for wheat enhancement due to its resilience against various pests and diseases affecting wheat (Mahelka et al. 2011). Since the early studies of the 1980s, extensive breeding programmes have commenced. Many of these programmes have utilized advancements in molecular technologies to expedite the enhancement of IWG (Bajgain et al. 2022). The utilization of the plant for dual purposes (grain and feed) has distinctly influenced the breeding processes and objectives. Furthermore, the recognition of perennial cereal as a potential candidate for domestication and enhancement as a biofuel crop, alongside its applications as fodder, has redirected the aforementioned breeding programmes.

4 Intermediate Wheatgrass Breeding for Forage

Forage crops are essential elements of the grassland and pasture agro-ecosystem. They constitute the primary source of fodder and nutrients for livestock. As primary producers, they are transformed into secondary production within the food chain via livestock production. Forage breeding is a multifaceted process that encompasses the morpho-physiological characteristics of the plant (longevity, reproductive strategies, mating systems) and atypical plant–environment interactions that influence plant performance across diverse grassland conditions. The primary objective of forage breeding is to create varieties that produce high and sustainable grass yields across diverse management systems. It also encompasses the creation of varieties that positively impact ecosystem functions, as well as animal growth and health (Rauf et al. 2016).

Table 1 Some studies on intermediate wheatgrass

Characteristics analysed in the study	Plant material	Technical	Result	Citation
Seed yield Volume wt. of seed Seed quality score Hay yield Lodging Plant height	Line A LineB S-6815 Oahe Ree Chief	Mass selection	The study found a positive correlation between the selection of spaced plants and seed yields in seeded tests, possibly due to the use of well-established plants aged 2 and 3 years and the moderately creeping root habit of intermediate wheat grass.	Knowles (1977)
In vitro dry matter Digestibility (IVDMD) Percent protein Dry matter	36 strains of intermediate wheatgrass	Selection	During the initial stages of a breeding program, simultaneous selection for first-cut yield and IVDMD is feasible due to positive and significant phenotypic correlation coefficients between these traits at both locations.	Vogel et al. (1986)
Spike emergence Plant height Leaf/stem (dry weight) NDF N concentration Digestible organic matter (IVDOM) For whole plant, stems and leaves	Intermediate wheatgrass	Selection	The study found moderately high narrow-sense heritability estimates for whole-plant IVDOM, NDF, and nitrogen content in stem and leaf fractions, IVDOM. Variance components for general combining ability (GCA) were generally superior to specific combining ability (SCA). Year effects for IVDOMan and NDF were considerable, but interactions with GCA and SCA were insignificant. Selection for increased whole-plant IVDOM in specific environments is expected to improve intermediate wheatgrass hay digestibility.	Berdahl et al. (1994)
Dry matter yield Crude protein NDF In vitro dry matter Herbage mass	Four intermediate wheatgrass	–	The results indicate the superior quality of intermediate wheatgrass for grazing livestock and the significant effect that slight enhancements in forage quality can exert on animal performance.	Moore et al. (1995)

Forage species with elevated nutritional value can enhance farm profitability by improving feed characteristics (Thomas et al. 2021). The primary characteristics that define the nutritional value of a specific feed are its chemical composition, intake, digestibility, and utilization efficiency. This pertains specifically to the fibre, lignin, and protein content of feed tissues and their digestibility characteristics (Cherney 2000). The topics examined in forage crop breeding are summarized in Table 2.

Table 2 Some traits examined in forage crop breeding studies

Yield	Quality

Plant height Tillering number Tilting rate Number of sprouting days Seed yield Green and dry forage yield	Crude protein rate NDF and ADF ratio ADL ratio In vitro dry matter digestibilities Macro and micronutrient content Relative feed value Dry matter intake

Prior investigations concerning intermediate wheatgrass have concentrated on its application as forage, grazing, and biomass cultivation (Lee et al. 2009). Several IWG cultivars have demonstrated enhancements in particular characteristics, including elevated forage yield and quality, stand persistence under grazing (Ogle et al. 2011).

Intermediate wheatgrass, despite its high yield and quality, has been restricted due to its low persistence and competitiveness with certain crops (Holt and Jefferson 1999). In some studies, it was reported to be subject to severe winter damage 4 years after establishment (Heinrichs and Clark 1961). Campbell (1961) advised against using intermediate wheatgrass for arid pastures with a lifespan of more than 3 years. Lawrence and Ashford (1966) noted that Intermediate wheatgrass could not withstand heavy leaf growth and suggested using it as hay instead of pasture. In other studies, mechanical leaf removal for forage harvesting has been reported to increase light intensity in the plant canopy and stimulate tillering production in grasses by increasing the red/far-red light ratio (Deregibus et al. 1983; Aamlid et al. 1997; Ugarte et al. 2010). Leaf removal can improve grass density and help maintain optimum leaf arrangement for photosynthesis (Youngner 1972). Consequently, properly timed leaf removal can improve forage quality by keeping tillering vegetative, while increasing total forage yield within the growing season. It can therefore ensure that the plant has a better plant.

5 Intermediate Wheatgrass Breeding for Grain

Annual crops occupy the predominant portion of the global agricultural land (Monfreda et al. 2008). Annual cereal crops, accounting for approximately 50% of global human caloric consumption (Kearney 2010), are linked to environmental issues including soil degradation, nutrient pollution in aquatic systems, and diminished habitats (Tilman 1999). Initiatives aimed at mitigating the detrimental environmental impacts of agriculture encompass the advancement of perennial crops (Glover et al. 2010). Perennial species do not require annual replanting, offering numerous advantages over annual crops, including diminished fertilizer and tillage usage, as well as the absence of irrigation needs in regions with ample rainfall (Crews et al. 2016). Moreover, perennial crops can enhance soil carbon and nitrogen reserves, mitigate soil erosion, and decrease nutrient loss due to leaching and runoff (Smith et al. 2013). A comprehensive root system can enhance soil carbon reserves over time and also increase water use efficiency (Zeri et al. 2013; Abraha et al. 2016). A plant can allocate its carbon both above and below ground, and these components yield ecological and/or economic advantages. Intermediate wheatgrass, a member of the Triticeae tribe within the Poaceae family alongside wheat and rye, is the inaugural perennial cereal crop cultivated and marketed for human consumption (Bajgain et al. 2020). Kernza was the inaugural domesticated perennial cereal crop to be commercialized (DeHaan et al. 2018). It was introduced to the United States as a forage crop in the early twentieth century and was predominantly cultivated for forage in the Great Plains and intermountain West regions (Ogle et al. 2011). IWG has experienced two decades of cultivation and selection, initially at the Big Flats Plant Materials Centre (Wagoner 1995; DeHaan et al. 2018; Cattani and Asselin 2018).

Breeders of intermediate wheatgrass have enhanced domestication characteristics, including larger seed size, increased grain yield, decreased fragmentation, and reduced plant height (Zhang et al. 2016). Field tests indicate that enhanced IWG populations produce greater grain yields than IWG forage crops across diverse soil and climatic conditions (Jungers et al. 2019), and the grain from these improved IWG populations is appropriate for numerous food applications. The US Food and Drug Administration has classified intermediate wheatgrass grain as 'Generally Regarded as Safe', allowing food companies to incorporate IWG grains as a novel ingredient (DeHaan and Ismail 2017). Certain cereal-type IWG lines have preserved feed quality characteristics, such as elevated biomass yield and nutritional value; consequently, IWG has been assessed as a dual-purpose crop, offering enhanced economic advantages by supplying cereal straw during the mid-summer harvest (Hunter et al. 2020).

6 Breeding of Intermediate Wheatgrass for Biofuel

Optimal energy crops ought to be perennial, high-yielding, require minimal to moderate inputs, be readily adaptable without necessitating extensive new cultural practices, possess favourable raw material characteristics, and be well-suited to the biorefinery's location (Parrish and Fike 2010). Despite the high demand for crops such as sorghum, miscanthus, and millet grass [*Panicum virgatum* (L.)] as biofuel feedstocks, it is improbable that a single plant or the combination of all three will serve as a comprehensive feedstock solution on a global scale or across all geographic regions in the United States. For biofuels to be genuinely viable, diverse crop groups, whether as monocultures or in mixtures, must be cultivated across various geographical regions.

C_4 perennial grasses have garnered greater attention as biofuel feedstocks compared to C_3 perennial grasses. This results from the elevated resource utilization efficiency and substantial yield potential of C_4 plants (van der Weijde et al. 2013). C_3 perennial grasses have largely been overlooked in the biofuel industry, yet numerous potential benefits indicate they warrant further consideration. C_3 perennial grasses have traditionally been cultivated as livestock fodder owing to their enhanced digestibility and elevated nutritional content (Burns and Fisher 2010). Utilizing these crops as biomass feedstocks could capitalize on established crop development, agronomic knowledge, and production infrastructure. Intermediate wheatgrass has been studied for these purposes in recent years (Payne et al. 2017). IWG is being cultivated as a perennial grain crop with potential applications as a biofuel feedstock. IWG biomass is abundant in feruloylated arabinoxylans and cellulose, rendering it appropriate for biofuel production (Dräger et al. 2024). The grain and biomass yields of IWG exhibit a favourable response to nitrogen fertilization, with grain yields reaching a maximum of 961 kg/ha and biomass yields varying between 9.2 and 12.3 Mg/ha at optimal fertilization rates (Jungers et al. 2017). Nonetheless, grain yields diminish with the age of the stand, and the ideal seeding dates for optimizing both grain and biomass yields remain under investigation (Jungers et al. 2022). The potential of IWG as a dual-use crop for grain and biofuel production is promising; however, additional genetic enhancement and agronomic research are required to improve its economic viability.

7 Intermediate Wheatgrass Breeding Targets

7.1 Yield Components

The primary aim of selection in IWG breeding programmes has been the enhancement of grain yield and seed size (Wagoner 1990). Breeders comprehend the correlations between yield, a multifaceted trait, and its constituent traits. Positive and frequently substantial correlations have been documented between seed size and weight in enhanced IWG germplasm (Larson et al. 2019). Multiple studies have indicated robust correlations between biological traits, including plant height, spike length, flag leaf area, and reproductive tiller count (Bajgain and Anderson 2021). This relationship has primarily been demonstrated to be advantageous for the cultivation of dual-purpose forage and grain IWG; however, it may also present certain drawbacks, including an elevated risk of lodging linked to taller plants (DeHaan et al. 2018). Conversely, characteristics like floret site utilization and the percentage of fully developed flowers significantly influence yield (Altendorf et al. 2021a). IWG generates a surplus of flowers compared to those utilized. This leads to low estimates of floret site utilization, varying from 12% to 60% (Altendorf et al. 2021b). This trait exhibits low heritability. The breeding gains from phenotypic selection are deemed insufficient. Hunter et al. (2020) also discovered that IWG grasses generated a greater number of non-reproductive siblings as they aged. This underscores the necessity of comprehending tillering dynamics and its influence on grain yield selection. Alongside research assessing IWG performance, it is essential to investigate the degree to which G × E interactions influence yield. G × E interaction is a recognized factor in plant breeding and agricultural production. As IWG production increases, breeding programs will persist in their efforts to comprehend and leverage G × E interactions for optimal yield, both regionally and at more localized scales.

8 Resistance to Biotic Stress

8.1 Fusarium Head Blight

It is a disease that can be observed under conditions of high humidity and air temperature of 20–25 °C, especially during pollination and grain filling (Manstretta and Rossi 2016). The causal agent is *F. graminearum* (Oliver et al. 2005; Turner et al. 2013). This causal agent causes the production of mitotoxin that can reduce the value of the grain. Infection can occur in the spikelets and causes a whitened appearance of the spikelets before senescence. Improving resistance to this disease and limiting the effects of its contamination by improving post-harvest practices is crucial for the development and widespread use of IWG.

8.2 Bacterial Leaf Streak

The etiological agents of bacterial leaf streak in IWG are the pathogenic bacteria *Xanthomonas translucens* pv. *undulosa* (*Xtu*) and *X. translucens* pv. *cerealis* (Mohan et al. 2001; Curland et al. 2021) The disease manifests as elongated lesions on leaves, ranging from light brown to dark brown in colour. Under extreme conditions, the lesions undergo necrosis, resulting in the formation of black streaks on the scales (Curland et al. 2018). Enhancing the prevalence of resistance alleles via genomic selection may serve as an efficacious approach to address bacterial leaf streak and other diseases in IWG breeding programs (Jannink et al. 2010).

8.3 Ergot

Ergot is a fungal affliction of cereal and forage crops induced by various species of the genus *Claviceps.* Although ergot can reduce cereal yields by as much as 10% (Wegulo and Carlson 2011), its more significant threat lies in its toxicity to human and animal health. A comprehensive understanding of guest resistance to ergot in cereal crops remains insufficient; however, flowering timing and pollen fertility are believed to be intertwined with guest resistance (Kodisch et al. 2020). Research aimed at identifying ergot-resistant rye genotypes indicates that disease incidence is diminished in germplasm, including local varieties and traditional cultivars with increased pollen production, while it is elevated in germplasm, such as modern synthetics and hybrid cultivars with reduced pollen production (Mirdita and Miedaner 2009).

9 Resistance to Abiotic Stress

In international breeding programs, studies are carried out with very different climatic conditions for the plant. Especially since climate change has the potential to increase drought and heat stress (Kukal and Irmak 2018), the topics studied for IWG are expected to be in parallel with this. In some studies, IWG has been reported as a source of abiotic resistance for wheat improvement (Zhang et al. 2021). With the increase in IWG cultivation area, environmental tests for abiotic stresses are likely to increase. Many issues such as cold acclimation, disruption of acclimation, winter survival of plants, light and temperature or their interactive effects are important for plants (Ergon 2017).

10 Modern Breeding Tools for Wheatgrass Breeding

Modern breeding tools used in intermediate wheatgrass are analysed below under the headings utilization of genomic resources and new approaches in trait improvement (Bajgain et al. 2022).

11 Utilization of Genomic Resources

11.1 IWG Genome Sequencing and Map Construction

DNA sequencing of the next generation and bioinformatics techniques have been the driving forces behind significant advancements in the field of molecular research (Schuster 2008). The discovery of polymorphic markers and population structure has been made possible by high-throughput and multiple genotyping methods such as genetic background screening (GBS) (Poland et al. 2012).

Intermediate wheatgrass (IWG) is a perennial grass undergoing domestication as a grain crop. Recent studies have utilized advanced genomic techniques to expedite the breeding process. Genotyping-by-sequencing (GBS) has been employed to construct genetic maps and identify markers throughout IWG's 21 chromosomes (Zhang et al. 2016; Altendorf et al. 2021a). Nested association mapping and linkage mapping have elucidated the intricate genetic architecture of flowering time in IWG (Altendorf et al. 2021b). Genomic selection (GS) models utilizing GBS-derived markers have demonstrated significant predictive capabilities for multiple agronomic traits. GS has facilitated the achievement of one breeding cycle annually, potentially augmenting the rate of genetic improvement by up to 2.6 times relative to phenotypic selection (Crain et al. 2021a). The application of GS in IWG breeding programs has yielded encouraging outcomes, with observed selection differentials ranging from 11% to 23% and anticipated genetic advancements of 6–14% annually for essential traits (Crain et al. 2021a).

11.2 Genetic Mapping of Yield and Domestication Traits

Until today, important studies have been carried out to create linkage maps (Altendorf et al. 2021b). In a study conducted to identify QTL and possible genes controlling agronomic and domestication traits in IWG, seed traits, such as seed length, seed width, seed weight, yield traits, such as plant height, plant weight, and domestication traits, such as flowering time and disease resistance, were examined.

Zhang et al. (2017) performed the inaugural quantitative trait loci (QTL) study within the International Genetic Web (IWG) at the University of Minnesota. GBS markers were employed to examine length, width, seed mass, and area within a diverse breeding population. The research identified 33 QTL via association mapping and 38 QTL through interval mapping, with 23 QTL common to both methodologies. The varied breeding population exhibited the highest logarithm-of-odds value for seed width. Nonetheless, merely 4731 QTL were employed for association mapping owing to the absence of a genome reference sequence assembly. Larson et al. (2019) conducted a study on domestication characteristics in the M26 × M35 lineage using clonally replicated field experiments in Kansas and Utah. The M26 parent, an F1 hybrid of C3–3471 × C3–3941, was a free-threshing, non-shattering plant in the Land Institute breeding program. The M26 × M35 family was evaluated for threshability, seed shattering, and essential spike yield characteristics. The study identified 11 QTL for 17 traits, with a minimum of two located on each linkage group. The QTL showed elevated LoD scores for seed length, seed shattering, and spike length. Nine key markers accounted for 46.8% of the variation in threshability over 5 years, while broad-sense heritability and QTL were maximized for early spike emergence. Seven QTL markers explained up to 37% of the variation in seed length, and five markers accounted for 20.8% of the variation in seed mass related to seed size traits.

Bajgain and Anderson (2021) conducted a GWAS study targeting multi-allelic and single markers to identify genomic regions influencing domestication traits in IWG. Specifically, measured, width, plant height, shattering, threshability, seed length, and mass utilizing the fourth-cycle IWG breeding population at the University of Minnesota, which comprised 637 genes. The population was evaluated in 2020 and 2019 at two sites in Minnesota, United States. Genome-wide association studies utilized 5379 haplotype blocks and 25,909 individual SNP markers. An examination of 31 SNP markers and 17 haplotype blocks revealed significant correlations with domestication traits across 14 and 11 chromosomes, respectively. A modest effect was noted in all marker–trait associations, with significant markers exhibiting a PVE of 4–6%. A single nucleotide polymorphism (SNP) marker on Chromosome 8 was identified in both single-marker and haplotype analyses, associated with seed length and mass. Divergent allelic frequencies and linkage disequilibrium with the quantitative trait locus (QTL) can result in variations in QTL detection between individual markers and haplotype blocks (Slatkin 2008; Hamblin and Jannink 2011). Bajgain and Anderson (2021) proposed utilizing both marker types in GWAS to enhance the understanding of the genetic architecture of complex quantitative traits in IWG.

11.3 Genomic Selection

Molecular techniques, such as GBS and alignment to a draft genome, have empowered the IWG breeding community to employ GS. In GS, markers are utilized to cover the entire genome, ideally ensuring that each gene is in linkage disequilibrium with a marker, thus facilitating the establishment of genomic-estimated breeding value (GEBV) for each individual (Meuwissen et al. 2001). Zhang et al. (2016) performed the first assessment of genomic selection (GS) in interspecific wheat germplasm at the University of Minnesota, suggesting that GS could be an effective selection strategy. Following the initial study, numerous supplementary studies have been conducted within the UMN and TLI breeding programmes (Bajgain et al. 2020; Crain et al. 2021a). A multitude of studies have evaluated GS in Triticeae species, as reviewed by Rutkoski et al. (2017). The application of GS in intermediate wheatgrass (IWG) has led researchers to investigate several factors affecting GS accuracy, including the choice of statistical models, the number of markers, and the configuration of training populations.

Research indicates that the selection of a statistical model substantially affects the accuracy of genomic selection (GS). Multiple models have been assessed utilizing GS, with the Gaussian kernel and random forest yielding the most favourable results. Nonetheless, models, such as RR-BLUP, Bayes A, B, Cpi, LASSO, ridge regression, and RKHS, demonstrated similar efficacy. The choice of a suitable statistical model is crucial for accurate results and genetic enhancements. The precision of genomic selection (GS) improves with increased marker density, and augmenting the quantity of markers can elevate genomic selection accuracy across various interspecific wheat germplasm (IWG) populations. Nonetheless, considerable discrepancies and increased variability were noted with fewer than 2500 markers in both studies. Notwithstanding considerable missing data in GBS data sets, the precision of GS models enhanced with the incorporation of additional markers, even amidst heightened missing data (Davey et al. 2011; Poland and Rife 2012; Crain et al. 2021b). Improvements in genotyping technologies, including whole-genome skim-sequencing methods, may increase the number of available markers for genomic selection. Previous research on training population design has demonstrated that enhancing the training population is more critical than improving family distribution within it. Although historical data may lead to diminished GS accuracy for wheat, the findings from the current TLI breeding programme indicate that its incorporation into models is beneficial.

12 Recent Approaches for Trait Enhancement

12.1 Genome Editing/EcoTILLING

Targeted Induced Local Lesions in Genomes (TILLING) and EcoTILLING (Ecotype TILLING) employ sophisticated DNA sequencing technologies to identify and detect induced and natural mutations in specific genomic regions of interest

(Barkley and Wang 2008). A multitude of essential domestication genes has been identified in cereal crop species related to characteristics such as seed size, shattering, and free threshing (Gross and Olsen 2010; Lenser and Theißen 2013; Swinnen et al. 2016; Purugganan 2019), and specific traits may hold potential significance in IWG (Larson et al. 2019; DeHaan et al. 2020). In addition to targeted mutagenesis of specific genes or traits, species such as wheat can now utilize base editors that do not necessitate double-strand breaks or exogenous DNA, as well as DNA-free genome-editing systems (Zong et al. 2018). These targeted mutagenesis techniques present potential for IWG, albeit necessitating labour-intensive and resource-intensive tissue culture processes. A refined plant regeneration system employing embryonic suspension cultures has been established for IWG and analogous species (Wang et al. 2003). Concurrently, alternative gene-editing methodologies utilizing viruses (Jackson and Li 2016; Hu et al. 2019) or nanoparticles (Cunningham et al. 2018; Demirer et al. 2019) to bypass tissue culture requirements are being advanced for various plant species, including wheat. Techniques for the direct delivery of plant growth regulators and gene-editing agents into somatic cells capable of generating gene-edited shoots, bypassing tissue culture, have been established for particular dicots (Maher et al. 2020). Diverse methodologies for targeted mutagenesis possess potential applications for IWG (DeHaan et al. 2020).

12.2 Genotyping Approaches

GBS are commonly used for GWAS and GS studies of IWG (Zhang et al. 2016; Crain et al. 2021a). However, WGS genotyping offers a cost-effective and high-throughput method for screening larger breeding populations, resulting in higher marker density and efficiency. To implement this technology in heterozygous IWG, it is essential to impute missing alleles. Recombinant inbred lines (RILs) or doubled-haploid (DH) populations are ideal for WGS strategies due to their homozygous nature and ease of imputation of missing alleles. Imputation is less accurate in out-crossing populations with high heterozygosity (IWG) due to low sequencing coverage misclassifying heterozygous alleles as homozygous (Scheben et al. 2017). Several software packages, including Beagle, STITCH, AlphaImpute, AlphaPeel, and magic impute, do not consider population structure for genotype calling and imputation (Browning and Browning 2009; Davies et al. 2016; Hickey et al. 2011; Whalen et al. 2018; Zheng et al. 2018). Current research explores using imputation programmes with the reference genome to achieve accurate imputations, even with low sequence coverage. WGS may become the preferred marker platform, but it has a cost and accuracy issue, particularly at low sequence coverage levels.

12.3 Phenomic Selection

The IWG is assessing multiple approaches to develop predictive breeding value models to expedite domestication and enhance efficiency. Despite GS reducing breeding cycles to 1 year and achieving notable advancements, the expense of genotyping species such as IWG, characterized by a complex genome, continues to exceed $10 per plant. TLI has collaborated with the Danforth Center, KSU, and USDA researchers to explore the application of phenomics-estimated breeding values (PEBVs) in crop selection to expedite enhancement. Recent studies demonstrate that a kinship matrix can forecast breeding values for traits such as yield, akin to genomic selection, provided there is adequate phenotypic dimensionality for traits like seedling growth and spectral signatures (Segura et al. 2014; Rincent et al. 2018; Parmley et al. 2019; Lane et al. 2020; Moreira et al. 2020). Perennial species can utilize phenomics for precise seedling selection of advanced traits. Following validation, phenomic relatedness may supplant genomic relatedness in breeding applications. Eliminating the necessity for genotyping breeding populations would decrease breeding expenses and facilitate larger, more diverse populations, thereby expediting progress (Van Tassel et al. 2022).

12.4 Introgression from Wheat

Annual wheat is the most extensively cultivated crop globally and possesses numerous advantageous qualities as a food source for humans. Variation for numerous essential traits may exist in IWG; however, identifying the genomic variants and breeding for the improvement of these traits may require several years. Specific characteristics, including kernel width, plumpness, large seed size, and semi-dwarf growth habit, have not been observed in IWG. Due to the feasibility of hybridization between annual wheat and IWG, TLI is incorporating small chromosomal segments from wheat into IWG via wide crosses, subsequently followed by backcrossing to IWG (Bajgain et al. 2022).

12.5 Mutation Breeding

Mutation breeding is an effective method in contemporary plant enhancement, employing induced genetic alterations to create novel crop varieties with advantageous characteristics (Yali and Mitiku 2022). This technique has been instrumental in crop evolution and domestication, providing benefits over conventional breeding methods by facilitating the swift enhancement of elite cultivars and the introduction of novel traits (Pathirana 2021). Both physical and chemical mutagens are utilized, with physical mutagens being more prevalent (Pandit et al. 2021). Mutation breeding has resulted in the development of over 3300 mutant varieties across more than 170 plant species worldwide (Pathirana 2021; Pandit et al. 2021). The amalgamation of mutation techniques with molecular methodologies is creating novel opportunities for crop improvement, tackling issues such as climate change adaptation, yield augmentation, and nutritional enhancement (Ali and Suryakant 2024; Pandit et al. 2021).

While IWG germplasm exhibits considerable variation for numerous traits relevant to domestication as a grain crop, certain variations appear to be deficient. To date, breeders have not discovered dwarfing alleles that significantly diminish height while enhancing grain yield. Another notable characteristic is seed plumpness. Wild plants possess exceedingly slender seeds, and a mutation resulting in more spherical seeds would constitute a significant enhancement.

Mutations considerably influence the genetic diversity of crop species (Mba 2013); however, identifying and utilizing induced mutations in outcrossing polyploid species such as IWG poses difficulties. Ukai (2010) has developed a model for identifying and applying mutated alleles in outcrossing species. Chemical mutagenesis has demonstrated effectiveness in wheat (Chen et al. 2012), and an EMS protocol for IWG was developed at TLI. A 0.8% EMS solution produced a 40% germination rate in wheat, but did not influence IWG. The conclusive IWG method entailed immersing seeds in a 0.18% EMS solution for 16 hours, followed by rinsing and storage. In the autumn of 2014, approximately 2400 EMS-treated seedlings were planted at TLI. Mutagenized plants were evaluated for dominant mutations to uncover new recessive mutants obscured by dominant alleles. Owing to the prevalence of dominant dwarfing genes in wheat (Sun et al. 2019), the objective was to identify short yet resilient semi-dwarf mutants. Regrettably, no novel mutations were identified that influence any traits of interest. All plants exhibited natural variation without significant deviation. A TILLING methodology is probably required to differentiate novel mutations from background variation (Chen et al. 2012). Homolog mutations across the three IWG subgroups can be pyramided using MAS.

13 Future Direction

The interest in sustainable food systems is increasing due to heightened awareness of environmental issues related to soil erosion, irrigation runoff, and greenhouse gas emissions. The cumulative effects of these environmental issues have significantly affected the ecosystem over decades and will intensify the challenges of food production. It is imperative to tackle these challenges with sustainable solutions and food supply. The cultivation of perennial food crops will provide environmental benefits, present alternatives to the declining food supply, motivate farmers to grow these crops, and meet the increasing consumer demand for sustainable food systems. In contrast, multifunctional products will significantly enhance the economic advantages for producers. Therefore, agronomic or breeding research for plants, such as intermediate wheatgrass cohosh, which display many of the previously mentioned characteristics, is of great importance.

References

Aamlid TS, Heide OM, Christie BR, McGraw RL (1997) Reproductive development and the establishment of potential seed yield in grasses and legumes. In: Fairey DT, Hampton JG (eds) Forage seed production, vol 1. CAB International, Wallingford, pp 9–44

Aasim M, Şahin Demirbağ N, Khawar KM, Kendir H, Özcan S (2011) Direct axillary shoot regeneration from the mature seed explant of the hairy vetch (Vicia villosa Roth). Arch Biol Sci 63:757–762

Abraha M, Gelfand I, Hamilton SK, Shao C, Su YJ, Robertson GP, Chen J (2016) Ecosystem water-use efficiency of annual corn and perennial grasslands: contributions from land-use history and species composition. Ecosystems 19:1001–1012

Adewara OA, Adebayo-Olajide TC, Ayedun JS, Kotun BC, Adeleke AJ, Brown AD et al (2024) Soil formation, soil health and soil biodiversity. In: Prospects for soil regeneration and its impact on environmental protection. Cham, Springer Nature, pp 95–121

Alderson J, Sharp WC (1994) Grass varieties in the United States, Agricultural handbook no. 170. Soil Conservation Service, USDA. US Government Printing Office, Washington, DC

Ali S, Suryakant TN (2024) Mutation breeding and its importance in modern plant breeding: A review. J Exp Agric Int 46(7):264–275. https://doi.org/10.9734/jeai/2024/v46i72581

Altendorf KR, DeHaan LR, Heineck GC, Zhang X, Anderson JA (2021a) Floret site utilization and reproductive tiller number are primary components of grain yield in intermediate wheatgrass spaced plants. Crop Sci 61(2):1073–1088

Altendorf KR, Larson SR, DeHaan LR, Crain J, Neyhart J, Dorn KM, Anderson JA (2021b) Nested association mapping reveals the genetic architecture of spike emergence and anthesis timing in intermediate wheatgrass. G3 Genes, Genom Genet 11(3):jkab025

Altın M, Gökkuş A, Koç A (2021) Çayır Mera Islahı. Palme Yayınevi

Asay KH (1995) Wheatgrasses and wildryes: the perennial Triticeae. Forages 1:373–394

Asay KH, Jensen KB (1996) Cool-season forage grasses. In: Moser LE et al (eds) Wheatgrass. Agronomy monograph no. 34. ASA-CSSA-SSSA, Madison, pp 691–724

Asbjornsen H, Hernandez-Santana V, Liebman M, Bayala J, Chen J, Helmers M et al (2013) Targeting perennial vegetation in agricultural landscapes for enhancing ecosystem services. Renew Agric Food Syst 29:101–125. https://doi.org/10.1017/S1742170512000385

Ashworth AJ, Katuwal S, Moore PA, Adams T, Anderson K, Owens PR (2022) Perenniality drives multifunctional forage-biomass filter strips' ability to improve water quality. Crop Sci 63:1–13. https://doi.org/10.1002/csc2.20878

Bajgain P, Anderson JA (2021) Multi-allelic haplotype-based association analysis identifies genomic regions controlling domestication traits in intermediate wheatgrass. Agriculture 11(7):667

Bajgain P, Zhang X, Jungers JM, DeHaan LR, Heim B, Sheaffer CC et al (2020) 'MN-Clearwater', the first food-grade intermediate wheatgrass (Kernza perennial grain) cultivar. J Plant Regist 14(3):288–297

Bajgain P, Crain JL, Cattani DJ, Larson SR, Altendorf KR, Anderson JA et al (2022) Breeding intermediate wheatgrass for grain production. Plant Breed Rev 46:119–217

Barkley NA, Wang ML (2008) Application of TILLING and EcoTILLING as reverse genetic approaches to elucidate the function of genes in plants and animals. Curr Genomics 9(4):212–226

Barkworth ME (2007) Flora of North America Vol. 24: Magnoliophyta: Commelinidae (in part): Poaceae, part 1. In: Barkworth ME et al (eds) Thinopyrum Á Löve. Oxford University Press, Oxford, pp 373–378

Berdahl JD, Karn JF, Dara ST (1994) Quantitative inheritance of forage quality traits in intermediate wheatgrass. Crop Sci 34(2):423–427

Browning BL, Browning SR (2009) A unified approach to genotype imputation and haplotype-phase inference for large data sets of trios and unrelated individuals. Am J Hum Genet 84(2):210–223

Burns JC, Fisher DS (2010) Intake and digestibility of improved selections of tall fescue and Orchardgrass hays. Crop Sci 50(1):419–426. https://doi.org/10.2135/cropsci2009.03.0157

Campbell JB (1961) Continuous versus repeated-seasonal grazing of grass-alfalfa mixtures at swift current, Saskatchewan. J Range Manag 14:72–77

Cattani DJ, Asselin SR (2018) Has selection for grain yieldaltered intermediate wheatgrass? Sustainability (Switzerland) 10:688

Cen Y, Zhang K, Zhang M, Zhao W, Pan X, Peng Y, Rubinato M (2024) A calculation model of the mean flow velocity of overland flow considering a variety of grass covers and raindrop's characteristics. Soil Tillage Res 242:106145

Chen L, Huang L, Min D, Phillips A, Wang S, Madgwick PJ et al (2012) Development and characterization of a new TILLING population of common bread wheat (*Triticum aestivum* L.). PLoS One 7(7):e41570

Cherney DJR (2000) Characterization of forages by chemical analysis. In: Givens D, Owen E, Axford R, Omed H (eds) Forage evaluation in ruminant nutrition. CAB Publishing, Wallingford, Oxon, pp 1–480

Crain J, DeHaan L, Poland J (2021a) Genomic prediction enables rapid selection of high-performing genets in an intermediate wheatgrass breeding program. Plant Genome 14(2):e20080

Crain J, Haghighattalab A, DeHaan L, Poland J (2021b) Development of whole-genome prediction models to increase the rate of genetic gain in intermediate wheatgrass (*Thinopyrum intermedium*) breeding. Plant Genome 14(2):e20089

Crews TE, Blesh J, Culman SW, Hayes RC, Jensen ES, Mack MC et al (2016) Going where no grains have gone before: from early to mid-succession. Agric Ecosyst Environ 223:223–238

Crews TE, Kemp L, Bowden JH, Murrell EG (2022) How the nitrogen economy of a perennial cereal-legume intercrop affects productivity: can synchrony be achieved? Front Sustain Food Syst 6:755548

Cunningham FJ, Goh NS, Demirer GS, Matos JL, Landry MP (2018) Nanoparticle-mediated delivery towards advancing plant genetic engineering. Trends Biotechnol 36(9):882–897

Curland RD, Gao L, Bull CT, Vinatzer BA, Dill-Macky R, Van Eck L, Ishimaru CA (2018) Genetic diversity and virulence of wheat and barley strains of *Xanthomonas translucens* from the upper midwestern United States. Phytopathology 108(4):443–453

Curland RD, Hallada KR, Ledman KE, Dill-Macky R (2021) First report of bacterial leaf streak caused by *Xanthomonas translucens* pv. *Undulosa* on cultivated wild rice (*Zizania palustris*) in Minnesota. Plant Dis 105(9):2711

Davey JW, Hohenlohe PA, Etter PD, Boone JQ, Catchen JM, Blaxter ML (2011) Genome-wide genetic marker discovery and genotyping using next-generation sequencing. Nat Rev Genet 12(7):499–510

Davies RW, Flint J, Myers S, Mott R (2016) Rapid genotype imputation from sequence without reference panels. Nat Genet 48(8):965–969

DeHaan LR, Ismail BP (2017) Perennial cereals provide ecosystem benefits. Cereal Foods World 62:278–281. https://doi.org/10.1094/CFW-62-6-0278

DeHaan L, Christians M, Crain J, Poland J (2018) Development and evolution of an intermediate wheatgrass domestication program. Sustainability 10:1499

DeHaan L, Larson S, López-Marqués RL, Wenkel S, Gao C, Palmgren M (2020) Roadmap for accelerated domestication of an emerging perennial grain crop. Trends Plant Sci 25(6):525–537

Demirer GS, Zhang H, Matos JL, Goh NS, Cunningham FJ, Sung Y et al (2019) High aspect ratio nanomaterials enable delivery of functional genetic material without DNA integration in mature plants. Nat Nanotechnol 14(5):456–464

Deregibus VA, Sanchez RA, Casal JJ (1983) Effects of light quality on tiller production in Lolium spp. Plant Physiol 72:900–902. https://doi.org/10.1104/pp.72.3.900

Devkota K (2020) Beekeeping: sustainable livelihoods and agriculture production in Nepal. Mod Beekeeping-Bases Sustain Prod 26:1–11

Dewey DR (1962) The genome structure of intermediate wheatgrass. J Hered 53:282–290

Dewey DR (1978) Intermediate wheatgrasses of Iran. Crop Sci 18:43–48

Dietzel R, Jarchow ME, Liebman M (2015) Above- and belowground growth, biomass, and nitrogen use in maize and reconstructed prairie cropping systems. Crop Sci 55(2):910–923. https://doi.org/10.2135/cropsci2014.08.0572

Dillman AC (1946) The beginnings of crested wheatgrass in North America 1. Agron J 38(3):237–250

Dräger H, Mobley J, Kamali P, Dorrani M, Lynn B, DeHaan L, Schendel RR (2024) Lignin, extractives and structural carbohydrate characteristics of *Thinopyrum intermedium* biomass reveal additional valorization opportunities for dual-crop utilization. J Sci Food Agric 104(15):9451–9461

Ergon Å (2017) Optimal regulation of the balance between productivity and overwintering of perennial grasses in a warmer climate. Agronomy 7(1):19

Gavrilescu M (2021) Water, soil, and plants interactions in a threatened environment. Water 13(19):2746

Glover JD, Reganold JP, Bell LW, Borevitz J, Brummer EC, Buckler ES et al (2010) Increased food and ecosystem security via perennial grains. Science 328(5986):1638–1639

Götzenberger L, de Bello F, Bråthen KA, Davison J, Dubuis A, Guisan A et al (2012) Ecological assembly rules in plant communities-approaches, patterns and prospects. Biol Rev 87(1):111–127

Gross BL, Olsen KM (2010) Genetic perspectives on crop domestication. Trends Plant Sci 15(9):529–537

Hamblin MT, Jannink JL (2011) Factors affecting the power of haplotype markers in association studies. Plant Genome 4(2):145

Heinrichs DH, Clark KW (1961) Clipping frequency and fertilizer effects on productivity and longevity of five grasses. Can J Plant Sci 41:98–108

Hickey JM, Kinghorn BP, Tier B, Wilson JF, Dunstan N, van der Werf JH (2011) A combined long-range phasing and long haplotype imputation method to impute phase for SNP genotypes. Genet Sel Evol 43:1–13

Hitchcock AS (1951) Manual of grasses of the United States, USDA miscellaneous publication 200, 2nd edn. U.S. Government Printing Office, Washington, DC

Holechek JL, Pieper RD, Herbel CH (1995) Range management: principles and practices, 2nd edn. Upper Saddle River, Prentice Hall, p 526

Holt NW, Jefferson PG (1999) Productivity and sustainability of four grazed grass–alfalfa mixtures. Can J Anim Sci 79(1):83–89

Hu J, Li S, Li Z, Li H, Song W, Zhao H et al (2019) A barley stripe mosaic virus-based guide RNA delivery system for targeted mutagenesis in wheat and maize. Mol Plant Pathol 20(10):1463–1474

Huddell A, Ernfors M, Crews T, Vico G, Menge DNL (2022) Nitrate leaching losses and the fate of 15N fertilizer in perennial intermediate wheatgrass and annual wheat-a field study. Sci Total Environ 857:159255. https://doi.org/10.1016/j.scitotenv.2022.159255

Hunter MC, Sheaffer CC, Culman SW, Jungers JM (2020) Effects of defoliation and row spacing on intermediate wheatgrass I: grain production. Agron J 112(3):1748–1763. https://doi.org/10.1002/agj2.20128

Jackson AO, Li Z (2016) Developments in plant negative-strand RNA virus reverse genetics. Annu Rev Phytopathol 54(1):469–498

Jannink JL, Lorenz AJ, Iwata H (2010) Genomic selection in plant breeding: from theory to practice. Brief Funct Genomics 9(2):166–177

Jensen KB, Anderson WF (2014) Rangeland and warm-season forage grasses. In: Smith S et al (eds) Yield gains in major U.S. field crops, CSSA special publication, vol 33. American Society of Agronomy, Inc, pp 219–266

Jungers JM, DeHaan LR, Betts KJ, Sheaffer CC, Wyse DL (2017) Intermediate wheatgrass grain and forage yield responses to nitrogen fertilization. Agron J 109(2):462–472

Jungers JM, Frahm CS, Tautges NE, Ehlke NJ, Wells MS, Wyse DL, Sheaffer CC (2018) Growth, development, and biomass partitioning of the perennial grain crop Thinopyrum intermedium. Ann Appl Biol 172(3):346–354

Jungers JM, DeHaan LH, Mulla DJ, Sheaffer CC, Wyse DL (2019) Reduced nitrate leaching in a perennial grain crop compared to maize in the upper Midwest, USA. Agric Ecosyst Environ 272:63–73. https://doi.org/10.1016/j.agee.2018.11.007

Jungers JM, Schiffner S, Sheaffer C, Ehlke NJ, DeHaan L, Torrion J et al (2022) Effects of seeding date on grain and biomass yield of intermediate wheatgrass. Agron J 114(4):2342–2351

Kantar MB, Tyl CE, Dorn KM, Zhang X, Jungers JM, Kaser JM et al (2016) Perennial grain and oilseed crops. Annu Rev Plant Biol 67(1):703–729. https://doi.org/10.1146/annurev-arplant-043015-112311

Kearney J (2010) Food consumption trends and drivers. Philos Trans R Soc B Biol Sci 365(1554):2793–2807

Kendir H, Sahin-Demirbag N, Aasim M, Khawar KM (2009) In vitro plant regeneration from Turkish Narbon Vetch (Vicia narbonensis L. var. narbonensis L.). Afr J Biotechnol 8(4):614–618

Knowles RP (1977) Recurrent mass selection for improved seed yields in intermediate wheatgrass 1. Crop Sci 17(1):51–54

Koc A, Kaya A, Gullap MK, Erkovan HI, Macit M, Karaoglu M (2014) The effect of supplemental concentrate feed on live weight gain of yearling heifers over grazing season in subirrigate drange lands of East Anatolia. Turk J Vete Anim Sci 38:278–284

Koc A, Schacht WH, Erkovan HI (2015) The history and current direction of rangeland management in Turkey. Rangelands 37(1):39–46

Kodisch A, Wilde P, Schmiedchen B, Fromme FJ, Rodemann B, Tratwal A et al (2020) Ergot infection in winter rye hybrids shows differential contribution of male and female genotypes and environment. Euphytica 216:1–14

Kukal MS, Irmak S (2018) Climate-driven crop yield and yield variability and climate change impacts on the US Great Plains agricultural production. Sci Rep 8(1):1–18

Lane HM, Murray SC, Montesinos-López OA, Montesinos-López A, Crossa J, Rooney DK, Barrero-Farfan ID, Fuente GNDL, Morgan CLS (2020) Phenomic selection and prediction of maize grain yield from near-infrared reflectance spectroscopy of kernels. Plant Phenome J 3:e20002

Larson S, DeHaan L, Poland J, Zhang X, Dorn K, Kantarski T et al (2019) Genome mapping of quantitative trait loci (QTL) controlling domestication traits of intermediate wheatgrass (*Thinopyrum intermedium*). Theor Appl Genet 132(8):2325–2351

Law EP, Wayman S, Pelzer CJ, DiTommaso A, Ryan MR (2021) Intercropping red clover with intermediate wheatgrass suppresses weeds without reducing grain yield. Agron J 114:1–17. https://doi.org/10.1002/agj2.20914

Lawrence T, Ashford R (1966) The productivity of intermediate wheatgrass as affected by initial harvest dates and recovery periods. Can J Plant Sci 46:9–15. https://doi.org/10.4141/cjps66-002

Lee D, Owens VN, Boe AR, Koo BC (2009) Biomass and seed yields of big bluestem, switchgrass, and intermediate wheatgrass in response to manure and harvest timing at two topographic positions. Glob Change Biol Bioenergy 1:171–179. https://doi.org/10.1111/j.1757-1707.2009.01008.x

Lenser T, Theißen G (2013) Molecular mechanisms involved in convergent crop domestication. Trends Plant Sci 18(12):704–714

Mahelka V, Kopecký D, Paštová L (2011) On the genome constitution and evolution of intermediate wheatgrass (*Thinopyrum intermedium*: Poaceae, Triticeae). BMC Evol Biol 11:1–17

Maher MF, Nasti RA, Vollbrecht M, Starker CG, Clark MD, Voytas DF (2020) Plant gene editing through de novo induction of meristems. Nat Biotechnol 38(1):84–89

Manstretta V, Rossi V (2016) Effects of temperature and moisture on development of fusarium graminearum perithecia in maize stalk residues. Appl Environ Microbiol 82(1):184–191

Mba C (2013) Induced mutations unleash the potentials of plant genetic resources for food and agriculture. Agronomy 3(1):200–231

Meuwissen TH, Hayes BJ, Goddard M (2001) Prediction of total genetic value using genome-wide dense marker maps. Genetics 157(4):1819–1829

Mirdita V, Miedaner T (2009) Resistance to ergot in self-incompatible germplasm resources of winter rye. J Phytopathol 157(6):350–355

Mohan SK, Bijman VP, St. John L (2001) Bacterial leaf stripe caused by *Xanthomonas translucens* pv. *Cerealis* on intermediate wheatgrass in Idaho. Plant Dis 85(8):921–921

Monfreda C, Ramankutty N, Foley JA (2008) Farming the planet: 2. Geographic distribution of crop areas, yields, physiological types, and net primary production in the year 2000. Glob Biogeochem Cycles 22(1):GB002947

Moore KJ, Vogel KP, Klopfenstein TJ, Masters RA, Anderson BE (1995) Evaluation of four intermediate wheatgrass populations under grazing. Agron J 87(4):744–747

Moreira FF, Oliveira HR, Volenec JJ, Rainey KM, Brito LF (2020) Integrating high-throughput phenotyping and statistical genomic methods to genetically improve longitudinal traits in crops. Front Plant Sci 11:681

Ning Z, Li Y, Zhao X, Han D, Zhan J (2022) Comparison of leaf and fine root traits between annuals and perennials, implicating the mechanism of species changes in desertified grasslands. Front Plant Sci 12:778547

Ogle D, St. John L, Tober D, Jensen K (2011) Plant guide for intermediate wheatgrass (*Thinopyrum intermedium*). USDA-Natural Resources Conservation Service, Idaho and North Dakota Plant Materials Centers. https://plants.usda.gov/plantguide/pdf/pg_thin6.pdf. Accessed 27 July 2016

Oliver RE, Cai X, Xu SS, Chen X, Stack RW (2005) Wheat-alien species derivatives: a novel source of resistance to fusarium head blight in wheat. Crop Sci 45(4):1353–1360

Oztas T, Koc A, Comakli B (2003) Changes in vegetation and soil properties along a slope on overgrazed and eroded rangelands. J Arid Environ 55(1):93–100

Pandit R, Bhusal B, Regmi R, Neupane P, Bhattarai K, Maharjan B et al (2021) Mutation breeding for crop improvement: a review. Rev Food Agric 2(1):31–35

Parmley K, Nagasubramanian K, Sarkar S, Ganapathysubramanian B, Singh AK (2019) Development of optimized phenomic predictors for efficient plant breeding decisions using phenomic-assisted selection in soybean. Plant Phenom 2019:5809404

Parrish DJ, Fike JH (2010) Growth and production of herbaceous energy crops. In: Soils, plant growth and crop production-Vol III. EOLSS Publication, Oxford, pp 16–43

Pathirana R (2021) Mutations in plant evolution, crop domestication and breeding. Trop Agric Res Ext 24(3):124–157

Payne C, Wolfrum EJ, Nagle N, Brummer JE, Hansen N (2017) Evaluation of fifteen cultivars of cool-season perennial grasses as biofuel feedstocks using near-infrared. Agron J 109(5):1923–1934

Poland JA, Rife TW (2012) Genotyping-by-sequencing for plant breeding and genetics. Plant Genome 5(3):92

Poland J, Endelman J, Dawson J, Rutkoski J, Wu S, Manes Y et al (2012) Genomic selection in wheat breeding using genotyping-by-sequencing. Plant Genome 5(3):103

Purugganan MD (2019) Evolutionary insights into the nature of plant domestication. Curr Biol 29(14):R705–R714

Rakkar M, Jungers JM, Sheaffer C, Bergquist G, Grossman J, Li F, Gutknecht JL (2023) Soil health improvements from using a novel perennial grain during the transition to organic production. Agric Ecosyst Environ 341:108164. https://doi.org/10.1016/j.agee.2022.108164

Rauf S, Sienkiewicz-Paderewska D, Malinowski DP, Hussain MM, Niazi IAK, Kausar M (2016) Forages: ecology, breeding objectives and procedures. In: Advances in plant breeding strategies: agronomic, abiotic and biotic stress traits. Springer, Cham, pp 149–201

Raven P, Wackernagel M (2020) Maintaining biodiversity will define our long-term success. Plant Diver 42(4):211–220

Rincent R, Charpentier JP, Faivre-Rampant P, Paux E, Le Gouis J, Bastien C, Segura V (2018) Phenomic selection is a low-cost and high-throughput method based on indirect predictions: proof of concept on wheat and poplar. G3 Genes, Genom Genet 8(12):3961–3972

Rutkoski JE, Crain J, Poland J, Sorrells ME (2017) Genomic selection for small grain improvement. In: Genomic selection for crop improvement: new molecular breeding strategies for crop improvement. Springer, Cham, pp 99–130

Sanderson MA, Goslee SC, Soder KJ, Skinner RH, Tracy BF, Deak A (2007) Plant species diversity, ecosystem function, and pasture management-a perspective. Can J Plant Sci 87(3):479–487

Scheben A, Batley J, Edwards D (2017) Genotyping-by-sequencing approaches to characterize crop genomes: choosing the right tool for the right application. Plant Biotechnol J 15(2):149–161

Schuster SC (2008) Next-generation sequencing transforms today's biology. Nat Methods 5(1):16–18

Segura V, Charpentier JPJP, Ader K, El Malki R, Guet J, Jorge VV et al (2014) The concept of phenomic selection: using Near InfraRed Spectroscopy (NIRS) to predict quantitative phenotypes. In: International poplar symposium (IPS-VI). International Union of Forest Research Organisations (IUFRO)

Slatkin M (2008) Linkage disequilibrium-understanding the evolutionary past and mapping the medical future. Nat Rev Genet 9(6):477–485

Smith CM, David MB, Mitchell CA, Masters MD, Anderson-Teixeira KJ, Bernacchi CJ, DeLucia EH (2013) Reduced nitrogen losses after conversion of row crop agriculture to perennial biofuel crops. J Environ Qual 42(1):219–228

Sun L, Yang W, Li Y, Shan Q, Ye X, Wang D, Yu K, Lu W, Xin P, Pei Z, Guo X, Liu D, Sun J, Zhan K, Chu J, Zhang A (2019) A wheat dominant dwarfing line with Rht12, which reduces stem cell length and affects gibberellic acid synthesis, is a 5AL terminal deletion line. Plant J 97:887–900

Swinnen G, Goossens A, Pauwels L (2016) Lessons from domestication: targeting cis-regulatory elements for crop improvement. Trends Plant Sci 21:506–515

Thomas DT, Flohr BM, Monjardino M, Loi A, Llewellyn RS, Lawes RA, Norman HC (2021) Selecting higher nutritive value annual pasture legumes increases the profitability of sheep production. Agric Syst 194:103272. https://doi.org/10.1016/j.agsy.2021.103272

Tilman D (1999) The ecological consequences of changes in biodiversity: a search for general principles. Ecology 80(5):1455–1474

Turner NC (2004) Sustainable production of crops and pastures under drought in a mediterranean environment. Ann Appl Biol 144(2):139–147

Turner MK, DeHaan LR, Jin Y, Anderson JA (2013) Wheatgrass-wheat partial amphiploids as a novel source of stem rust and *fusarium* head blight resistance. Crop Sci 53(5):1994–2005

Ugarte CC, Trupkin SA, Ghiglione H, Slafer G, Casal JJ (2010) Low red/far-red ratios delay spike and stem growth in wheat. J Exp Bot 61:3151–3162. https://doi.org/10.1093/jxb/erq140

Ukai Y (2010) Theoretical studies on population development design in mutation breeding of allogamous plants. Breed Sci 60(3):267–278

van der Weijde T, Alvim Kamei CL, Torres AF, Vermerris W, Dolstra O, Visser RGF, Trindade LM (2013) The potential of C4 grasses for cellulosic biofuel production. Front Plant Sci 4:107. https://doi.org/10.3389/fpls.2013.00107

Van Tassel DL, DeHaan LR, Diaz-Garcia L, Hershberger J, Rubin MJ, Schlautman B et al (2022) Re-imagining crop domestication in the era of high throughput phenomics. Curr Opin Plant Biol 65:102150

Vogel KP, Reece PE, Lamb JFS (1986) Genotype and genotype × environment interaction effects for forage yield and quality of intermediate wheatgrass 1. Crop Sci 26(4):653–658

Wagoner P (1990) Perennial grain new use for intermediate wheatgrass. J Soil Water Conserv 45:81–82

Wagoner P (1995) Intermediate wheatgrass (*Thinopyrum intermedium*). In: Development of a perennial grain crop. Cereals and pseudocereals. Chapman & Hall, London

Wang ZY, Bell J, Hopkins A (2003) Establishment of a plant regeneration system for wheatgrasses (*Thinopyrum*, *Agropyron* and *Pascopyrum*). Plant Cell Tissue Organ Cult 73:265–273

Wegulo SN, Carlson MP (2011) Ergot of small grain cereals and grasses and its health effects on humans and livestock. University of Nebraska–Lincoln Extension, Lincoln, Nebraska

Wezel A, Stöckli S, Tasser E, Nitsch H, Vincent A (2021) Good pastures, good meadows: mountain farmers' assessment, perceptions on ecosystem services, and proposals for biodiversity management. Sustainability 13(10):5609

Whalen A, Ros-Freixedes R, Wilson DL, Gorjanc G, Hickey JM (2018) Hybrid peeling for fast and accurate calling, phasing, and imputation with sequence data of any coverage in pedigrees. Genet Sel Evol 50:1–15

Wróbel B, Zielewicz W, Staniak M (2023) Challenges of pasture feeding systems-opportunities and constraints. Agriculture 13(5):974

Yali W, Mitiku T (2022) Mutation breeding and its importance in modern plant breeding. J Plant Sci 10(2):64–70

Youngner VB (1972) Physiology of defoliation and regrowth. In: Youngner VB, McKell CM (eds) The biology and utilization of grasses. Academic, New York, pp 292–303

Zeri M, Hussain MZ, Anderson-Teixeira KJ, DeLucia E, Bernacchi CJ (2013) Water use efficiency of perennial and annual bioenergy crops in Central Illinois. J Geophys Res Biogeo 118(2):581–589

Zhang X, Sallam A, Gao L et al (2016) Establishment and optimization of genomic selection to accelerate the domestication and improvement of intermediate wheatgrass. Plant Genome 9:1–18. https://doi.org/10.3835/plantgenome2015.07.0059

Zhang X, Larson SR, Gao L, Teh SL, DeHaan LR, Fraser M et al (2017) Uncovering the genetic architecture of seed weight and size in intermediate wheatgrass through linkage and association mapping. Plant Genome 10(3):plantgenome2017-03

Zhang X, Cui C, Bao Y, Wang H, Li X (2021) Molecular cytogenetic characterization of a novel wheat-Thinopyrum intermedium introgression line tolerant to phosphorus deficiency. Crop J 9(4):816–822

Zheng C, Boer MP, van Eeuwijk FA (2018) Accurate genotype imputation in multiparental populations from low-coverage sequence. Genetics 210(1):71–82

Zong Y, Song Q, Li C, Jin S, Zhang D, Wang Y et al (2018) Efficient C-to-T base editing in plants using a fusion of nCas9 and human APOBEC3A. Nat Biotechnol 36(10):950–953

Recent Advancements in Breeding and Genomics of Orchardgrass

Muhammed Tatar, Meliha Feryal Sarıkaya, Amjad Ali, İlker Yüce, Sedat Severoglu, Muhammad Tanveer Altaf, Mahmut Kaplan, and Kağan Kökten

Abstract

Grassland pastures are essential for livestock production, supporting animal health, productivity, and rural economies. Wheat fodder crops, such as orchardgrass, are vital because of their versatility, substantial biomass yield, and nutritional advantages. These are well-known cold climate–growing crops. Orchardgrass is a moisture-loving species, it yields more in moist conditions, needs less water, and plays a crucial role in enhancing ecological stability and soil health. Its roots are soil which plays a major role in its stability and helps to prevent soil erosion, improve water retention capacity of soil, and facilitate carbon sequestration. It is known as a key plant for solving abiotic constraints. Due to its extensive genetic variety availability, it has been successfully improved by employing of breeding tools that make use of cutting-edge molecular technology. Recent advancement in genomic methodologies, including marker-assisted selection, genome-wide association studies (GWAS), and Next generation Sequencing (NGS), which have helped in the breeding of fast, better yielded, and resistance and more tolerant plants. These developed plants have been able to meet the requirements of shifting agricultural systems and climatic circum-

M. Tatar (✉) · M. F. Sarıkaya · A. Ali · İ. Yüce · K. Kökten
Faculty of Agricultural Sciences and Technologies, Sivas University of Science and Technology, Sivas, Turkey
e-mail: mtatar@sivas.edu.tr

S. Severoglu
Department of Field Crops, Faculty of Agriculture, University of Atatürk, Erzurum, Turkey

M. T. Altaf
Department of Field Crops, Faculty of Agriculture and Natural Sciences, Recep Tayyip Erdoğan University, Pazar, Rize, Turkey

M. Kaplan
Department of Field Crops, Faculty of Agriculture, University of Erciyes, Kayseri, Turkey

F. S. Baloch et al. (eds.), *Forage Crops in the Omics Era*,
https://doi.org/10.1007/978-981-95-9249-4_15

stances. The stability and livestock feed choice for better food make it more popular to the orchardgrass for rapid breeding to develop more and more biomass. This chapter fully defines all the characteristics related to breeding and further advances the research on orchardgrass to develop better varieties.

Keywords

Wheatgrass · *Dactylis glomerata* L. · Molecular breeding · Next generation sequencing · Sustainable agriculture production

1 Introduction

Grassland and pasture areas are crucial feed resources for livestock-raising societies, enhancing animal health and productivity (Koc et al. 2014). Meadows and pastures are crucial ecosystems providing food and serving environmental functions like directing water cycles, regulating surface runoff, and increasing soil percolation (Cen et al. 2024). Due to their extensive root systems, they have the ability to store carbon, making them valuable for sustainable environmental management. Rangelands, covering large areas, store 50% more carbon than forests and may increase their carbon absorbing role due to their resilience to higher temperatures (Reinermann et al. 2020).

Pastures are the primary food source for livestock, with *Dactylis glomerata* L. (Orchardgrass) being a crucial member of the wheatgrass family used extensively in grassland pastures (Bajgain et al. 2019). Orchardgrass is a crucial forage crop in cold climates, efficiently growing biomass, reducing soil erosion, and contributing to environmental balance by trapping carbon in roots (Sakiroglu et al. 2020). Meadow-pasture ecosystems impact livestock productivity, with high biomass production positively correlated with animal food content. Genetic diversity of cereal forage crops develops resistant, productive crops (Cen et al. 2024).

Orchardgrass, a plant native to stony and rocky terrains, is utilized for ecological restoration in desertified lands due to its ability to grow in difficult-to-till areas. Grassland-pasture ecosystems have a constantly changing and dynamic structure. Changes in environmental factors directly affect these ecosystems and provide diversity in vegetation (Wezel et al. 2021). In this context, adaptation of meadow pastures to climate changes offers important opportunities to increase livestock productivity. In these areas, different plant communities are formed depending on changes in climatic factors such as precipitation and temperature. Pastures are grasslands with a very wide distribution in the world. One of the climates in which grasslands occur is the Mediterranean climate and another is the steppe (temperate zone continental) climate (Turner 2004). Shrub pastures are seen in areas with Mediterranean climate. In the steppe climate, convectional precipitation occurs when the sun heats the water masses and the air rises. Due to the low and short-term precipitation, wheatgrass density is observed in these pastures (Fig. 1) (Reinermann et al. 2020).

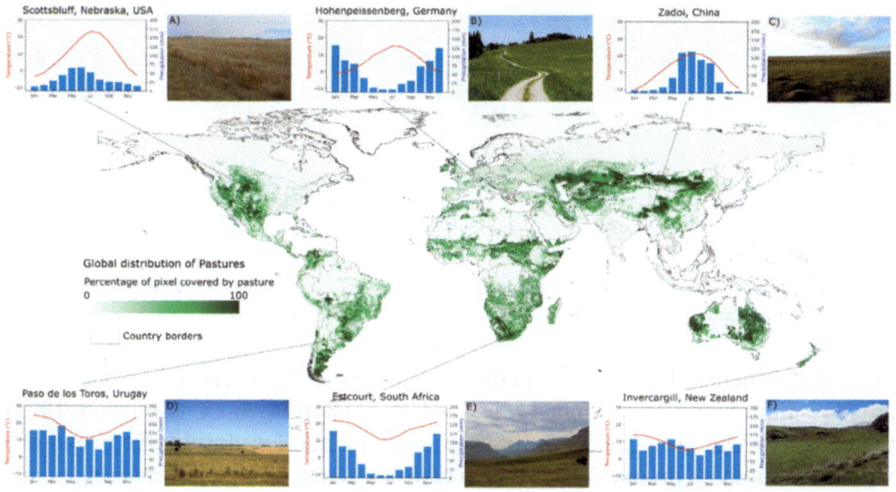

Fig. 1 The global distribution of rangelands, average temperature and precipitation between 1960s and early 2000s, and high climate diversity enable the creation of diverse rangeland biomes (Reinermann et al. 2020)

Orchardgrass, a perennial plant in the wheatgrass family, is a crucial feed source for livestock farming communities, offering high nutritional value and yield (Bajgain et al. 2019). Orchardgrass is being improved through breeding efforts to enhance yield, disease resistance, and environmental tolerance, using molecular techniques and genetic diversity to create sustainable, durable livestock forage crops (Dietzel et al. 2015). The breeding and adaptation of orchardgrass to environmental conditions is crucial for addressing future ecological and economic challenges.

2 Morphological Characteristics of Orchardgrass

Orchardgrass is a perennial wheatgrass species extensively farmed globally, yielding substantial production during chilly seasons. Plant height fluctuates between 60 and 200 cm, contingent upon environmental conditions (Fig. 2) (Jung and Baker 1985).

This plant is used as a meadow-pasture and fodder plant, especially due to its high nutritional value, coarse structure, and tall leaves (Gökçe and Avcı 2019; Huang et al. 2020; Altaf et al. 2024). There are few leaves on the stems. However, the base of the plant has abundant leaves. The cross-section of the leaf blade is "V" shaped. The leaf blade is flat and keeled. The collar is membranous and 3–10 mm long. There is no auricle. The leaf blade is 2–12 mm wide and up to 1 m long. Leaf colour varies from light green to dark blue–green. Leaf characters vary considerably depending on environmental conditions and cultivars. The inflorescence is a panicle 8–15 cm long. The lower branches of the panicle are longer and more branched than

Fig. 2 Orchardgrass plant. (**a**) General appearance. (**b**) Vegetative stage of the plant. (**c**) Physiologically mature plant. (**d**) Physiologically mature seed. (Original by Tatar)

the upper branches. The spikelets in the panicle have 2–5 flowers each. Spiked fruits are 4–7 mm long, 1–1.5 mm wide, and 0.2–0.8 mm thick (Gençkan 1983). Orchardgrass can grow successfully even on stony and low-fertility soils and provides high biomass production. It is also resistant to environmental challenges such as drought, cold, and shade due to its strong adaptability (Sandage and Wixidham 2006). Although its nutritional value is similar to alfalfa, its nutritional value decreases to half of alfalfa during the flowering period (Gökçe and Avcı 2019). Orchardgrass has the ability to adapt to different climatic and soil conditions and is found in naturally temperate regions of Europe, North Africa, and Asia. Its varieties are divided into three main groups: winter hardy, summer dormant types, and winter active types (Manga et al. 2002). However, it stands out as the fourth most common grass species with an annual seed harvest of 14,000 tons worldwide and is widely used especially in dry pastures (Huang et al. 2020). In addition to being used for silage and grazing by livestock, meadow grass has an important role in grassland restoration and ecological improvement (Mao et al. 2023).

Orchardgrass pastures, when managed correctly, lead to significant productivity increases in livestock production. The leaf/stalk ratio is higher than that of reed fescue pasture, and animals grazing on this pasture, especially in early spring and summer, consume more feed and gain more live weight compared to other pastures (Hatipoğlu and Kökten 2009).

3 Orchardgrass Cultivation

Orchardgrass is a cool-season plant that grows best when daytime temperatures are 22 °C and nighttime temperatures are 12 °C (Sandage and Wixidham 2006). When temperatures rise above 28 °C, growth and tillering are reduced. According to research, orchardgrass germinates in the range of 15–25 °C and germination time varies between 5 and 14 days (Gökçe and Avcı 2019). This plant produces most of its biomass in the spring, while this production is lower in the fall. Orchardgrass is moderately resistant to winter cold and shade tolerant. With these characteristics, it can grow in conditions where it receives 33% of full illumination for 3 years without a decrease in productivity (Jung and Baker 1985).

Orchardgrass is a plant that can grow in different soil conditions, but it cannot grow efficiently in saline soils and areas with high groundwater. The optimum soil pH value is between 5.8 and 7 (Bush 2006). Nitrogen application has a positive effect on the development of orchardgrass, and it is recommended to grow this plant either pure or mixed with legumes. Orchardgrass sown in a well seedbed should be supplemented with fertilizers such as nitrogen (2.5–3 kg/ha), phosphorus (5–15 kg/ha), and potassium (5–7.5 kg/ha). Sowing can be done with 1 kg/ha of seed depending on soil preparation and sowing depth should be 1–1.5 cm. Orchardgrass is a suitable plant for grazing, providing very nutritious grasses, especially in spring. This plant, which is suitable for rotational grazing, may weaken if grazed continuously. Therefore, the duration and intensity of grazing should be carefully controlled. To obtain high yields, grazing should be done when the plant reaches 20–25 cm in height. When orchardgrass is grown mixed with legumes, the yield of the mixture is higher, and the quality is better. However, the proportion of legumes in this mixture should not exceed 30%. In addition, orchardgrass plant is a foreign pollinated tetraploid and when sown for seed production, 70–100 cm row spacing is left, and 0.7–0.9 kg/da seed is used. The seed should be dried before ripening and stored after harvest.

Orchardgrass is also susceptible to various pests. Brown streak disease caused by *Scolecotrichum graminis*, leaf blight disease caused by *Rhynchosporium orthosporum*, rust and powdery mildew diseases damage the plant. In addition, pests such as Japanese black beetle and green June beetle feed on roots and above-ground organs (Hatipoğlu and Kökten 2009).

4 Systematics, Botanical Origin, Distribution, Domestication, and Evolution

The monospecific genus *Dactylis* L. belongs to the family Poaceae, subfamily Pooideae and is characterized by a base chromosome number $x = 7$ (Watson et al. 1986). *Dactylis* is an allogamous and variable genus, especially in the Poaceae family, containing diploid ($2n = 14$), tetraploid ($2n = 28$), and hexaploid ($2n = 42$) populations. The genus consists of at least 20 different subspecies, of which the single species *Dactylis glomerata* is known as "Orchardgrass" or "cocksfoot" (Table 1) (Lindner et al. 2004; Anonymous 2024).

Table 1 Subspecies of Orchardgrass

Diploids		***Tetraploids***
altaica	*lusitanica*	*glomerata*
aschersoniana	*metlesicsii*	*marina*
ibizensis = nestorii	*mairei*	*hylodes*
castellata	*reichenbachii*	*slovenica*
hyrcana	*parthiana*	*oceanica*
himalayensis	*santai*	*hispanica*
judaica	*smithii*	
izcoi = Galician	*sinensis*	***Hexaploid***
juncinella	*woronowii*	*hispanica*

Orchardgrass is native to Europe, North Africa, and some parts of Asia and was introduced to the United States in the 1750s. In 1763, seeds were shipped from Virginia to England and spread to Western Europe, South America, Australia, New Zealand, and Japan (Balasko and Nelson 2003). Orchardgrass is especially notable for its tetraploid forms. These species were derived from their original diploid ancestors by chromosome pairing, and these polyploids have spread over larger geographical areas (Fiasson et al. 1987). Tetraploids constitute about 95% of the genus *Dactylis*, showing a great morphological diversity. Although most subspecies are diploid, tetraploids are self-incompatible, outcrossing and highly heterozygous species that maintain a high level of genetic variation. These characteristics cause cross-pollinated species to have more genetic diversity than self-pollinated species (Larson et al. 2004). This heterozygosity makes it possible for orchardgrass, which is particularly tolerant to cold climates and shade, to spread worldwide (Jiang et al. 2014). Orchardgrass is now widely used as a forage crop in all temperate regions.

The evolutionary process of the genus *Dactylis* is closely linked to the glacial period. Migration to glacial refugia led to the interaction and hybridization of different forms, allowing the stronger offspring to colonize new areas as glacial areas retreated (Stebbins 1971). During the last glacial period, the northern range of *Dactylis* retreated to a dispersed refuge in southern Europe. Glaciation also allowed *Dactylis* to spread southwards into the North African grasslands. Here, it has been observed to be quite widespread, but there is no evidence of any spread into the high tropical regions of Africa. It has also spread to the Canary Islands and the Cape Verde Islands, probably through bird migrations (Lumaret and Borrill 1988).

Orchardgrass is cultivated as an economically important and widespread forage crop in China. Molecular studies have revealed that the diploid *sinensis* species in China is closer to the *aschersoniana* species in Europe. This suggests that the *sinensis* species may have migrated from Europe to China or spread along the historic "silk road." Orchardgrass, a high-yielding perennial forage grass, is a crucial part of the livestock industry in North America, Oceania, Europe, and Asia due to its adaptability and high yield (Jiang et al. 2014). *Dactylis*' adaptive radiation exemplifies polyploid evolution, starting with migration to glacial refugia and hybridizing populations, allowing them to colonize cooling climates (Stewart and Ellison 2010). *Dactylis* spread was influenced by human migration and trade, with orchardgrass possibly migrating from Europe to China or via the Silk Road (Jiang et al. 2014).

5 Germplasm and Genetic Resources of Orchardgrass

Orchardgrass, a popular cool-season forage crop in North Africa, Europe, and Asia, is cultivated in North America, Europe, and Japan due to its high nutritional value and productivity. At elevations ranging from 1000 to 3600 m, this plant can be found growing naturally in the southwestern and northwestern regions of China. It can be found growing on forest edges, shrublands, and semi-alpine meadows (Peng et al. 2008). Orchardgrass plays a crucial role in providing essential ecosystem services in both agricultural areas and natural ecosystems.

Stewart and Ellison (2010) have classified the genetic resources available for *Dactylis* breeding into four distinct gene pools. These gene pools are divided into the following categories: primary, secondary, tertiary, and quaternary. A more comprehensive understanding of the genetic variety of plants can be gained via the use of this classification. The individuals that are employed directly in breeding are included in the primary gene pool. These people are also capable of easily hybridizing within the species itself. In addition to cultivars and elite breeding lines that have been suited for use in agriculture, this pool contains genetic resources that are essential for direct breeding, such as orchardgrass. Species that are able to exchange genes with the primary gene pool are included in the secondary gene pool. These species, however, exhibit hybridization processes that are more difficult to achieve and provide alternative genetic variation. Furthermore, the tertiary gene pool is comprised of populations from which beneficial gene transfer can be accomplished through the application of biotechnological techniques. This pool of genes comprises more distant relatives. There are genetic resources of more distant relatives of the species that are contained inside the quaternary gene pool. These resources can be accessed through the application of modern genetic engineering techniques. In both natural ecosystems and agricultural settings, orchardgrass is a significant contributor to the ecosystem service that it provides.

Over the course of the past few years, significant efforts have been made to broaden the selection of orchardgrass genetic material. Research on *Dactylis* accessions is carried out by international organizations such as the European Cooperative Plant Genetic Resources Programme (ECPGR), the National Plant Genetic Resources System (NPGS) of the United States Department of Agriculture, and the International Organization for Plant Information (IOPI). These organizations are responsible for conducting the most comprehensive studies on those accessions. These endeavours are geared towards ensuring the collection and preservation of the genetic resources that are associated with the species. By conducting an exhaustive investigation on the karyotypes of *Dactylis* species, Wetschnig (1991) was able to uncover the genetic variety of forms that have varying levels of ploidy ($2n = 14, 28, 42$). On the other hand, these findings highlight the significance of doing a more in-depth investigation of genetic characteristics, particularly an investigation into addition and substitution lines.

In a number of research, the size of the genome of orchardgrass has been investigated, and it has been discovered that the size of the genome varies substantially between various populations. A reduction in genome size of up to 30% was observed by Reeves et al. (1998) when the elevation of the organism increased. On the other hand, Vilhar et al. (2002) did not discover any evidence that supported the existence of this connection. According to the findings of Tuna et al. (2007), diploid populations that originate from Mediterranean climates have a lower DNA mass in comparison to forms that originate from Northern Europe. Such investigations make it possible to conduct a more in-depth investigation of the connections that exist between the size of the genome and the environmental factors. The breeding of orchardgrass has been accomplished by gathering material from vast geographical areas. The objective of this breeding is to increase the plant's capacity to adapt to climatic change while simultaneously boosting its productivity. As an illustration, the mountainous ecotypes that are situated in southern France exhibit a higher tolerance to snow cover. Some types exhibited distinct variances in their ability to tolerate drought and winter dormancy, according to experiments carried out in Australia. According to Sanada et al. (2007), these kinds demonstrate that genetic material is utilized extensively and plays a significant part in the production of varieties that are able to adapt to a variety of environmental circumstances. According to the findings of these studies, breeders of orchardgrass have successfully employed genetic material, which has enabled the production of novel cultivars that have the capacity to adapt to a wide range of environments.

A chromosome-scale diploid meadowgrass genome was estimated to be roughly 1.84 gigabytes in size was presented by Huang et al. (2020) after they conducted an analysis of the genome of orchardgrass, which is a worldwide significant forage crop. It is estimated that there are 40,088 protein-coding genes in the genome, the majority of which are mobile elements. Many of these elements are retrotransposons, particularly long terminal repeats (LTRs), which are a type of superposition. Within the scope of this study, the genetic networks of significant characteristics, such as blooming time, were uncovered, and four major candidate genes that govern these characteristics were found. Additionally, during the process of re-sequencing the genomes of 76 different species of orchardgrass from throughout the world, it was discovered that germplasm from Northern Europe and East Asia are brought together through integrated trading channels. In addition, this study demonstrates the evolutionary connections that exist between orchardgrass and eight other grass species that are related to it.

6 Orchardgrass Breeding

In regions that are classified as temperate, orchardgrass is a significant crop that is utilized for fodder. It is hardy and long-lived, and it is subject to diseases, pests, and environmental challenges. It is cultivated to boost yield and quality despite its hardiness and longevity. As a result, the breeding of orchardgrass is done with the intention of enhancing significant agronomic characteristics such as increased production, resistance to drought, and cold tolerance. In addition, the development of this species is being sped up by contemporary breeding techniques such as molecular markers and genetic engineering. According to Stebbins and Zohary (1959), orchardgrass is a prime example of adaptive radiation and polyploid evolution, both of which are extremely significant in the process of adaptation and evolution. According to Sahuquillo and Lumaret (1999), research conducted through taxonomic, morphological, and genetic investigations has shed light on the evolutionary history of the species and showed the function that tetraploidy plays in the process of evolution.

During the breeding process, both genotypic and phenotypic selection approaches are utilized extensively. All throughout the world, particularly in the United States of America, Canada, Europe, and New Zealand, these techniques have been successfully implemented (Casler et al. 2000, 2002). Clonal selection techniques and the collection of genetic material from huge populations are the two methods that are utilized in the process of synthetic varietal breeding. Techniques such as RAPD, AFLP, ISSR, and SSR are often utilized in successful breeding programmes (Xie et al. 2011, 2012). Molecular marker is an important tool that may be used to determine the genetic diversity of a population. According to Cömertpay and Özpınar (2019), these markers contribute to the comprehension of population structure and the identification of breeding materials.

6.1 Classical/Conventional Breeding

The work that has been done in breeding has primarily been established based on traditional methods of population breeding. Phenotypic selection, hybridization, and recombination are some of the processes that are included in these strategies, which are typically utilized on natural and cultivated populations of orchardgrass (Gauthier and Bernard 1998). The process of phenotypic selection and hybridization provides breeders with the ability to pick plants that possess particular characteristics and to increase the genetic diversity of these vegetation. In addition, the introduction of novel genetic variants into populations can be a means by which crossing populations might enhance the adaptability of plants. Phenotypic selection is a key factor in the process by which plants alter their characteristics in response to changes in their environment, particularly in various ecoregions. Through ecological selection, these mechanisms make it possible for plants to adapt to the environment in which they are found. It has been demonstrated that the combination of ecological selection and phenotypic selection can improve adaptation (Nakayama et al. 1997).

Classical breeding programmes often have the goals of enhancing forage yield, disease and insect resistance, and tolerance to abiotic conditions, particularly drought and cold weather. These are all common objectives. According to Acharya et al. (2006), the winter hardiness of orchardgrass is lower than that of other species, such as timothy grass (*Phleum pratense* L.) and meadow fescue (*Festuca pratensis* Huds.), which are typically utilized in parts of the country that are typically located further north. Consequently, it is of utmost significance to make efforts to enhance the winter hardiness, snow mould resistance, and frost tolerance of orchardgrass, particularly in places that are located in the greater north.

"Wasemidori" is an example of a synthetic variety that was created by natural selection for winter hardiness and resistance to snow mould (Terada et al. 1991). This variety is comprised of eight clones and is a synthetic variation (Tajimi 1974). Research indicates that characteristics such as winter hardiness in cold areas, water tolerance in desert places, and productivity in high fertility soils are particularly essential focal points of breeding. Studies have shown that these characteristics are particularly relevant. According to research conducted in France (Nakayama et al. 1997), special selections have been done in order to develop plant varieties of orchardgrass that are resistant to the harsh winter conditions. Breeders have significant chances to supply genetic diversity and generate new genotypes, particularly as a result of the studies that have been conducted to adapt to cold climes. According to Tosun and Sagsoz (1994), researchers are conducting studies to ascertain the physical and biological properties of orchardgrass plants that are growing in various ecological zones. Additionally, they aim to enhance the potential for these species to adapt to their respective places. In addition, the tetraploidy of orchardgrass, which means that it possesses four sets of chromosomes, is a significant genetic component that can be used in breeding research to help boost production and resistance. Tetraploid plants are typically larger and more robust than other types of plants, contributing to their increased resistance to environmental challenges. In addition, environmental factors such as the acidity of the soil might have an effect on the productivity of orchardgrass grass. The persistence of this plant typically reduces when it is grown in soil conditions that are acidic. Several studies (Poozesh et al. 2010) have demonstrated that the use of chemical fertilizers and other types of fertilizer have an impact on the acidity of the soil and encourage the growth of plants. However, among the elements that influence the recovery of orchardgrass in acidic soils, a significant role has also been found as being played by various types of fertilizers (Kakihara and Ogura 2022).

Further expansion of the genetic diversity of orchardgrass is possible by the utilization of genetic material derived from local wild sources. According to Peng et al. (2008), these materials can be utilized in genetic breeding research, which enables the generation of varieties that are better suited to the conditions of the environment. Studies have also been conducted to discover the physical properties of these species as well as the parameters under which these genotypes are able to flourish in the most effective manner (Nabhan et al. 2021). At the end of the day, the breeding of orchardgrass is done with the intention of not only increasing the amount of fodder it produces but also improving its resistance to the conditions of the environment. In this respect, it is believed that breeding research will make a significant contribution to the acquisition of varieties that are more resistant to environmental factors, more productive, and more environmentally compatible.

6.2 Molecular Breeding

Rapid advancements in biotechnology have made major contributions in recent years, particularly in areas such as genetic resources, genetic diversity, characterization, breeding, conservation, and variety development (Aasim et al. 2022; Yalinkiliç et al. 2024). These contributions have been particularly substantial in the past few years. The preservation and expansion of genetic diversity are two of the most important factors in ensuring the long-term viability of agricultural practices. It is necessary to establish priorities for the selection and preservation of genotypes in order to ensure the conservation of genetic resources. The application of molecular techniques in genetic breeding has resulted in significant advancements, particularly in the field of feed crops like orchardgrass. According to Stewart and Ellison (2010), genetic markers are utilized extensively in the process of analysing the genetic structure of plants and selecting individuals that possess the necessary characteristics. When measuring genetic diversity, molecular markers, which play a key role in understanding the genetic basis of phenotypic features, make it easier to differentiate between populations since they assist the process of discrimination. The incorporation of molecular genetic approaches into plant breeding through the use of marker-assisted selection (MAS) has resulted in an increase in the effectiveness of artificial selection between phenotypes in recent years. These methods make it possible for genetic markers to determine the correlations that exist between environmental interactions and variation, particularly using procedures such as quantitative trait loci mapping (Marone et al. 2012).

Cutting-edge methods, such as the DArT technology, make it possible to conduct polymorphism analysis at thousands of loci, which in turn makes it possible to derive fingerprints from the entire genome. According to Mace et al. (2008), these technologies will bring about substantial advancements in the field of plant breeding as well as the identification of genetic diversity. In addition, the technique known as new generation sequencing (NGS) has accelerated the process of genome-wide mapping and characterization, which has led to an increase in the effectiveness of markers that are utilized in plant breeding. Nevertheless, there are not much studies done on forage crops, and thus there is a need for greater research in this area.

Molecular markers can be categorized into three primary types: morphological, biochemical, and DNA-based. Each marker type demonstrates varying levels of polymorphism and repeatability (Agarwal et al. 2008). These markers are utilized in several applications, including biodiversity analysis, phylogenetic studies, germplasm management, and variety identification. Marker systems like RAPD, AFLP, ISSR, SRAP, EST-SSR, and SSR are extensively utilized, particularly in biodiversity assessment. Moreover, SCoT (Starter Code Targeted) polymorphism has garnered interest as a more efficient and dependable genetic marker technique in plant breeding (Guo et al. 2012). Molecular approaches and marker-assisted breeding have been established to enhance the efficiency and speed of plant breeding processes in the future. AFLP analysis is reliable and demonstrates a significant quantity of reproducible polymorphic bands with few primer combinations. Both strategies are rapid, cost-effective, and do not necessitate prior sequence knowledge (Agarwal et al. 2008). ISSR markers contain several highly informative multi-allelic loci. These provide highly discriminative information with good reproducibility and are relatively abundant. Of these three techniques, AFLP is the most labour-intensive and time-consuming, but also the most reliable (Costa et al. 2016). Recent molecular studies on orchardgrass have been characterized by marker type and number of primers, collection site/location/total plant material, results and references are summarized in Table 2.

In breeding studies on orchardgrass using genetic diversity and molecular technologies, orchardgrass is one of the most widely used perennial grass species in Europe, along with *Lolium perenne* and *Festuca arundinacea*. This plant species attracts attention with its high productivity, rich protein content, and resistance to drought, cold, and shade conditions, especially in pure cultivation. Biotechnological advancements have enhanced the study of orchardgrass genetic diversity, utilizing techniques like genetic mapping and molecular markers for plant breeding. DNA

Table 2 Primers used type for Orchardgrass genetic diversity and results

No	Marker type and primer number	Collection place/location/total plant material	Conclusion	References
1	15 SSR	53 genotypes collected from 5 continents	127 alleles and the number of alleles per locus ranged from 5 to 12, with a polymorphic rate (P) of 95.21%, showing high genetic diversity in the Asian (90.55%) and European (86.61%) groups.	Xie et al. (2009)
2	21 SSR	China (11), Russia (1), Kazakhstan (1), India (2) and Japan (1) Total 16 genotypes	A total of 143 polymorphic allelic bands with an average of 6.8 alleles per locus and a range of 313 alleles, with an average polymorphic ratio (P) of 90.7% for the species, showed greater genetic variation among accessions (65.75%) and between geographic regions (75.58%).	Xie et al. (2010b)
3	15 SSR	Baoxing, Anba, Bote, Kaimo varieties and *Dactylis glomerata* ssp. *woronowii* subsp. and improved breeding line YA02–116 Total 6 genotypes	The average of 7.6 alleles per locus and the polymorphic rate among 120 individuals was 100%, indicating a high degree of genetic diversity. The highest genetic diversity among the six genotype accessions was observed in Kaimo ($H = 0.2518$; $I = 0.3916$; $P = 87.3\%$) and the lowest level of genetic diversity was observed in breeding line YA02–116 ($H = 0.1806$; $I = 0.2788$; $P = 58.73\%$), indicating that the used accessions have the potential to provide a rich genetic resource in breeding programs for Chinese Orchardgrass varieties.	Xie et al. (2012)
4	29 SSR	21 countries (Russia, Romania, Switzerland, Poland, Hungary, Lithuania, England, Canada, Australia, New Zealand, Denmark, Italy, Spain, Slovakia, Luxembourg, France, Czech Republic and Netherlands countries 1 variety, Germany 3, USA 2 and China 8 varieties) Total 32 genotypes	A total of 229 bands with an average of 7.9 bands per marker, with an average polymorphic rate of 92.1% for the species and Polymorphism information content (PIC) between 0.771 and 0.893, 11 bands showing stable and reproducible SSR patterns were obtained by primer pairs A01E14, A01K14 and D02K13.	Jiang et al. (2013)

5	29 SSR	1861 plant genotypes from 59 natural and semi-natural populations in Switzerland (20), Bulgaria (20) and Norway (19)	The average heterozygosity (HE, C) ranges between 0.44 and 0.59, with the highest heterozygosity in individuals obtained in the Norwegian region, the Bulgarian population presents a high genetic diversity, while the population in the Norwegian region reveals a high amount of rare, localized alleles.	Last et al. (2013)
6	14 SSR	Turkey 11 provinces (Bolu (20), Bursa (20), Çanakkale (20), Gaziantep (20), Isparta (20), İzmir (20), Kars (20), Kırşehir (20), Ordu (20), Rize (20), and Zonguldak (20)) 220 plant genotypes in 11 populations	A total of 72 alleles, average number of alleles per locus was 5.14, all primers used in the study were polymorphic, populations were grouped under four groups in the dendrogram.	Demirkol and Aşçı (2017)
7	24 SSR	32 genotypes	SSR primers produced 126 alleles in total, the number of alleles ranged between 3 and 7, the average number of alleles per locus was 5.25, the allele sizes ranged between 101 bp and 354 bp, and the polymorphism rate was 100% for each primer, and it was reported that SSR technique is a highly suitable and effective technique for molecular identification of Orchardgrass genotypes.	Cömertpay and Özpınar (2019)
8	14 SSR	9 natural populations including 2 *Dactylis glomerata* subspecies	Among the studied populations, an average of 5.27 alleles with a polymorphism information content (PIC) of 0.69 polymorphism per locus set, the average polymorphic rate for the species was 86.08%, indicating a high degree of genetic diversity.	Chtourou-Ghorbel et al. (2024)
9	12 ISSR	50 genotypes from China and other countries	There were 101 polymorphic bands with an average of 8.41 bands in each primer pair, the average percentage of polymorphic bands was 86.3.8%, the Orchardgrass accessions were divided into 5 groups and the samples from the same continent were classified in the same group, while the Chinese and United States accessions were in different groups.	Zeng et al. (2006)

(continued)

Table 2 (continued)

No	Marker type and primer number	Collection place/location/total plant material	Conclusion	References
10	22 ISSR	Total 34 genotypes	A total of 424 loci, 34 loci were monomorphic and 390 loci were polymorphic among all genotypes studied, and the average percentage of polymorphisms was 70.39%.	Azar et al. (2022)
11	12 RAPD	6 genotypes belonging to each population	More than 40 polymorphic bands were obtained to determine the ploidy level of the population and its diversity among populations, indicating a high level of gene flow and homogeneous distribution among Orchardgrass populations.	Tuna et al. (2004)
12	21 SRAP	China (26), UK (2), Denmark (9), Netherlands (2), Switzerland (8), Germany (2), Australia (2), Canada (2) and USA (7) Total 60 genotypes	480 bands, of which 405 (84.38%) were polymorphic, Orchardgrass from China and the United States had higher genetic diversity, the genetic diversity of Orchardgrass from Australia was different from other collections in terms of cluster analysis method, the information provided by SRAP markers was consistent with morphological variability and karyotype, and the SRAP marker system can be used effectively in the study of Orchardgrass genetic variability and evolutionary history.	Zeng et al. (2008)
13	24 SCoT	Denmark (7), Russia (1), Switzerland (8), United Kingdom (2), France (1), Italy (1), Luxembourg (1), Netherlands (3), Romania (1), Spain (1), Czech Republic (1), Germany (4), Hungary, Lithuania (1), Poland, Slovakia (2), Canada and United States (10) Total 95 genotypes	In total, 273 polymorphic bands were obtained, the average number of bands per primer was 11.4, the average polymorphic rate for the species was 83.4%, 69.13% variation within groups and 30.87% variation between groups, and a significant geographical difference between North America and Oceania.	Jiang et al. (2014)

14	9 AFLP	32 wild genotypes	400 DNA fragments were amplified, of which 322 (80.50%) were polymorphic. For the 34 accessions, genetic similarity coefficients (GSC) ranged from 0.69 to 0.93, with an average of 0.81, with ploidy levels (87.98%) and geographic regions (85.65%) showing greater rates of variability.	Peng et al. (2008)
15	89 RAPD and 47 ISSR	Total 100 genotypes	Analysis of the RAPD dendrogram showed that all genotypes could be separated, while the number of ISSR markers were not sufficient to separate all genotypes.	Costa et al. (2005)
16	15 EST-SSR and 15 SSR	China (13), Iran (3), Turkey (4), India (4), Pakistan (2), Denmark (2), Finland (1), Norway (1), Switzerland (2), Ukraine (1), Russia (2), Romania (2), Bulgaria (2), Greece (2), Italy (2), Portugal (2), Spain (5), Australia (2), New Zealand (2), USA (5), Canada (2), Chile (1), Colombia (1), Morocco (4), Egypt (2), Tunisia (3) and Algeria (2) 74 genotypes consisting of cultivars, wild materials, advanced lines and subspecies from 27 countries and 9 geographical regions	190 polymorphic bands with an average of 6.3 alleles per SSR locus, polymorphic ratio (*P*) 84.63%, proportion of variance explained by within- and between- geographical groups diversity 74.87% and 25.13%, respectively, among the nine geographical regions, the genetic diversity of weeds in Asia (*P*: 94.44%), North Africa (*P*: 88.89%) and Europe (*P*: 87.30%) was higher, while that of weeds in Oceania (*P*: 40.48%), Eastern Europe (*P*: 30.16%) and Northern Europe (*P*: 48.40%), SSR data proved to be useful in determining genetic relationships among diverse accession collections, and most accessions clustered in accordance with pedigree relationships, adaptation zones or geographic origin.	Xie et al. (2010a)
17	7 SSR and 6 ISSR	From northern, central and southern Greece Total 72 genotypes	There is a high level of genetic diversity between populations with ISSRs 38% and SSRs 17% and within populations 62% and 83% respectively, and it was revealed that SSRs are more informative about genetic diversity within populations and ISSRs are more informative about genetic diversity among populations.	Madesis et al. (2014)

(continued)

Table 2 (continued)

No	Marker type and primer number	Collection place/location/total plant material	Conclusion	References
18	26 RAPD, 22 ISSR and 4 AFLP	100 genotypes	RAPD produced 97 bands (40 were polymorphic (41.2%)), ISSR 91 bands (54 were polymorphic (59.3%)) and AFLP 100 bands (92 were polymorphic (92%)). It was concluded that RAPD, ISSR, and AFLP molecular markers were useful in discriminating the genetic diversity of the present collection of Orchardgrass and that AFLP was the best due to its high degree of polymorphism (91%), while the polymorphism levels revealed by RAPD and ISSR techniques were lower than AFLP (41% and 59%), although these techniques are easier and less costly to implement.	Costa et al. (2016)
19	18 EST-SSR and 21 SCoT	United States (2), Australia (3), China (14), Japan (2), Spain (4), Turkey (2), Israel (2), Portugal (3), Russian Federation (2), Former Soviet Union (2), Italy (1), Poland (2), Finland (2), Bulgaria (1), Ireland (2), Iran (3), India (2), Pakistan (1), Tunisia (2), Egypt (1), Chile (1), Netherlands (2), Canada (1), Denmark (2), Morocco (3), France (2), Greece (2), Romania (1), Serbia and Montenegro (1), Libya (1), Hungary (1), Belgium (1), Argentina (1), Afghanistan (1), Algeria (1) and Colombia (1) Total 75 genotypes	A total of 164 bands for EST-SSR markers and 289 bands for SCoT, 148 (90.24%) for EST-SSR and 272 (94.12%) for SCoT were polymorphic. When EST-SSR and SCoT marker efficiencies were compared, it was found that the efficiency of SCoT (12.43%) was higher than EST-SSR (4.82%).	Yan et al. (2016)

20	20 SSR and 40 EST-SSR	5 Korean varieties (Ordione, Onnuri, Onnuri 2ho, LuckyOne, and LuckyOne 2ho) 2 non-Korean (Potomac and Amba) Total 7 populations, 12 individuals from each population Total 84 individuals	A total of 607 alleles, the number of alleles ranged from 3 to 27, with an average of 10.1. polymorphism informative content (PIC) values per locus ranged from 0.127 to 0.919 in 84 individuals (12 individuals from each of the seven cultivars), with an average of 0.585.	Nam et al. (2024)

polymorphisms are particularly useful for detecting genetic differences, accelerating genetic mapping and breeding processes (Qiao et al. 2020).

Orchardgrass breeding uses breeding methods and phenotypic, morphological, and cytological studies to develop high-yielding cultivars resistant to environmental stresses. Phenological studies understand plant development and adaptation strategies, while morphological analyses reveal plant structure changes (Piao et al. 2019). Cytological analyses help understand plant structure, ploidy level, and productivity effects. Biochemical markers study genetic diversity and environmental adaptation, forming basis for plant breeding and developing resistant varieties for increased agricultural productivity.

Next-generation sequencing (NGS) technologies offer efficient marker identification through techniques like genome-wide association mapping and QTL mapping (Nadeem et al. 2018). Systems like DArTseq and SNP markers analyse thousands of loci with high efficiency. RNA sequencing technology (RNA-Seq) detects gene expression accurately. The Illumina HiSeq 2000 platform is fast and cost-effective for plant transcriptome sequencing (Qiao et al. 2020; Altaf et al. 2025). RNA-Seq has become an important method for understanding the effects of environmental factors on plants, especially drought, waterlogging, and other abiotic stresses.

Orchardgrass serves as a crucial model organism for studying genetic responses to abiotic stresses like drought and flooding, using RNA-Seq technology to identify gene expression changes. Ji et al. (2018) study found significant differences in the expression of 41 miRNAs in leaves and roots of drought-resistant Baoxing variety after 18-day water deficit treatments. One miRNA, miRNA ancestor-miR164c-3p, was associated with heat shock protein A (HSP70), suggesting it may play a crucial role in orchardgrass adaptation. Similarly, Qiao et al. (2020) study on orchardgrass revealed that oxidoreductase activity and oxidation-reduction processes are crucial for adaptation to flooding, highlighting the genetic control mechanisms involved. This study underscores the importance of transcriptomic studies in understanding plant genetic adaptation to abiotic stresses like waterlogging.

Molecular studies on orchardgrass have significantly improved our understanding of their genetic adaptation to abiotic stresses like drought and waterlogging. RNA-Seq and other NGS-based techniques have enhanced our understanding of these plants' stress responses, which is crucial for plant breeding. The use of NGS-based genetic markers and molecular markers has improved agricultural productivity, and a better understanding of their role in plant adaptation will enhance future breeding.

Research on heat stress, disease resistance, and environmental adaptability using genetic and molecular methods widely cultivated in temperate regions, orchardgrass is a high-yielding fodder crop that can withstand environmental stressors. We now have a better understanding of this plant's stress responses and environmental adaptation, thanks to a number of molecular studies. Huang et al. (2015) investigated the differential expression of genes in response to heat stress using Baoxing and 01998 cultivars. According to the findings of high-throughput sequencing using the Illumina HiSeq 2000 platform, these cultivars exposed to heat stress at 30 °C and

38 °C had a total of 3527 and 2649 differentially expressed genes (DEGs). This work represented a significant step in comprehending the genetic pathways regulating orchardgrass's heat tolerance, as the Baoxing cultivar displayed higher DEGs during heat stress (Huang et al. 2015). In a similar vein, Feng et al. (2017) conducted a thorough transcriptome study to comprehend the rules governing flowering and vernalization. Twelve thousand nine hundred sixty seven DEGs were found during the flowering phases and 4689 DEGs during vernalization in this study on the cultivar DONATA. This study demonstrated the significance of transcription factors in the vernalization process by demonstrating the impact of photoperiod, circadian rhythm, and other metabolic processes on blooming (Feng et al. 2017).

According to Ren et al. (2020), who studied rust resistance mechanisms in orchardgrass, the plant's defence against pathogens is provided by NBS (nucleotide binding site) resistance genes. The majority of the 413 NBS genes found in the study are expressed in various tissues, including roots, stems, and leaves. It was discovered that NBS resistance genes contribute to both biotic and abiotic stress resistance (Ren et al. 2020). Lastly, a study by Jin et al. (2024) looked at how orchardgrass was affected by *Puccinia graminis* rust disease. A genome-wide association study (GWAS) examination of 333 distinct orchardgrass accessions revealed 91 genes that may be resistant to rust disease. It was discovered that this resistance is mostly caused by the glutathione S-transferase (GST) gene family (Jin et al. 2024).

The research reveals orchardgrass's genetic mechanisms against environmental stresses and diseases, promoting molecular breeding studies. This will enhance agricultural productivity and adapt to environmental changes. The integration of classical and modern breeding techniques is crucial for developing resistant varieties.

7 Use of Biotechnological Methods in Orchardgrass Breeding Programs

Orchardgrass plants have been genetically modified using various techniques to enhance genetic diversity and environmental resistance. These methods include electroporation from protoplasts, polyethylene glycol application, and micro injection bombardment. Protoplast culture, a technique where plant cells are treated without a cell wall, is crucial for integrating genetic material into the plant cell. Successful transgenic plants have been obtained in orchardgrass using electroporation and PEG treatments. In these methods, foreign genes can be integrated into the plant cell by temporarily making the cell membrane permeable (Horn et al. 1988). Electroporation is particularly favoured as a high-throughput method for genetic transformation.

Another important genetic engineering technique is micro injection bombardment. In this method, microscopic particles carrying the genetic material are directed at high speed into plant tissues, thereby delivering the genetic material directly into the cells. In studies on orchardgrass, a successful genetic transformation was achieved by applying micro injection bombardment to young leaf tissues (Denchev

Table 3 Successful transformation of orchardgrass (Denchev et al. 2000)

Explant and/or target tissue	Vector/method	Observations
Mesophyll cells in cultivated foliar segments	Plasmid pAHC25 harbours the /lidA gene, which encodes for /3-glucuronidase (GUS), as well as the bar gene that imparts resistance to phosphinothricin-based herbicides, such as Basta. Both genes were regulated by the Ubi 1 promoter. Plasmid was introduced into mesophyll cells using microprojectile bombardment.	Elevated efficiency for transient GUS expression in leaf segments and for the recovery of Basta-tolerant plants. Leaf tissue from Bastatolerant plants generated somatic embryos that had positive GUS staining. Transgenic plants exhibit phenotypic normality.
Protoplasts generated from suspension culture	Plasmid pCIB709 harbours the APH IV gene, which imparts resistance to hygromycin. Direct uptake of DNA plasmid by protoplasts produced from suspension culture using the 35S promoter	Suboptimal efficiency, with only a limited number of validated transgenic plants produced. Certain plants exhibited weakness and were both male and female infertile.

et al. 1997). In another study by Denchev et al. (2000), direct somatic embryogenesis was obtained from mesophyll cells by microprojectile bombardment of orchardgrass leaf tissue (Table 3). This method has been used as an effective tool to increase genetic diversity and integrate desirable traits in plants.

Agrobacterium-mediated genetic transformation is a widely used technique to create permanent genetic changes in plants. In orchardgrass plants, *Agrobacterium*-mediated genetic transformation was successfully performed using seed-derived callus tissues. This method enables the production of transgenic plants by transferring genetic material directly into the plant cell (Lee et al. 2006). One of the most important advantages of *Agrobacterium*-mediated genetic transformation is the ability to create permanent genetic changes in plants. In orchardgrass, an effective genetic transformation system was developed by microprojection bombardment of green tissues with high regeneration capacity. These green tissues derived from mature seeds of Rapido cultivar were used for genetic modification by microprojection bombardment and successful transgenic plants were obtained (Cho et al. 2001). Increasing regeneration capacity is a critical factor for the successful application of genetic engineering.

The transfer of heat shock protein (HSP) gene into plants is an important study to increase the resistance of orchardgrass to environmental stresses, especially high temperatures. For this purpose, HSP gene was transferred to embryogenic callus tissue via *Agrobacterium*, and the transgenic plants obtained were determined to be heat tolerant up to 60 °C by leaf disk evaluations. This temperature is lethal for non-transgenic plants (Kim et al. 2008). This genetic modification was successful in increasing the tolerance of orchardgrass to temperature stress.

These genetic engineering studies serve many important purposes such as making orchardgrass resistant to environmental stresses, increasing their yields and improving their resistance to diseases. In particular, heat tolerance and resistance to other environmental stresses provided by transgenic plants can enable these plants to grow more efficiently in different climatic conditions. Such biotechnological applications will continue to play an important role in orchardgrass breeding and future genetic engineering studies.

8 Future Perspective and Conservation Strategies for Orchardgrass

The future development of important agricultural species such as orchardgrass requires the maintenance of pragmatic field breeding programmes to conserve genetic diversity and effectively implement breeding efforts. However, this species currently has limited investments, and the genetic material of its diploid forms is threatened by habitat destruction and climate change. This situation is particularly complicated by the fact that many forms are underrepresented in genebanks. It is therefore crucial that a wide range of these forms are collected and conserved. Sustaining large-scale breeding programmes at the international level is a critical requirement to enable the collection of genetic material, transfer of wild genetic material, and application of molecular resources.

The development of orchardgrass is a complex process that requires the integration of in situ (in natural habitats) and ex situ (storage of genetic material) methods. In this process, it is important to consider biotechnology and environmental factors, as well as sustainable agricultural practices. In situ conservation ensures that species are preserved in their natural habitats and native plant species, such as orchardgrass, support biodiversity in ecosystems, maintaining environmental balance. However, factors such as habitat loss and climate change threaten the existence of these species. In this context, it is necessary to protect their natural habitats, monitor environmental conditions, and increase their adaptability. Natural populations of orchardgrass have many adaptive abilities that develop resilience to environmental challenges, for example, tolerance to cold and arid conditions. These traits can be used as resources in future breeding efforts. However, in situ conservation protects genetic diversity and ecosystem services, allowing species to adapt to environmental changes through natural selection processes (Lumaret et al. 1989).

Ex situ conservation is the storage of genetic material away from its natural habitat, preserving genetic diversity. It is crucial for plants like orchardgrass to ensure long-term preservation and inclusion in breeding programs. However, existing collections are often incomplete and underrepresented, limiting their effective use and success. Ex situ collections allow for genetic improvement using biotechnological tools and can be monitored through molecular markers. However, limited funding and capacity issues pose challenges in maintaining the genetic integrity of these collections (Sanada et al. 2007).

Molecular biotechnology is crucial for the future development of orchardgrass, enabling better understanding of genetic diversity and targeted breeding methods. This technology can be used to modify traits quickly and effectively, while considering ethical and environmental impacts. Climate change is a significant concern for crops sensitive to water stress and temperature changes, making it essential to develop resistant varieties for sustainable agriculture. Harmonized application of in situ and ex situ methods will enhance genetic resource conservation and adaptation potential, requiring international cooperation, technological advances, and adequate funding.

9 Conclusion

Orchardgrass is a crucial forage crop for livestock production, particularly in temperate climates, due to its high biomass yield, carbohydrate level, and resistance to abiotic stresses. It is used in milk and meat production and contributes to ecosystem services. Fescue, a low-cost, high-yielding, and sustainable forage crop, is also essential for long-term livestock production. Genome-wide research is enabling efficient genetic analyses to improve fescue productivity. Next-generation sequencing technologies and techniques like genome-wide association mapping can develop more productive and resistant species under different conditions and breeding systems.

The molecular analysis of meadow grass genome will advance functional genetic research and accelerate plant breeding processes, leading to increased productivity and resistant varieties. Fescue breeding is of strategic importance, focusing on biotechnological tools and modern breeding methods. Genome-based breeding of fescue is becoming a key element of sustainable animal husbandry, crucial for food security and environmental sustainability. Deepening genetic diversity and analysis of fescue will provide a solid foundation for future livestock breeding systems and sustainable solutions to increase forage crop productivity.

References

Aasim M, Katirci R, Baloch FS, Mustafa Z, Bakhsh A, Nadeem MA et al (2022) Innovation in the breeding of common bean through a combined approach of in vitro regeneration and machine learning algorithms. Front Genet 13:897696

Acharya SN, Kastelic JP, Beauchemin KA, Messenger DF (2006) A review of research progress on cicer milkvetch (*Astragalus cicer* L.). Can J Plant Sci 86(1):49–62. https://doi.org/10.4141/P04-174

Agarwal M, Shrivastava N, Padh H (2008) Advances in molecular marker techniques and their applications in plant sciences. Plant Cell Rep 27:617–631. https://doi.org/10.1007/s00299-008-0507-z

Altaf MT, Cavagnaro PF, Kokten K, Ali A, Morales A, Tatar M, Baloch FS (2024) Genotyping-by-sequencing derived SNP markers reveal genetic diversity and population structure of *Dactylis glomerata* germplasm. Front Plant Sci 16:1530585. https://doi.org/10.3389/fpls.2025.1530585

Altaf MT, Cavagnaro PF, Kokten K, Ali A, Morales A, Tatar M et al (2025) Genotyping-by-sequencing derived SNP markers reveal genetic diversity and population structure of Dactylis glomerata germplasm. Front Plant Sci 16:1530585

Anonymous (2024). https://en.wikipedia.org/wiki/Dactylis_glomerata. Accessed Dec 24 2024

Azar SS, Mohammadi R, Nouraein M (2022) Evaluation of genetic diversity of native and non-native *Dactylis glomerata* L. ecotypes using ISSR molecular markers. Shahid Bahonar Univ Kerman Iranian Biotechnol Soc 14(2):133–154. https://doi.org/10.22103/jab.2022.18454.1352

Bajgain P, Zhang X, Anderson JA (2019) Genome-wide association study of yield component traits in intermediate wheatgrass and implications in genomic selection and breeding. G3 Genes Genom Genet 9(8):2429–2439. https://doi.org/10.1534/g3.119.400073

Balasko JA, Nelson CJ (2003) Grasses for northern areas. Forages 1:125–148

Bush T (2006) Orchardgrass (*Dactylis glomerata* L.). Plant fact slieet. USDA-NRC. www.plants.usda.gov/fachsheet/doc/fs-dagl.doc

Casler MD, Fales SL, McElroy AR, Hall MH, Hoffman LD, Leath KT (2000) Genetic progress from 40 years of orchardgrass breeding in North America measured under hay management. Crop Sci 40(4):1019–1025. https://doi.org/10.2135/cropsci2000.4041019x

Casler MD, Fales SL, McElroy AR, Hall MH, Hoffman LD, Undersander DJ, Leath KT (2002) Half-sib family selection for forage yield in orchardgrass. Plant Breed 121(1):43–48. https://doi.org/10.1046/j.1439-0523.2002.00665.x

Cen Y, Zhang K, Zhang M, Zhao W, Pan X, Peng Y, Rubinato M (2024) A calculation model of the mean flow velocity of overland flow considering a variety of grass covers and raindrop's characteristics. Soil Tillage Res 242:106145. https://doi.org/10.1016/j.still.2024.106145

Cho MJ, Choi HW, Lemaux PG (2001) Transformed T 0 orchardgrass (*Dactylis glomerata* L.) plants produced from highly regenerative tissues derived from mature seeds. Plant Cell Rep 20:318–324. https://doi.org/10.1007/s002990100330

Chtourou-Ghorbel N, Guenni K, Bedoui M, Chadded H, Sai-Kachout S, Trifi-Farah N (2024) Genetic and phenological diversity of Tunisian natural populations of *Dactylis glomerata* L. Genet Resour Crop Evol 71:3471–3484. https://doi.org/10.1007/s10722-024-01869-9

Cömertpay G, Özpınar H (2019) Determination of genetic diversity in elite orchardgrass (*Dactylis glomerata* L.) genotypes using SSR markers. Turk J Agric Food Sci Technol 7(1):127–133. https://doi.org/10.24925/turjaf.v7i1.127-133.2313

Costa R, Pereira G, Vicente C, de Sousa MT (2005) The use of molecular markers in genetic variability analysis of a collection of orchardgrass L. In: Molecular breeding for the genetic improvement of forage crops and turf. Wageningen Academic, Wageningen, pp 254–255. https://doi.org/10.3920/9789086865550_132

Costa R, Pereira G, Garrido I, Tavares-de-Sousa MM, Espinosa F (2016) Comparison of RAPD, ISSR, and AFLP molecular markers to reveal and classify orchardgrass (*Dactylis glomerata* L.) germplasm variations. PLoS One 11(4):e0152972. https://doi.org/10.1371/journal.pone.0152972

Demirkol G, Aşçı ÖÖ (2017) The evaluation of genetic diversity in orchardgrass (*Dactylis glomerata* L.) populations. Ordu Univ J Sci Technol 7(2):289–294

Denchev PD, Songstad DD, McDaniel JK, Conger BV (1997) Transgenic orchardgrass (*Dactylis glomerata*) plants by direct embryogenesis from micro projecticle bombarded leaf cells. Plant Cell Rep 16:813–819. https://doi.org/10.1007/s002990050326

Denchev PD, McDaniel JK, Conger BV (2000) Transgenic orchardgrass (*Dactylis glomerata* L.). In: Transgenic crops I. Springer Berlin Heidelberg, Berlin, Heidelberg, pp 151–163. https://doi.org/10.1007/978-3-642-59612-4_10

Dietzel R, Jarchow ME, Liebman M (2015) Above-and belowground growth, biomass, and nitrogen use in maize and reconstructed prairie cropping systems. Crop Sci 55(2):910–923. https://doi.org/10.2135/cropsci2014.08.0572

Feng G, Huang L, Li J, Wang J, Xu L, Pan L et al (2017) Comprehensive transcriptome analysis reveals distinct regulatory programs during vernalization and floral bud development of orchardgrass (*Dactylis glomerata* L.). BMC Plant Biol 17:1–19. https://doi.org/10.1186/s12870-017-1170-8

Fiasson JL, Ardouin P, Jay M (1987) A phylogenetic ground plan of the specific complex *Dactylis glomerata*. Biochem Syst Ecol 15(2):225–229

Gauthier L, Bernard J (1998) Morphological and agronomic characteristics of orchardgrass and their relationship to environmental adaptation. Agric Res Rev 11:45–58

Gençkan MS (1983) Yembitkileri Tarımı. Ege Üniversitesi Ziraat Fakültesi yayınları no: 467, Ege Üniversitesi Matbaası, İzmir

Gökçe N, Avcı MA (2019) The effects of different temperatures on germination in (Orchardgrass L.) genotypes determined by some plant characteristics. J Bahri Dagdas Crop Res 8(2):306–314

Guo DL, Zhang JY, Liu CH (2012) Genetic diversity in some grape varieties revealed by SCoT analyses. Mol Biol Rep 39:5307–5313. https://doi.org/10.1007/s11033-011-1329-6

Hatipoğlu R, Kökten K (2009) Domuz ayrığı (*Dactylis* sp. L.). In: Avcıoğlu R, Hatipoğlu R, Karadağ Y (eds) Yem Bitkileri Buğdaygil ve Diğer Familyalardan Yembitkileri, Cilt III, T.C. Tarım ve Köyişleri Bakanlığı, Tarımsal Üretim ve Geliştirme Genel Müdürlüğü, pp 611–616

Horn ME, Shillito RD, Conger BV, Harms CT (1988) Transgenic plants of orchardgrass (*Dactylis glomerata* L.) from protoplasts. Plant Cell Rep 7:469–472. https://doi.org/10.1007/BF00272733

Huang LK, Yan HD, Zhao XX, Zhang XQ, Wang J, Frazier T et al (2015) Identifying differentially expressed genes under heat stress and developing molecular markers in orchardgrass (*Dactylis glomerata* L.) through transcriptome analysis. Mol Ecol Resour 15(6):1497–1509. https://doi.org/10.1111/1755-0998.12418

Huang L, Feng G, Yan H, Zhang Z, Bushman BS, Wang J et al (2020) Genome assembly provides insights into the genome evolution and flowering regulation of orchardgrass. Plant Biotechnol J 18(2):373–388. https://doi.org/10.1111/pbi.13205

Ji Y, Chen P, Chen J, Pennerman KK, Liang X, Yan H et al (2018) Combinations of small RNA, RNA, and degradome sequencing uncovers the expression pattern of microRNA–mRNA pairs adapting to drought stress in leaf and root of *Dactylis glomerata* L. Int J Mol Sci 19(10):3114. https://doi.org/10.3390/ijms19103114

Jiang LF, Zhang XQ, Ma X, Huang LK, Xie WG, Ma YM, Zhao YF (2013) Identification of orchardgrass (*Dactylis glomerata* L.) cultivars by using simple sequence repeat markers. Genet Mol Res 12(4):5111–5123. https://doi.org/10.4238/2013.october.29.5

Jiang LF, Qi X, Zhang XQ, Huang LK, Ma X, Xie WG (2014) Analysis of diversity and relationships among orchardgrass (*Dactylis glomerata* L.) accessions using start codon-targeted markers. Genet Mol Res 13(2):4406–4418

Jin Y, Feng G, Luo J, Yan H, Sun M, Jing T et al (2024) Combined genome-wide association study and transcriptome analysis reveal candidate genes for resistance to rust (*Puccinia graminis*) in *Dactylis glomerata*. Plant Dis 108:PDIS-02. https://doi.org/10.1094/PDIS-02-24-0360-RE

Jung GA, Baker BS (1985) Orchardgrass. In: Heath ME, Bames RF, Metcalfe DS (eds) Forages: the science of grassland agriculture. Iowa State University Pres, Iowa, pp 224–232

Kakihara H, Ogura SI (2022) Effect of soil acidification on regrowth of orchardgrass (*Dactylis glomerata*) under application of grazing cattle dung, cattle manure compost, and chemical fertilizer. Grassl Sci 68(3):255–262. https://doi.org/10.1111/grs.12361

Kim KY, Jang YS, Cha JY, Son D, Choi GJ, Seo S, Lee SJ (2008) Acquisition of thermotolerance in transgenic orchardgrass plants with DgHSP17. 2 gene. Asian Australas J Anim Sci 21(5):657–662. https://doi.org/10.5713/ajas.2008.60725

Koc A, Kaya A, Güllap MK, Erkovan HI, Macit M, Karaoğlu M (2014) The effect of supplemental concentrate feed on live weight gain of yearling heifers over grazing season in subirrigated rangelands of East Anatolia. Turk J Vet Anim Sci 38(3):278–284. https://doi.org/10.3906/vet-1309-56

Larson SR, Jones TA, Jensen KB (2004) Population structure in *Pseudoroegneria spicata* (Poaceae: Triticeae) modeled by Bayesian clustering of AFLP genotypes. Am J Bot 91(11):1789–1801. https://doi.org/10.3732/ajb.91.11.1789

Last L, Widmer F, Fjellstad W, Stoyanova S, Kölliker R (2013) Genetic diversity of natural orchardgrass (*Dactylis glomerata* L.) populations in three regions in Europe. BMC Genet 14:1–14. https://doi.org/10.1186/1471-2156-14-102

Lee SH, Lee DG, Woo HS, Lee KW, Kim DH, Kwak SS et al (2006) Agrobacterium-mediated transformation of Orchardgrass via seed-derived callus. Plant Biotechnol Rep 1(4):218–222. https://doi.org/10.1016/j.plantsci.2006.05.006

Lindner R, Lema M, Garcia A (2004) Extended genetic resources of *Dactylis glomerata* subsp. *izcoi* in Galicia (northwest Spain). Genet Resour Crop Evol 51:437–442. https://doi.org/10.1023/B:GRES.0000023457.30193.63

Lumaret R, Borrill M (1988) Cytology, genetics, and evolution in the genus Dactylis. Crit Rev Plant Sci 7(1):55–91

Lumaret R, Barrientos A, Muir A (1989) The role of unreduced gametes in the polyploid evolution of Orchardgrass. Evol Ecol 4(3):298–305

Mace ES, Xia L, Jordan DR, Halloran K, Parh DK, Huttner E et al (2008) DArT markers: diversity analyses and mapping in Sorghum bicolor. BMC Genomics 9:1–11. https://doi.org/10.1186/1471-2164-9-26

Madesis P, Abraham EM, Kalivas A, Ganopoulos I, Tsaftaris A (2014) Genetic diversity and structure of natural *Dactylis glomerata* L. populations revealed by morphological and microsatellite-based (SSR/ISSR) markers. Genet Mol Res 13(2):4226–4240. https://doi.org/10.4238/2014.June.9.8

Manga İ, Acar Z, Ayan İ (2002) Buğdaygil Yem Bitkileri. Ondokuz Mayıs Üniversitesi Ziraat Fakültesi Ders Kitabı no: 6, Yayın

Mao P, Ou C, Jia Z, Hong L, Ma W (2023) Research progress for seed production technology of herbage and turfgrass in China. Chin J Grassl 45:1–11

Marone D, Laido G, Gadaleta A, Colasuonno P, Ficco DB, Giancaspro A et al (2012) A high-density consensus map of A and B wheat genomes. Theor Appl Genet 125:1619–1638. https://doi.org/10.1007/s00122-012-1939-y

Nabhan A, Arvas Ö, Lakmes A (2021) Scanning morphological traits of orchardgrass (*Dactylis glomerata* L.) genotypes in the natural rangelands of eastern Anatolia. Yuzuncu Yıl Univ J Agric Sci 31(1):111–120. https://doi.org/10.29133/yyutbd.809169

Nadeem MA, Nawaz MA, Shahid MQ, Doğan Y, Comertpay G, Yıldız M, ... & Baloch FS (2018) DNA molecular markers in plant breeding: current status and recent advancements in genomic selection and genome editing. Biotechnology & Biotechnological Equipment, 32(2), 261–285

Nakayama K, Kato M, Kobayashi M (1997) Winter hardiness and morphological characteristics of Orchardgrass cultivars in Japan. Euphytica 94(3):311–317

Nam DG, Baek ES, Hwang EB, Kim SK, Gwak SC, Lee YH et al (2024) Genetic diversity patterns within and among varieties of Korean orchardgrass (*Dactylis glomerata* L.) based on SSR and EST-SSR analysis. J Crop Sci Biotechnol 27:449–461. https://doi.org/10.1007/s12892-024-00241-8

Peng YAN, Zhang X, Deng Y, Ma X (2008) Evaluation of genetic diversity in wild orchardgrass (*Dactylis glomerata* L.) based on AFLP markers. Hereditas 145(4):174–181. https://doi.org/10.1111/j.0018-0661.2008.02038.x

Piao S, Liu Q, Chen A, Janssens IA, Fu Y, Dai J et al (2019) Plant phenology and global climate change: current progresses and challenges. Glob Chang Biol 25(6):1922–1940. https://doi.org/10.1111/gcb.14619

Poozesh V, Castillon P, Cruz P, Bertoni G (2010) Re-evaluation of the liming-fertilization interaction in grasslands on poor and acid soils. Grass Forage Sci 65(2):260–272. https://doi.org/10.1111/j.1365-2494.2010.00744.x

Qiao D, Zhang Y, Xiong X, Li M, Cai K, Luo H, Zeng B (2020) Transcriptome analysis on responses of orchardgrass (*Dactylis glomerata* L.) leaves to a short term flooding. Hereditas 157:1–16. https://doi.org/10.1186/s41065-020-00134-0

Reeves G, Francis D, Davies MS, Rogers HJ, Hodkinson TR (1998) Genome size is negatively correlated with altitude in natural populations of *Dactylis glomerata*. Ann Bot 82(suppl_1):99–105. https://doi.org/10.1006/anbo.1998.0751

Reinermann S, Asam S, Kuenzer C (2020) Remote sensing of grassland production and management-a review. Remote Sens 12(12):1949. https://doi.org/10.3390/rs12121949

Ren S, Sun M, Yan H, Wu B, Jing T, Huang L, Zeng B (2020) Identification and distribution of NBS-encoding resistance genes of *Dactylis glomerata* L. and its expression under abiotic and biotic stress. Biochem Genet 58:824–847. https://doi.org/10.1007/s10528-020-09977-8

Sahuquillo E, Lumaret R (1999) Chloroplast DNA variation in *Dactylis glomerata* L. taxa endemic to the Macaronesian islands. Mol Ecol 8(11):1797–1803. https://doi.org/10.1046/j.1365-294x.1999.00755.x

Sakiroglu M, Dong C, Hall MB, Jungers J, Picasso V (2020) How does nitrogen and forage harvest affect belowground biomass and nonstructural carbohydrates in dual-use Kernza intermediate wheatgrass? Crop Sci 60(5):2562–2573. https://doi.org/10.1002/csc2.20239

Sanada Y, Takai T, Yamada T (2007) Ecotypic variation of water-soluble carbohydrate concentration and winter hardiness in cocksfoot (*Dactylis glomerata* L.). Euphytica 153:267–280. https://doi.org/10.1007/s10681-006-9262-9

Sandage LJ, Windham TE (2006) Orchard grass. Cool season perennial grasses series. University of Arkansas Cooperative Extension Service. Available at: https://www.uaex.uada.edu/publications/

Stebbins GL (1971) Chromosomal evolution in higher plants. Edward Arnold, London

Stebbins GL, Zohary D (1959) Cytogenetic and evolutionary studies in the genus Dactyis: 2 plates. In: Morphology, distribution, and interrelationships of the diploid subspecies. University of California Press

Stewart AV, Ellison NW (2010) The genus Dactylis; wealth of wild species: role in plant genome elucidation and improvement. Springer, New York

Tajimi A (1974) Tetrasomic inheritance of stem rust resistance in orchardgrass. Jpn J Grassl Sci 20(1):26–30. https://doi.org/10.14941/grass.20.26

Terada Y, Daido H, Ito K, Araki H, Hojito S, Sugita S, Kawabata S, Abe J, Saiga S, Suzuki S (1991) Breeding of 'Wasemidori' orchardgrass (*Dactylis glomerata* L.) and its characteristics. Res Bull Hokkaido Natl Agric Exp Stn 155:101–117

Tosun M, Sagsoz S (1994) Determination of morphological and biological parameters of naturally growing orchardgrass (*Dactylis glomerata* ssp *hispanica* (Roth) Nyman) in Erzurum region collected from different locations. In: Field crop symposium, Izmir, pp 39–43

Tuna M, Khadka DK, Shrestha MK, Arumuganathan K, Golan-Goldhirsh A (2004) Characterization of natural orchardgrass (*Dactylis glomerata* L.) populations of the Thrace Region of Turkey based on ploidy and DNA polymorphisms. Euphytica 135(1):39–46. https://doi.org/10.1023/B:EUPH.0000009537.08697.4e

Tuna M, Teykin E, Buyukbaser A, Budak H (2007) Nuclear DNA variation in the grass genus *Dactylis* L. In: Poster presentation, 5th international symposium on the molecular breeding of forage and turf, 1–6 July 2007, Sapporo, Japan

Turner NC (2004) Sustainable production of crops and pastures under drought in a Mediterranean environment. Ann Appl Biol 144(2):139–147. https://doi.org/10.1111/j.1744-7348.2004.tb00327.x

Vilhar B, Vidic T, Jogan N, Dermastia M (2002) Genome size and the nucleolar number as estimators of ploidy level in *Dactylis glomerata* in the Slovenian Alps. Plant Syst Evol 234:1–13. https://doi.org/10.1007/s00606-002-0186-0

Watson L, Dallwitz MJ, Johnston CR (1986) Grass genera of the world: 728 detailed descriptions from an automated database. Aust J Bot 34(2):223–230. https://doi.org/10.1071/BT9860223

Wetschnig W (1991) Karyotype Morphology and Some Diploid Subspecies of Orchardgrass L. (Poaceae). Phyton Horn 31:35–55

Wezel A, Stöckli S, Tasser E, Nitsch H, Vincent A (2021) Good pastures, good meadows: Mountain farmers' assessment, perceptions on ecosystem services, and proposals for biodiversity management. Sustainability 13(10):5609. https://doi.org/10.3390/su13105609

Xie WG, Zhang XQ, Ma X, Peng Y, Huang LK (2009) Genetic variation and relationship in orchardgrass (*Dactylis glomerata* L.) germplasm detected by SSR markers. Hereditas 31(6):654–662. https://doi.org/10.3724/sp.j.1005.2009.00654

Xie WG, Zhang XQ, Cai HW, Liu W, Peng Y (2010a) Genetic diversity analysis and transferability of cereal EST-SSR markers to orchardgrass (*Dactylis glomerata* L.). Biochem Syst Ecol 38(4):740–749. https://doi.org/10.1016/j.bse.2010.06.009

Xie WG, Zhang XQ, Ma X, Cai HW, Huang LK, Peng Y, Zeng B (2010b) Diversity comparison and phylogenetic relationships of cocksfoot (*Dactylis glomerata* L.) germplasm as revealed by SSR markers. Can J Plant Sci 90(1):13–21. https://doi.org/10.4141/CJPS08192

Xie W, Zhang X, Cai H, Huang L, Peng Y, MaX. (2011) Genetic maps of SSR and SRAP markers in diploid orchardgrass (*Dactylis glomerata* L.) using the pseudo-testcross strategy. Genome 54:212–221. https://doi.org/10.1139/G10-111

Xie WG, Lu XF, Zhang XQ, Huang LK, Cheng L (2012) Genetic variation and comparison of orchardgrass (*Dactylis glomerata* L.) cultivars and wild accessions as revealed by SSR markers. Genet Mol Res 11:425–433. https://doi.org/10.4238/2012.February.24.1

Yalinkiliç NA, Başbağ S, Altaf MT, Ali A, Nadeem MA, Baloch FS (2024) Applicability of SCoT markers in unraveling genetic variation and population structure among sugar beet (*Beta vulgaris* L.) germplasm. Mol Biol Rep 51(1):584. https://doi.org/10.1007/s11033-024-09526-1

Yan H, Zhang Y, Zeng B, Yin G, Zhang X, Ji Y et al (2016) Genetic diversity and association of EST-SSR and SCoT markers with rust traits in orchardgrass (*Dactylis glomerata* L.). Molecules 21(1):66. https://doi.org/10.3390/molecules21010066

Zeng B, Zhang XQ, Fan Y, Lan Y, Ma X, Peng Y, Liu W (2006) Genetic diversity of *Dactylis glomerata* germplasm resources detected by Inter-simple Sequence Repeats (ISSRs) molecular markers. Hereditas 28(9):1093–1100. https://doi.org/10.1360/yc-006-1093

Zeng B, Zhang XQ, Lan Y, Yang WY (2008) Evaluation of genetic diversity and relationships in orchardgrass (*Dactylis glomerata* L.) germplasm based on SRAP markers. Can J Plant Sci 88(1):53–60. https://doi.org/10.1360/yc-006-1093

Recent Advancements in Breeding and Genomics of Ryegrasses

Wahid Ullah, Arshad Ali, Zahid Ullah, Ahmad Ali, Javed Iqbal, Naveed Alam, Sidratul Muntaha, Banzeer Ahsan Abbasi, Madiha Imtiaz, and Shumaila Ijaz

Abstract

Recent advancements in ryegrass genomics and breeding have significantly enhanced our understanding and improvement of this vital forage species. The development of a high-quality, chromosome-scale genome association for perennial ryegrass (*Lolium perenne* L.) has been a cornerstone of these efforts. This assembly, achieved through the integration of PacBio long-read sequencing, BioNano optical mapping, Illumina short-read polishing, and Hi-C scaffolding, covers approximately 2.55 Gb of the genome organized into seven pseudo-chromosomes. It includes annotations for over 54,000 high-confidence gene models and various non-coding RNAs, providing a comprehensive resource for comparative genomic studies and breeding applications. In parallel, the implementation of genomic selection (GS) strategies has accelerated genetic gains in ryegrass breeding programs. By leveraging genome-wide marker data, GS

W. Ullah · A. Ali · Z. Ullah · A. Ali · S. Muntaha
Center for Plant Sciences and Biodiversity, University of Swat, Charbagh Swat, Pakistan
e-mail: zahidtaxon@uswat.edu.pk; naveedalam@uswat.edu.pk

J. Iqbal (✉) · M. Imtiaz
Department of Botany, Bacha Khan University, Charsadda, Khyber Pakhtunkhwa, Pakistan

N. Alam
Institute of Forest Sciences, University of Swat, Charbagh Swat, Pakistan

B. A. Abbasi
Department of Botany, Rawalpindi Women University, Rawalpindi, Pakistan

S. Ijaz
Department of Burn and Plastic Surgery, Shenzhen Institute of Translational Medicine, The First Affiliated Hospital of Shenzhen University, School of Biomedical Engineering, Shenzhen University Medical School, Shenzhen University, Shenzhen, China

F. S. Baloch et al. (eds.), *Forage Crops in the Omics Era*,
https://doi.org/10.1007/978-981-95-9249-4_16

enables the prediction of breeding values, facilitating the selection of superior genotypes for traits including biomass yield and disease resistance. Studies have demonstrated that GS can double or even triple the rate of genetic gain compared to traditional phenotypic selection methods. Moreover, the application of CRISPR/Cas9 genome editing technologies has opened new avenues for precise genetic modifications in ryegrass. Recent research utilizing an Agrobacterium-mediated delivery system with a single promoter-based CRISPR/Cas9-sgRNA construct achieved a 29% editing efficiency in the Phytoene Desaturase (PDS) gene of perennial ryegrass, highlighting the potential of this approach for functional genomics studies and trait improvement. Collectively, these advancements in genomic resources, selection methodologies, and genome editing tools have the potential to greatly improve the efficiency and effectiveness of ryegrass breeding programs, leading to the development of enhanced cultivars with desirable agronomic traits.

Keywords

Ryegrass · Breeding programs · CRISPR/Cas9 · Desirable agronomic traits

1 Introduction

The ryegrass (*Lolium*) is placed in the family Poaceae, being distinct from *Festuca*, and has undergone differentiation into distinct species (Charmet et al. 1997). The natural and artificial hybrids between ryegrass and European *Festuca* species have been reported (Jenkin 1993). Due to the close resemblance of the two genera, they are collectively called the *Lolium-Festuca* group or ryegrasses (Hand et al. 2010). The genus *Lolium* consists of ten species which are distributed in Northern Africa, Europe and Asia. Moreover, it is being cultivated and distributed in temperate areas of the world (Zhang et al. 2020). *Lolium multiflorum* and *Lolium perrene* are the economically important species of the genus and have also been used for pasture and lawn cover purposes. Perennial ryegrass (*Lolium perenne* L.) is an allogamous species having genetic gametophytic self-incompatibility. It has a significant role as the cultivated forage species in northwest America, Europe, Australia, South Africa, Japan, and New Zealand (Humphreys et al. 2009). The breeding program of this species started in the 1920s and the breeders were depending on the recurrent selection of phenotypic superior individual (Hayes et al. 2013). This method has resulted in significant advancement in many attributes in the past decades such as spring growth and rust resistance. Development of better cultivars of perennial ryegrass as grazers has had a long history in Australia and New Zealand (Stewart 2006), and for grazing system in Europe (Wilkins and Humphreys, 2003). Particularly in dry matter yield, there has been an improvement in productive character (McDonagh et al. 2016) but also in nutritive value and ground cover (Chaves et al. 2009).

Many grass species were sown by farmers and agriculturists for short-term grassland breeding for many years. The perennial ryegrass species seeds were sold in Europe earlier the Second World War; however, these seeds were developed from lines of appropriate seed yield avoid taking foraging performance (Wilkins and Humphreys, 2003). During that time, farmers would frequently sow the grass seeds they had harvested in the surrounding natural area. The ryegrass breeding program firts time started in the UK in 1919, and by 1931 the first ever cultivar of the blended grass was obtained (Humphreys et al. 2006). Individual research, education, and breeding organizations in Europe, North America, Australia, and New Zealand started serious research on breeding perennial forage grasses about 1950 to 1960. The large diversity of natural population and landscape are the primary reasons to breeding program. Many grass cultivars were developed and put on the market to sell until 1970, which was the result of numerous crosses of plants that had been chosen in landraces and ecotypes (Humphreys et al. 2009).

Perennial ryegrass is considered as the most important and mostly used grass member for cultivating the grasslands in temperate regions and has resulted in major efforts of breeding globally, especially in European countries where they represent about 50% of the marketed grass seeds (Humphreys et al. 2009). Due to their leafy and highly tillering manner, it is considered high herbage yields of good feedings for animals feeding for many decades. Especially because the ryegrass is more resilient to grazing-induced defoliation and animal tramping than other temperate grass species. Ryegrasses have spike emergence dates longer than 7 weeks, indicating a wide range of variability. Regarding the rapid development of spikes, breeders have produced early intermediate cultivars that are more practical for grazing or cutting the regrowth as green forage for animals and the cultivation process, as well as late cultivars that are suited to a long spring grazing season. The European National Lists typically divide perennial ryegrass varieties into numerous earliness groups based on the spike's emerging timing for accessibility considerations.

Perennial ryegrass is well modified to a broad range of environment in which dairy farming is carried out. Over the past two decades, concern has been mounting over the persistence of newly sown ryegrass-based pastures on dairy farms, particularly in the upper North Island of New Zealand (Tozer et al. 2011). Data for the longevity of perennial ryegrass pastures on commercial farms are sparse, but 10 years is often taken as an expectation. Sanderson and Webster (2009) estimated that 8% of pasture area used for dairy grazing is renewed annually in New Zealand, implying an average lifespan of 12 years. Using data reported by Kelly et al. (2011) from a survey of dairy farmer ($n = 776$, in Waikato and Bay of Plenty) experiences with pasture renewal, it can be calculated that, in 2009–10, average renewal rates were approximately 21% per annum, inferring that pastures were lasting only 5 years on average in these important upper North Island dairy regions. In some situations, ryegrass-based pastures have been reported to fail within 2–4 years, e.g., Lee et al. (2018), McCahon et al. (2021). A high rate of pasture renewal adds substantial costs to farm operating expenses associated with maintaining feed supply (Brazendale et al. 2011), and increases the risks of adverse environmental impacts because soils are disturbed by cultivation more frequently (Betteridge et al. 2011; Wall et al. 2021). Plant breeding and/or management interventions to reduce these impacts would help secure the economic and environmental sustainability of pasture-based farming systems in areas currently experiencing persistence problems, and in areas where future climate change can be expected to create similar problems (Sarıkaya and Tatar 2023; Beukes et al. 2021).

2 Ryegrass Breeding Strategies

Lolium and *Festuca* (the ryegrasses) are highly self-incompatible and primarily have a cross-pollination mechanism; the ultimate synthetic cultivars are different from the self-pollinated hybrids. Descended from several carefully selected parents, ryegrass cultivars are genetically different but morphologically related populations. The breeding process of ryegrass requires millions of dollars of investment and usually spreads over 10–15 years from conceptualization to the release of a new cultivar. The plant breeders use a vast array of strategies for developing new cultivars. The process starts with screening of phenotypes in the field from several thousand genotypes over a period of several years and is subjected to a variety of stresses, grazing, and cutting. Thus, the cross-pollination of different ryegrasses (variable in a particular trait) produces the first generation of plants (F1), which gives rise to F2 plants by cross-pollination of F1 plants, including remaining culms of older plants and germplasm accessions.

Further steps in the breeding program involve more careful screening of the elite plants for additional characters and seed production of synthetic generation 1 (Syn 1), testing of progeny, and then production of the second synthetic generation (Syn 2), as well as on-farm experiments for 2–3 years, after repeated cutting and grazing. Based on the availability of seeds and the goal of the program, the breeding methods vary a lot. Often, there is a limited seed supply in the start of the breeding program; therefore, the experiments are confined to small plots or small replicated rows. With the passage of time and advancement of experiments, more seeds become available for cultivation on a larger scale and in different localities, with many replications. A significant number of these breeding lines often do not reach the market (Jensen et al. 2009). Details about the different breeding strategies are given in Fig. 1.

3 Hybridization and Introgression Breeding

Among the common plant breeding strategies, hybridization and introgression involve breeding between the progeny of genetically diverse parents, as it shows hybrid vigor with greater production. However, these qualities may reduce from generation to generation, during continuous production (McClain 1982). In hybridization, individual plants of two different species belonging to the same genus are crossed with each other. The F1 hybrids of this cross will contain the same number of genes of both species of the parents, although this factor will be influenced in the next successive generation based on what the hybrids will be crossed with later. In the case of introgression breeding, F1 hybrids from the parental cross of two different species are repeatedly backcrossed with one of the parent species, leading to a complex mixture of parental genes. Ryegrasses (*Lolium* and *Festuca*) possess

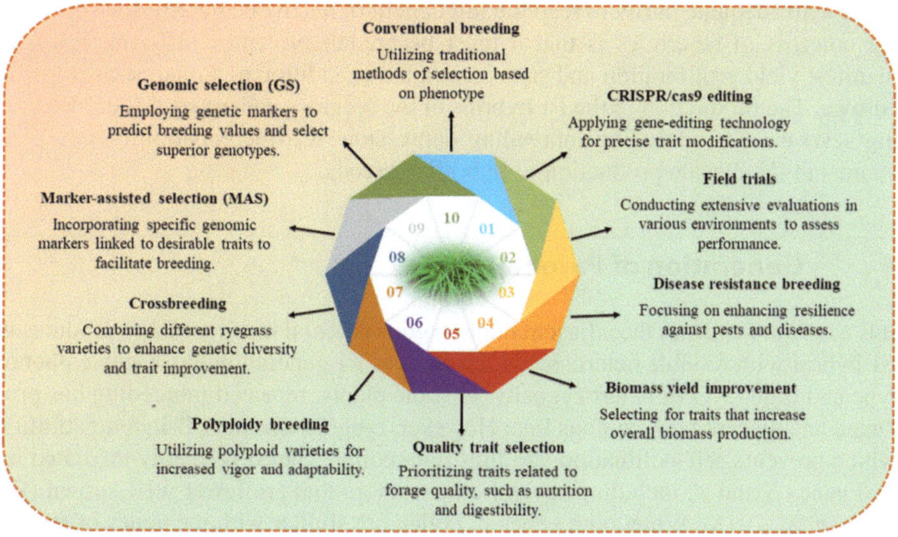

Fig. 1 Ryegrass breeding strategies

enormous genetic diversity and huge opportunities for hybridization and introgression breeding (Humphreys 2005).

Different ecotypes and various cultivars of the ryegrasses were used by Barrett et al. (2010) for the synthesis of semi-hybrid ryegrasses in perennial species. It was found that the mean yield of above-ground biomass from the hybrids was on average 1.6% greater than the best producing parents, although with variations for different cultivars. Remarkably, hybrid vigor was higher in crosses between cultivars with different heading dates than in crosses between cultivars with similar heading dates. This study suggests that more research should be done on semi-hybrid breeding methods. Farm production in New Zealand would probably shift as a result of the F1 hybrid ryegrass breeding, which has the added advantages of increased yield and stress tolerance (Clark et al. 2007). The commercial manufacture of F1 hybrid ryegrass has not been tried so far due to the small, bisexual flowers, which prevent large-scale manual outcrossing. Moreover, the absence of genetic resources inhibits further complex crossing strategies, which are further complicated by the failure to efficiently inbreed ryegrass because of self-incompatibility. However, the information has now been gathered from other similar grass species, where in addition to transcriptomics and genomic sequence for ryegrass, heterosis, and new techniques including transgenic pollination control, CRISPR gene editing have been used, thereby providing a stimulating opportunity to produce F1 hybrid ryegrass (Wu et al. 2016).

The production of F1 hybrid seed requires three fundamental basics to be successful. The first is the formation of different homogametic, closely neighboring homozygous parental lines, that usually accomplish through inbreeding. The second is that the parental lines have to be crossed effectively to produce F1 hybrid seeds with adequate purity to reap the full benefit of heterosis. Finally, the benefit of the outcross of heterotics is that it must be predictable and consistent, and also manifest yield proliferation and stress tolerance in addition to already existing elite cultivar. The production of the F1 hybrids of the ryegrass undergoes difficulties during several steps, including inbreeding depression, self-incompatibility, and constraints in large-scale production of hybridized seeds.

4 Generation of Parental Lines

It is necessary to cross the adjacent homozygous parental line in order to produce an F1 hybrid with reliable heterosis, which ensures F1 generation has a stable phenotype and high level of heterozygosity. In some plants, repeated inbreeding has produced an adjacent homozygous line. However, ryegrass shows self-incompatibility, which prevents self-pollination and thus inbreeding. The SI is mainly mediated by two genes *S* and *Z*, including several other factors that are fewer well known (Do Canto et al. 2018). When both genes in pollen are similar with the genes of female tissues, that is S and Z, then the pollen tube growth is stopped. If the *S* or *Z* genes are not similar to each other as a result, the growth of pollen tube continues, and fertilization can take place. To produce inbred ryegrass, the self-incompatibility must be reduced.

In ryegrass, self-incompatibility can be reduced by mutation or high temperature treatment (Do Canto et al. 2018). However, this SI can occur occasionally and because of this uncertainty, breeders have fewer options for which lines to inbreed. The current improvement in gene editing technology may aid by altering the S or Z genes in ryegrass, allowing SI to be overcome. The sequencing of two genes, namely, *DUF247*, underlying the *S* or *Z* loci is now recognized (Thorogood et al. 2017). It is possible to produce a knockout line by using CRISPR technology by targeting the coding regions of such genes. This will enable the breeder to produce readily self-fertile line in several varieties than other approaches (Ye et al. 2018).

In addition, inbreeding depression is a challenge in producing inbred parental lines (Yamada 2001). Upon repeated self-fertilization in inbreeding, depression causes the loss of vigor due to a lethal recessive allele becoming homozygous, often resulting in the formation of poor set of seed, which prevents commercial seed production and breeding. Inbreeding depression, through years of trailing, has occurred in maize to reduce the selection of the inbred line (Duvick 1999). The issue of ID could be reduced in ryegrass through the selection of line with better production of seeds. Such a selection would need enough quantity of self-cross breeding material with a huge range of allelic diversity between different loci that are responsible for seed set. This could be generated by utilizing the gene control approaches.

5 Crossing Genetically Distinct Parental Lines

When an appropriate inbred line has been selected and generated, such a line on large scale must be outcrossed. While SI usually ensures the outcrosses of ryegrass, the inbred line can be self-fertilized. Inside the hybrid seed, selfed progeny would lead to a low performing proportion of inbred seeds among the F1 hybrid seeds, which would greatly impact upon the entire crops production. Consequently, it is essential to prevent selfing for the production of F1 hybrid seeds. The hybrid production of maize was initially benefited due to the presence of unisexual flower in maize, which enabled the cultivars to detassel the female and cross with a distinct line. Ryegrass, like various other plants, has bisexual flowers, making it impossible to make hybrids in such a way.

One hybrid breeding method is to use the SI system in ryegrass, which has limited diversity in S and Z genes. When two populations with restricted distinct alleles at these loci are crossed, approximately 83% of the resulting offspring are F1 hybrids. This is due to the complementary genetic composition of the populations (Pembleton et al. 2015). Unlike self-fertile line, this strategy does not achieve the high level of homozygosity often desired in inbred line (Posselt 1993). The identification of genetic sequence of *S* and *Z* loci enables the use of marker to streamline the selection process (Thorogood et al. 2017). This makes the breeding process more targeted and efficient.

The cytoplasmic male sterility has been used in rice for the production F1 hybrid rice by three-way crossing system. The first CMS line that carries a mutation in a mitochondrial gene leads to the production of non-viable pollen. The second is the maintainer line the pollen; this line is used to propagate the CMS line, ensuring the mutation is retained since mitochondria are maternally inherited. The last one is restorer line, which contains a nuclear gene that offsets the CMS mutation, enabling the production of viable seeds for yield. A parallel system could be adapted for F1 hybrid ryegrass if a functional CMS line could be established. Unlike rice, ryegrass is primarily grown for its vegetative tissue in pastures; hence, a restorer allele might not be necessary. In fact, excluding the restorer could be advantageous in preventing unwanted pollen dispersal. Technologies like CRISPR could potentially modify mitochondrial or chloroplast genomes to induce CMS, face limitations. The inability to target guide RNAs to organelles effectively constrains the use of such techniques (Gammage et al. 2018).

Recent advancements in hybrid rice production have utilized environmentally male sterile line. EGMS lines are propagated in environments that do not induce sterility and later transferred to conditions that trigger sterility. In these sterility-inducing environments, the EGMS lines are crossed with a pollen donor line that lacks temperature sensitivity to produce F1 hybrids. This strategy has been highly effective in China, boosting rice yields while maintaining or even reducing the land required for cultivation (Yuan LongPing 2014). The genes and mutations responsible for EGMS in rice have been identified. If analogous genes exist in ryegrass, it may be possible to replicate these traits through gene-editing technologies (Li et al. 2016).

5.1 Predictable Heterosis in Ryegrass

Once an appropriate control system of pollination and inbred line are in place, the level of heterosis must be optimized in F1 hybrid. The F1 hybrid could be beneficial for farmers when the heterosis provides advantages to enhance elite cultivars and also should be reproducible and predictable. It is well recognized that different crossings provide varying levels of yield. It is impossible to predict which crossings will result in the most active offspring; the genetic distance appears to be associated with the yield enhancement (Cheres et al. 2000).

6 Recent Advancements in the Genomics of Ryegrasses

Ryegrasses are an economically significant forage grass; they are broadly cultivated in temperate areas globally. In the classification system, they are placed along with wheat, oats, barley, and *Brachypodium distachyon* in the Pooideae subfamily of the grass family, i.e., Poaceae (Byrne et al. 2015). They are the principal forage grasses contributing to diary invention in the temperate regions of the world. The genus *Lolium* contains nine distinct species having an evolutionary relationship with *Festuca*. Members of *Lolium* are moderately interfertile, forming a distinct ploidy series and integrating variation in agronomy, phenology, and certain adaptive characters (Humphreys et al. 2006). *Lolium perrene* commonly called perennial rye grass ($2n = 14$) is an important fodder grass in temperate regions and used as silage, pasture, and hay. They provide high yield, tolerance for grazing, long growing season, and high digestibility and palatability for ruminant animals. It is estimated that perennial ryegrass has more than 100 years old history, and it is realized that the genetic diversity in fodder grass is more modest than the diversity present in grain crops. Compared to genetic developments for grain production, forage yield has only seen moderate genetic gains (Humphreys 1997). Therefore, by applying molecular breeding and hybridizing approaches, it is considered a more effective way to increase the rate of genetic grow for perennial ryegrass (Arias Aguirre et al. 2012). Using genomic-based methods for breeding programs directly improve the accessibility and utility of reference genomes by generating high-density marker data (such as SNPs). It has consequences for agricultural production as well as for the importance of ryegrasses, which contribute to the ecosystem services of perennial grasslands, biodiversity reserves, water catchment, and carbon appropriation. Details about recent advancements in the genomics of ryegrasses are given in Fig. 2.

Members of the genus *Lolium* are composed of many perennial and annual species (Catalán et al. 1997). They are indigenous to Europe, North Africa, and temperate Asia also expanded to New Zealand, South America, southern Africa, Australia, and North America throughout the past 200 years (Humphreys et al. 2009). In tainted commercial crop seed and animal feed, they were primarily seeds of *Lolium* as weedy seeds are often found in commercial crop seed and animal feed, and they were primarily grown on new lands used for grazing animals, cover crops, and as turf. Their morphological characteristics are similar among different species;

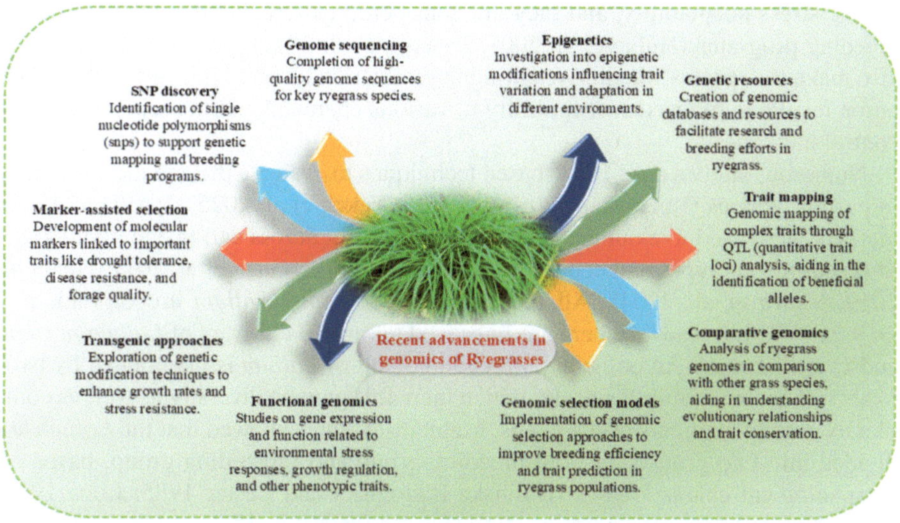

Fig. 2 Recent advancements in genomics of ryegrasses

distinction of species is detectable in the flowering stage by the flowers number, inflorescence structure, and shape (Terrell 1968). *Lolium perrene, L. rigidum,* and *L. perrene L. ssp. multiflorum* are distinctive members of the genus. The species are adaptable to varied climates, and their use patterns vary based on the place grown and end use. *Lolium perenne* is adapted to temperate areas and has stronger cold resistance than the other two species (Zhang et al. 2009). It is also adapted around the world because of its high feed value and perenniality. On the other hand, *Lolium multiflorum and Lolium rigidum* exhibit annual activity and are mostly used for grazing in hot places, including the Southern United States and Australia. Throughout the domestication process, natural and deliberate adaptations occurred, allowing *Lolium* species to develop into pastures. Hence, various cultivars have been cultivated for increased heat and cold tolerance. They are regarded prominent weeds of global significance in cropping systems, invading both agricultural and non-agricultural regions.

7 Self-Incompatibility and Genetic Variation of *Lolium*

Members of the genus *Lolium* possess self-incompatibility mechanisms, which minimize the possibility of successful pollen fertilization with unmatched loci. This phenomenon is regulated by numerous distinct multiallelic loci (Takayama and Isogai 2005). As a member of the grassland, the genetic variation in *Lolium* population was investigated for its impact on vegetative persistence and morphology (O'Connor et al. 2020; Katova and Vulchinkov 2019). Some members of the genus are also studied in regard to herbicide resistance mechanisms and genetic inheritance (Karn and Jasieniuk 2017). Their high genetic variability leads to biotic and abiotic stress adaptability, and they are considered extremely valuable qualities in breeding programs (Hulke et al. 2008; Pearson et al. 2011). So, these features can also make the species weedier, resulting in production losses. The high rate of adaptation results in reduced susceptibility to various chemical and non-chemical weed control methods.

Numerous studies used DNA-based techniques to evaluate the genetic variability in pasture grasses (Loera-Sánchez et al. 2019; Yeken et al. 2025). RAPD and SSR markers were used to examine genetic heterogeneity across and inside populations of breeding genotypes and ecotypes of Italian ryegrass (Guan et al. 2017; Nie et al. 2019; Vieira et al. 2004). All the members of genus *Lolium* are diploid, i.e., $2n = 2x = 14$, but many of them are improved tetraploid cultivars of *Lolium perrene* and *L. multiflorum.* An obligate outbreeding habit of *Lolium* is maintained by two-locus self-incompatibility system, which increases genetic diversity as well as complexity among members of the genus. Many researchers showed that the *Lolium* are divided into two groups, i.e., out breeding group and inbreeding group, based on morphological characteristics (Bulinska-Radomska and Lester 1985; Loos 1993; Bennett et al. 2000), seed proteins (Bulinska-Radomska and Lester 1985), interspecific hybridization and isozymes (Charmet and Balfourier 1994; Charmet et al. 1997), and molecular markers such as random amplified polymorphic DNA (RAPD) markers, restriction fragment length polymorphism (RFLP) markers, (Charmet et al. 1997), internal transcribed spacer (ITS) rDNA (Charmet et al. 1997; Gaut et al. 2000; Catalan et al. 2004), sequence-related amplified polymorphism (SRAP) markers (Cheng et al. 2016a), and chloroplast DNA (Balfourier et al. 2000; Cheng et al. 2016b). On the other hand, both *Lolium perenne* and *L. rigidum* are outcrossing species on the basis of a study involving 51 self-sustaining populations collected from Europe and Middle East. Majority of the accessions of *L. rigidum* formed cluster with *L. remotum* and *L. temulentum,* and those of *L. perenne* populations, formed two different clusters based on plastid DNA markers (Balfourier et al. 2000). Likewise, according to Cresswell et al. (2001), classification on the basis of improved fragment length polymorphism (AFLP) markers analyzed the population in *L. perenne*, four of *L. multiflorum*, and nine of their hybrids, *L. hybridum,* in Portugal location.

8 Conclusion

In conclusion, ryegrass breeding and genomics developments mark a paradigm change in our understanding and development of this important fodder plant. A strong basis for upcoming comparative genomic research and focused breeding initiatives is established by the high-quality, chromosome-scale genome assembly for perennial ryegrass and comprehensive gene annotations. By permitting more accurate selection based on anticipated breeding values, the incorporation of genomic selection techniques has dramatically accelerated the genetic advances in breeding operations. The advent of CRISPR/Cas9 genome editing expands the toolkit accessible to scientists and breeders by enabling creative and focused alterations that may result in notable enhancements in characteristics such as disease resistance and biomass production. In addition to improving the quality and sustainability of ryegrass cultivars, these combined tactics support the more general agricultural objectives of productivity and resilience. To meet the challenges presented by shifting environmental circumstances and rising forage quality demands, we will need to continue investing in these genomic breakthroughs.

Author Contributions W.U., A.A., Z.U., A.A., and J.I., designed and conceived the chapter idea. W.U, Z.U., B.A.A., and A.A. wrote the original draft. N.A., S.M., B.A.A., and S.I., reviewed and edited the manuscript and provided valuable insights. All authors have made valuable revisions, edited the manuscript, and approved the final version.

References

Arias Aguirre A, Studer B, Frei U, Lübberstedt T (2012) Prospects for hybrid breeding in bioenergy grasses. Bioenergy Res 5:10–19

Balfourier F, Imbert C, Charmet G (2000) Evidence for phylogeographic structure in Lolium species related to the spread of agriculture in Europe. A cpDNA study. Theor Appl Genet 101(1):131–138

Bennett SJ, Hayward MD, Marshall DF (2000) Morphological differentiation in four species of the genus Lolium. Genet Resour Crop Evol 47(3):247–255

Barrett BA, Turner MA, Lyons TB, Rolston MP, Easton HS (2010) Evaluation of semi-hybrid perennial ryegrass populations. Proc N Z Grassland Assoc:11–16. https://doi.org/10.33584/jnzg.2010.72.2810

Betteridge K, Crush J, Ledgard S, Barton MS (2011) Nitrogen leaching implications of poor pasture persistence. NZGA Res Pract Ser 15:79–84

Beukes P, Babylon A, Griffiths W, Woodward S, Kalaugher E, Sood A, Chapman D (2021) Modelling perennial ryegrass (Lolium perenne) persistence and productivity for the Upper North Island under current and future climate. NZGA Res Pract Ser 17:297–306

Brazendale R, Bryant JR, Lambert MG, Holmes CW, Fraser TJ (2011) Pasture persistence: how much is it worth? NZGA Res Pract Ser 15:3–6

Bulinska-Radomska Z, Lester RN (1985) Relationships between five species of Lolium (Poaceae). Plant Syst Evol 148(3):169–175

Byrne SL, Nagy I, Pfeifer M, Armstead I, Swain S, Studer B et al (2015) A synteny-based draft genome sequence of the forage grass Lolium perenne. Plant J 84(4):816–826

Catalán P, Kellogg EA, Olmstead RG (1997) Phylogeny of Poaceae subfamily Pooideae based on chloroplast ndhF gene sequences. Mol Phylogenet Evol 8(2):150–166

Catalán P, Torrecilla P, Rodrıguez JAL, Olmstead RG (2004) Phylogeny of the festucoid grasses of subtribe Loliinae and allies (Poeae, Pooideae) inferred from ITS and trnL–F sequences. Mol Phylogenet Evol 31(2):517–541

Charmet G, Balfourier F (1994) Influence of ecological factors on population differentiation in perennial ryegrass (Lolium perenne L.). Genet Resour Crop Evol 41(3):175–184

Charmet G, Ravel C, Balfourier F (1997) Phylogenetic analysis in the Festuca-Lolium complex using molecular markers and ITS rDNA. Theor Appl Genet 94:1038–1046

Chaves B, De Vliegher A, Van Waes J, Carlier L, Marynissen B (2009) Change in agronomic performance of Lolium perenne and Lolium multiflorum varieties in the past 40 years based on data from Belgian VCU trials. Plant Breed 128(6):680–690

Cheng L, Zhang N, Huang B (2016a). Effects of 1-aminocyclopropane-1-carboxylate-deaminase–producing bacteria on perennial ryegrass growth and physiological responses to salinity stress. J Amer Soc Hort Sci 141(3):233–241

Cheng Y, Zhou K, Humphreys MW, Harper JA, Ma X, Zhang X, Huang L (2016b) Phylogenetic relationships in the Festuca-Lolium complex (Loliinae; Poaceae): new insights from chloroplast sequences. Front Ecol Evol 4:89

Cheres MT, Miller JF, Crane JM, Knapp SJ (2000) Genetic distance as a predictor of heterosis and hybrid performance within and between heterotic groups in sunflower. Theor Appl Genet 100:889–894

Cresswell A, Sackville Hamilton NR, Roy AK, Viegas BMF (2001) Use of amplified fragment length polymorphism markers to assess genetic diversity of Lolium species from Portugal. Mol Ecol 10(1):229–241

Clark DA, Caradus JR, Monaghan RM, Sharp P, Thorrold BS (2007) Issues and options for future dairy farming in New Zealand. N Z J Agric Res 50(2):203–221

Do Canto J, Studer B, Frei U, Lübberstedt T (2018) Fine mapping a self-fertility locus in perennial ryegrass. Theor Appl Genet 131:817–827

Duvick DN (1999) Heterosis: feeding people and protecting natural resources. In: Genetics and exploitation of heterosis in crops. American Society of America, Madison, pp 19–29

Gammage PA, Moraes CT, Minczuk M (2018) Mitochondrial genome engineering: the revolution may not be CRISPR-Ized. Trends Genet 34(2):101–110

Gaut BS, Tredway LP, Kubik C, Gaut RL, Meyer W (2000) Phylogenetic relationships and genetic diversity among members of the Festuca-Lolium complex (Poaceae) based on ITS sequence data. Plant Syst Evol 224(1):33–53

Guan X, Yuyama N, Stewart A, Ding C, Xu N, Kiyoshi T, Cai H (2017) Genetic diversity and structure of Lolium species surveyed on nuclear simple sequence repeat and cytoplasmic markers. Front Plant Sci 8:584

Hand ML, Cogan NO, Stewart AV, Forster JW (2010) Evolutionary history of tall fescue morphotypes inferred from molecular phylogenetics of the Lolium-Festuca species complex. BMC Evol Biol 10:1–17

Hayes BJ, Cogan NO, Pembleton LW, Goddard ME, Wang J, Spangenberg GC, Forster JW (2013) Prospects for genomic selection in forage plant species. Plant Breed 132(2):133–143

Hulke BS, Watkins E, Wyse DL, Ehlke NJ (2008) Freezing tolerance of selected perennial ryegrass (Lolium perenne L.) accessions and its association with field winterhardiness and turf traits. Euphytica 163:131–141

Humphreys MO (1997) The contribution of conventional plant breeding to forage crop improvement. In: Proceedings of the 18th International Grassland Congress', Winnipeg and Saskatoon, Canada, pp 8–17

Humphreys MO (2005) Genetic improvement of forage crops–past, present and future. J Agric Sci 143(6):441–448

Humphreys MW, Yadav RS, Cairns AJ, Turner LB, Humphreys J, Skøt L (2006) A changing climate for grassland research. New Phytol 169(1):9–26

Humphreys M, Feuerstein ULF, Vandewalle M, Baert J (2009) Ryegrasses. In: Fodder crops and amenity grasses. Springer New York, New York, pp 211–260

Jensen JG, Popay AJ, Tapper BA (2009) Argentine stem weevil adults are affected by meadow fescue endophyte and its loline alkaloids. N Z Plant Prot 62:12–18

Karn E, Jasieniuk M (2017) Genetic diversity and structure of Lolium perenne ssp. multiflorum in California vineyards and orchards indicate potential for spread of herbicide resistance via gene flow. Evol Appl 10(6):616–629

Katova A, Vulchinkov J (2019) Variability of morphological characters of collection accessions of perennial ryegrass (Lolium perenne L.). Bulg J Agric Sci 25:1015–1023

Kelly S, Smith E, Brazendale R (2011) Pasture renewal in the Waikato and Bay of Plenty regions: An overview of farmer practice, experience and attitudes. Pasture Persistence–Grassland Research and Practice Series, 15:21–24

Lee J, Thom E, Waugh D, Bell N, McNeill M, Wilson D, Chapman D (2018) Trajectory and causes of decline in the botanical composition of dairy-grazed pasture in the Waikato. J N Z Grassl 79:89–96

Li Q, Zhang D, Chen M, Liang W, Wei J, Qi Y, Yuan Z (2016) Development of japonica photosensitive genic male sterile rice lines by editing carbon starved anther using CRISPR/Cas9. J Genet Genomics Yi Chuan Xue Bao 43(6):415–419

Loera-Sánchez M, Studer B, Kölliker R (2019) DNA-based assessment of genetic diversity in grassland plant species: challenges, approaches, and applications. Agronomy 9(12):881

Loos BP (1993) Morphological variation in Dutch perennial ryegrass (Lolium perenne L.) populations, in relation to environmental factors. Euphytica 74(1):97–107

McCahon K, McCahon A, Ussher G (2021) Diversified pastures at the front line of climate change in Northland: farmers experiences, new directions and wider implications for other parts of the country. NZGA Res Pract Ser 17:213–224

McClain EF (1982) Effects of generation and method of synthesis on performance of orchardgrass cultivars. Crop Sci 22(6):1129–1133

McDonagh J, O'Donovan M, McEvoy M, Gilliland TJ (2016) Genetic gain in perennial ryegrass (*Lolium perenne*) varieties 1973 to 2013. Euphytica 212(2):187–199

Nie G, Huang T, Ma X, Huang L, Peng Y, Yan Y et al (2019) Genetic variability evaluation and cultivar identification of tetraploid annual ryegrass using SSR markers. PeerJ 7:e7742

O'Connor JR, Jahufer MZZ, Lyons T (2020) Examining perennial ryegrass (Lolium perenne L.) persistence through comparative genetic analyses of two cultivars after nine years in the field. Euphytica 216(3):36

Pearson A, Cogan NO, Baillie RC, Hand ML, Bandaranayake CK, Erb S et al (2011) Identification of QTLs for morphological traits influencing waterlogging tolerance in perennial ryegrass (Lolium perenne L.). Theor Appl Genet 122:609–622

Pembleton LW, Shinozuka H, Wang J, Spangenberg GC, Forster JW, Cogan NO (2015) Design of an F1 hybrid breeding strategy for ryegrasses based on selection of self-incompatibility locus-specific alleles. Front Plant Sci 6:764

Posselt UK (1993) Hybrid production in Lolium perenne based on incompatibility. Euphytica 71:29–33

Sanderson K, Webster M (2009) Economic analysis of the value of pasture to the New Zealand economy. Report to Pasture Renewal Charitable Trust. Wellington: BERL, 42

Sarıkaya MF, Tatar M (2023) Global climate change and crop production. Uluslararası Sivas Bilim ve Teknoloji Üniversitesi Dergisi 2(2):65–78

Stewart AV (2006) Genetic origins of perennial ryegrass (Lolium perenne) for New Zealand pastures. pp. 11–20. In: Proceedings of the 13th Australasian Plant Breeding Conference

Takayama S, Isogai A (2005) Self-incompatibility in plants. Annu Rev Plant Biol 56(1):467–489

Terrell EE (1968) A taxonomic revision of the genus Lolium (No. 1392). Agricultural Research Service, US Department of Agriculture

Thorogood D, Yates S, Manzanares C, Skot L, Hegarty M, Blackmore T et al (2017) A novel multivariate approach to phenotyping and association mapping of multi-locus gametophytic self-incompatibility reveals S, Z, and other loci in a perennial ryegrass (Poaceae) population. Front Plant Sci 8:1331

Tozer K, Cameron C, Thom E (2011) Pasture persistence: farmer observations and field measurements. NZGA: Research and Practice Series 15:25–30

Vieira EA, Castro CM, Oliveira ACD, Carvalho FIFD, Zimmer PD, Martins LF (2004) Genetic structure of annual ryegrass (Lolium multiflorum) populations estimated by RAPD. Sci Agric 61:407–413

Wall AM, Goodrich JP, Schipper LA (2021) Importance of pasture resilience for New Zealand's agricultural carbon stocks. In: Resilient pastures symposium. Grasslands research and practice series, vol 17, pp 190–200. New Zealand Grassland Association (NZGA), Waikato, New Zealand

Wilkins PW, Humphreys MO (2003) Progress in breeding perennial forage grasses for temperate agriculture. J Agric Sci 140(2):129–150

Wu Y, Fox TW, Trimnell MR, Wang L, Xu RJ, Cigan AM et al (2016) Development of a novel recessive genetic male sterility system for hybrid seed production in maize and other cross-pollinating crops. Plant Biotechnol J 14(3):1046–1054

Yamada T (2001) Introduction of a self-compatible gene of Lolium temulentum L. to perennial ryegrass (*Lolium perenne* L.) for the purpose of the production of inbred lines of perennial ryegrass. Euphytica 122(2):213–217

Ye M, Peng Z, Tang D, Yang Z, Li D, Xu Y et al (2018) Generation of self-compatible diploid potato by knockout of S-RNase. Nat Plants 4(9):651–654

Yeken MZ, Tekin M, Ali A, Altaf MT, Qureshi SA, Çelik A et al (2025) Advancements in lentil breeding: harnessing molecular markers and omics approaches for resistance to biotic and abiotic stresses. Crop Pasture Sci 76(11):CP24350

Yuan LongPing YL (2014) Development of hybrid rice to ensure food security

Zhang C, Fei SZ, Warnke S, Li L, Hannapel D (2009) Transcriptome profiling in perennial ryegrass during cold acclimation by using expressed sequence tag analysis. J Plant Physiol 166:1436–1445

Zhang J, Li H, Jiang Y, Li H, Zhang Z, Xu Z et al (2020) Natural variation of physiological traits, molecular markers, and chlorophyll catabolic genes associated with heat tolerance in perennial ryegrass accessions. BMC Plant Biol 20:1–17

Future Perspectives of Forage Breeding for Sustainable Agriculture

Sadia Zafar, Javed Iqbal, Muhammad Arslan Ashraf, Rizwan Rasheed, Banzeer Ahsan Abbasi, Inam Mehdi Khan, Rehman Iqbal, Shoaib Khan, Shumaila Ijaz, Zakir Ullah, Ghulam Murtaza, Rashid Iqbal, Hassan Sher, Zahid Ullah, and Ahmad Ali

Abstract

Forage breeding plays a crucial role in advancing sustainable agriculture by addressing key challenges related to climate change, food security, and environmental degradation. This discussion highlights the potential of forage breeding,

S. Zafar (✉) · I. M. Khan · R. Iqbal
Department of Botany, Division of Science and Technology, University of Education Lahore, Punjab, Pakistan
e-mail: sadia.zafar@ue.edu.pk

J. Iqbal (✉)
Department of Botany, Bacha Khan University, Charsadda, Khyber Pakhtunkhwa, Pakistan

M. A. Ashraf · R. Rasheed
Department of Botany, Government College University, Faisalabad, Pakistan
e-mail: marslanashraf@gcuf.edu.pk; drrizwan@gcuf.edu.pk

B. A. Abbasi
Department of Botany, Rawalpindi Women University, Rawalpindi, Pakistan

S. Khan
Department of Chemistry, Abbottabad University of Science and Technology (AUST), Abbottabad, Pakistan

S. Ijaz
Department of Burn and Plastic Surgery, Shenzhen Institute of Translational Medicine, The First Affiliated Hospital of Shenzhen University, School of Biomedical Engineering, Shenzhen University Medical School, Shenzhen University, Shenzhen, China

Z. Ullah
Department of Plant Sciences, Faculty of Biological Sciences, Quaid-i-Azam University, Islamabad, Pakistan
e-mail: zakirullah@bs.qau.edu.pk

F. S. Baloch et al. (eds.), *Forage Crops in the Omics Era*,
https://doi.org/10.1007/978-981-95-9249-4_17

focusing on the synergy between pioneering technologies and sustainable agricultural practices. The development of forage crops is undergoing a significant transformation with the adoption of advanced tools like genomic selection, CRISPR gene editing, and high-throughput phenotyping. These technologies empower breeders to develop crop varieties that are more resilient to environmental stresses, have enhanced nutritional value, and exhibit greater resistance to diseases and pests. Moreover, integrating forages into agroecological practices such as crop rotation, agroforestry, and cover cropping presents considerable opportunities to improve soil health, biodiversity, and ecosystem services. The role of socioeconomic and regulatory frameworks in facilitating the adoption of sustainable forage systems is crucial, as is the application of digital technologies and precision agriculture in optimizing forage management. Furthermore, global cooperation and the transfer of technology are essential for spreading the benefits of forage innovations worldwide. The future of forage breeding is poised to make a significant contribution to sustainable agriculture by bolstering food security, enhancing rural livelihoods, and fostering environmental sustainability through the combination of technological advances, sustainable practices, and comprehensive socioeconomic strategies.

Keywords

Forage breeding · Sustainable agriculture · Genomic selection · CRISPR gene editing · Agroecology · Climate resilience

1 Forage

Forage crops are utilized to feed about 1.52 billion cattle, 1.21 billion sheep, 1.02 billion goats, and 0.21 billion buffalo worldwide in order to support the production of various dairy products, meat, and wool, as well as to give farmers, especially those in developing nations, a stable income and important ecosystem services (FAOSTAT 2020). Approximately 0.60 billion resource-poor smallholder farmers depend on livestock commodities for their livelihood, contributing roughly $1.4 trillion to the global economy and employing over 1.3 billion people (Lee 2018). The performance of the livestock sector is largely dependent on the availability of

G. Murtaza
School of Agriculture, Yunnan University, Kunming, Yunnan, China

R. Iqbal
Department of Agronomy, Faculty of Agriculture and Environment, The Islamia University of Bahawalpur, Bahawalpur, Pakistan

Department of Life Sciences, Western Caspian University, Baku, Azerbaijan
e-mail: rashid.iqbal@iub.edu.pk

H. Sher · Z. Ullah · A. Ali
Center for Plant Sciences and Biodiversity, University of Swat, Charbagh Swat, Pakistan

high-quality feed throughout the year (Maheswari et al. 2024). In contrast to improved forage supply, disease and pest resistance, breeding for fodder quality traits is considered an essential secondary activity. All things considered, the nutritional quality aspects of forage are undervalued and are not seen to be as important as forage biomass in influencing market prices (Chand et al. 2022). High levels of digestibility, non-structural carbohydrates, crude protein, mild tannins, elevated palatability, high sulphur amino acid, suitable minerals, and low levels of anti-nutritional factor can be found in high-quality fodder and feed. But the obvious truth is that cattle intake rate and foraging velocity would be determined by how palatably the forage crop was produced.

2 Importance of Forage in Sustainable Agriculture

Forage is important in sustainable agriculture for a number of reasons.

- *Soil health and erosion control*:

  The deep root systems of forage crops, such legumes and grasses, aid in stabilizing the soil and reducing erosion caused by water and wind.

  Additionally, these root systems maintain the soil's organic matter content and structure, which raises soil fertility and improves water-holding capacity (Deveci et al. 2016).

  A combination of grasses and legumes improves soil structure more than either one alone (Howard 2020; Aasim et al. 2011). The scientists discovered that the older, stronger root fibres in smooth brome, crested wheatgrass, and thin wheatgrass were more effective in enhancing structure than the slender, easily decomposing roots. Smooth brome and crested wheatgrass produced about twice as many root fibres with around double the tensil strength in 3–4 years as slender wheatgrass. There are numerous reports of higher yields when farmed crops are rotated with forages. The magnitude of production disparities observed when no manure or fertilizer is given to different rotations is demonstrated by data from Ohio (Suresh Kumar 2023).

  The Morrow plots in Urbana, Illinois, where corn was planted year after year for 55 years starting in 1888, offer another illustration of the role that forages play in corn rotation. Without the addition of manure, residues, or mineral nutrients, continuous maize averaged 1819 kg/ha. Under the same conditions and timeframe, corn yielded an average of approximately 11 kg/ha in a corn–oats–clover rotation (Bochalya et al. 2021).

- *Nutrition of cattle*:

  In the northeast and central regions, the establishment and advancement of animal husbandry were facilitated by the enduring, invasive types of feed grasses and legumes. On the other hand, a comparable rise in animal production in the southern states was hampered by a number of problems, including a dearth of high-quality fodder resources. The Great Plains' original grasses were used by domestic cattle, which led to a rapid increase in the country's massive livestock output. Numerous changes have occurred in the widespread use of natural and tame grasslands for the production of ruminant livestock, many of which are linked to increases in per capita disposable income and population expansion. The production and consumption of feedstuffs have gradually changed as a result of the increasing demand for animal protein and the corresponding need for more efficient animal production methods. A larger focus was on substituting high-energy grain crops for forages in order to meet the energy needs of dairy cows with high yields and ensure quick liveweight growth in feedlots. This shift was brought about in part by technological advancements that simplified the production and application of grain in animal feeding operations, in addition to the seeming difficulty of maintaining the quality and output of forages throughout the growing season.

3 Pest Control

Numerous accounts of forages suppressing weeds can be found in the scientific literature. Entz et al. (2002) included multiple research in their review study that showed a significant decrease in weed formation in rotations with fodder crops compared to those without. In one study, the amount of grain produced in forage-containing rotations comprising wild oats (Avena fatua L.) was less than 1%, but in continuous wheat or wheat-fallow systems, the amount was 15%. On-farm observations corroborate research indicating that adding forages into grain crop rotations suppresses weeds. According to a survey conducted by Entz et al. (1995), a majority of over 80% of commercial grain and forage growers in Manitoba and Saskatchewan reported that weed growth decreased after forage crops. In that same poll, 33% of participants said that the benefits of forages in suppressing weeds persisted for longer than 2 years after the growth of forages was stopped. Among the weeds that were controlled were perennials like Canada thistle and wild oat, as well as annuals like green foxtail. Forages have an effect on weeds and other pests as well. A 3-year hay crop incorporated among multiple rotations was the sole reason that Entz et al. (2002) found that common root rot was suppressed in another investigation.

4 Forage Quality

The degree to which a forage can elicit an animal response that the animal finds attractive can be used to evaluate its quality. The following are some factors that affect the quality of the forage.

- *Palatability*

  Taste, feel, and smell are the factors that influence an animal's choice of forage. Therefore, factors that affect texture, leafiness, fertilization, dung or pee patches, moisture level, insect infestation, or substances that give a forage a sweet, sour, or salty flavour may all affect how appetizing it is. Superior grade forages are usually quite tasty (Ayyadurai et al. 2013). The amount of fodder that is absorbed as it moves through an animal's digestive system is known as digestibility, and it varies widely. Compared to 80%–90% of juvenile, leafy plant tissues, less than 50% of mature, stemmy material is digested (Kaithwas et al. 2020).

- *Nutrient Content*

  Will the forage supply enough nutrients after it has been digested? Typically, 70%–90% of living forage plants are made of water. In order to standardize analyses, feed production and nutrient content are frequently expressed on a dry matter (DM) basis. Dry matter found in feed may be divided into two primary groups: (1) cell contents, which include non-structural plant tissue components including protein, sugar, and starch and (2) structural cell wall components, which include cellulose, hemicellulose, and lignin (Moore et al. 2022).

- *Anti-quality Factors*

  Forage may include a diversity of substances that weaken an animal's performance, make them ill, or even kill them. These substances include mycotoxins, cyanoglycosides, alkaloids, nitrates, and tannins (Katoch 2023). The plant species (including weeds) present, the season, the surrounding environment, and the sensitivity of the animals all affect the presence and/or intensity of these factors. Anti-quality components in dangerous concentrations cannot be found in high-quality forages (Katoch 2022).

- *Animal Performance*

  The best indicator of the quality of feed is animal performance, particularly when feed is given to animals freely and without restriction. The term "nutritive value" refers to the potential for providing nutrients, such as digestibility and nutrient content, as well as the quantity that animals would eat and any anti-quality elements that may be present. Numerous elements related to the plants or the animals themselves might have an effect on an animal's performance. Any of these aspects that are not properly taken into account may lower an animal's performance level, which lowers potential revenue (Dickson 2022).

5 Factors Affecting Forage Quality

Forage quality is affected by several things. The kind of fodder, the maturity level at harvest, and the methods for gathering and storing fodder are the three most important variables. Temperatures during forage development, diversity, and soil quality and fertilization are examples of secondary influences (Moore et al. 2022).

- *Species Differences*

Grass versus Legumes

Legumes typically produce feed that is of higher quality than grasses. This is due to the fact that legumes frequently have higher protein contents and lower fibre contents than grasses. An important advantage of cultivating legumes alongside grasses is better fodder quality. Figure 2 compares alfalfa and timothy from a mixed stand's second cut to show typical species variations in quality. When alfalfa was in its early bloom, its crude protein (CP) content was 16%, whereas Timothy's was 9.5%. On the other hand, grasses may achieve CP levels similar to legume feed if they get significant nitrogen fertilizer applications. Timothy exhibited much greater amounts of neutral detergent fibre (NDF) in the same comparison as alfalfa. When comparing grass forages to legumes, greater NDF (total fibre) levels and a slower rate of fibre (cell wall) digestion usually translate into lower voluntary intake. Increased forage consumption leads to an increase in nutrients due to faster digestion (Šidlauskaitė et al. 2022).

Cool-Season versus Warm-Season Grasses

The quality of fodder varies greatly among the grasses that are grown for feed in the United States. The two main types of forage grasses are warm season (tropical or subtropical locations) and cool season (temperate regions). Tall fescue, perennial and annual ryegrass, Kentucky bluegrass, and orchardgrass are examples of cool-season grasses. Warm-season grasses include dallisgrass, bahiagrass, maize, and Bermudagrass. Generally speaking, warm-season grasses are of lower quality than cool-season varieties. Compared to warm-season grasses, cool-season grass species have an average digestibility that is almost 9% greater. In addition, warm-season grasses have lower crude protein levels than cool-season grasses. Annual grasses are often of superior grade than perennials within each group. Warm-season grasses convert sunlight into food more effectively than cool-season grasses because of changes in leaf architecture (tissue arrangement or structure); nonetheless, the larger percentage of highly lignified, less digestible tissues in their leaves is a characteristic of warm-season grasses (Niyigena et al. 2021).

- *Temperature*

  High-temperature plants often provide lower quality fodder than lower temperature plants, while cool-season species are those that thrive best in the winter months. But, if grown in a warm area as opposed to a cold one, fodder of any species usually has a lesser quality. In one research, annual ryegrass, for instance, grew to 59% leaf material when cultivated at 50° to 59 °F, but only 36% leaf matter when grown at 68° to 77 °F (Melo et al. 2022).

- *Maturity Stage*

  The maturation stage of a species is the main factor that determines the quality of its feed upon harvest. As forage gets more mature, its quality declines. Cool-season grasses, for instance, frequently exhibit dry matter (DM) digestibilities greater than 80% in the first 2–3 weeks of growth in the spring. After then, digestibility starts to decrease by 1%–3% units every day until it falls below 50%. Animals intake of fodder is also influenced by maturity during harvest (Newman et al. 2024).

  Plant growth and increased fibrousness cause a sharp decline in feed consumption. DM digestibility and ingestion values for cool-season grass has gathered at several stages of maturity. Similar effects in a wide range of animals have been reported in other investigations. As plants become older, their intake capability declines, and their NDF concentration rises. This is because digesting non-fibre components of fodder is easier than digesting non-fibre components. Additionally, when plants become older, the rate at which fibre is digested decreases. As fodder matures, digestion therefore slows down significantly (Stokes and Prostko 2024).

6 Forage Nutritional Profiles

Forage legumes often perform better than grasses in terms of nutritional content. Temperature is an essential environmental factor that determines the digestibility of fodder crops because it influences the concentration of dry matter yield, mineral ash, and digestibility in those crops.

Sanz-Sáez et al. (2012) found that the rate of digestibility decline in temperate regions was higher (0.5 unit/day) compared to tropical environments (0.1 unit/day). In a manner similar to this, the growth habits of forage crops—tall versus dwarf and rapid versus slow—also impact the feed's quality by prioritising protection and architectural supports over leaf development (Lee 2018). As a result, the microenvironment and patterns of plant development greatly impact the nutritional value and calibre of fodder crops. Proteins, carbohydrates, and fats/amount define the nutritional status of forage crops. These carbon-based nutrients give animals energy in the form of vitamins and minerals, and the composition of each crop determines how digestible it is (Capstaff and Miller 2018). Additionally, plant secondary metabolites such as tannins and phenols affect feed digestibility and livestock productivity in addition to actively participating in defence against biotic and abiotic stressors (Zhang et al. 2020).

- *Carbohydrates*: Since carbohydrates make up the majority of the dry matter yield (DMY) of grasseous forage crops (between 50% and 80%), their reduced presence in leguminous forage feeds may be offset by providing the livestock with an extra source of grains or grasses (Hatfield and Kalscheur 2020). Forage crops often include three different types of carbohydrates: water-soluble, water-insoluble, and intermediate (Watts 2009). Both microbial digestion and herbivore breaking of the glycosidic link convert these carbohydrates into simple sugars. Higher lignin concentrations in fodder crops decrease digestibility via interacting with cell wall proteins and structural carbohydrates, according to Ruckle et al. (2017). According to Hatfield and Kalscheur (2020), carbohydrates are thus essential for animal health, reproduction, development, behaviour, and the improved quality of animal products like milk and meat.
- *Proteins*: RuBisCO is the most prevalent protein in plants and a major source of nitrogen (Jonker and Yu 2017). The type of the crop, the climate, the time of harvest, and management techniques all affect crude protein levels (Habermann et al. 2019). Studies by Aboagye et al. (2018) and Christensen et al. (2017) show that protein concentration also affects methane emission, microbial population, ruminant buffering capacity, nitrogen intake, and lowers the risk of acidosis. Fruits and vegetables are browned by enzymes called polyphenol oxidases

(PPOs) that contain copper. By delaying the breakdown of fatty acids and nitrogen in the rumen and silo, they also help fodder crops.

- After consuming feed, ruminants' high protein breakdown of forage crops causes them to excrete an excessive amount of nitrogen, which is a major source of diffuse pollution. But during the rumen fermentation and ensilage process, quinones interact with forage proteins to form phenol–protein complexes that are resistant to proteolytic enzymes (Parveen et al. 2010). According to reports, feeding ruminants red clover enhances the efficiency of N-use and the quality of the protein via the complexing of leaf protein and the creation of protein-bound phenols (Lee 2014). Moreover, feeding animals red clover silage boosted the accumulation of healthy polyunsaturated fatty acids (PUFA) in animal byproducts and decreased lipolysis (Lee et al. 2009). Plant proteases and lipases are deactivated by PPOs, protecting plant protein and glycerol-based polyunsaturated fatty acids. PPO catalyses the deactivation process by converting diphenols into quinones (Bittner 2006).

- *Lipids*: Seed and pollen are important sources of lipid buildup in plants and are the main sources of energy during germination (Mukhtiar et al. 2016). The primary cause of leaf lipid deposition, the cell membrane makes up around 3.5% of the lipid in dry matter (Barrett et al. 2015). PUFAs make up the majority of lipids in forage crops, ranging from 10 to 30 g/kg. Of these, α-linolenic acid, which accounts for 62% of total fat, is the most prevalent kind. The ratio and makeup of dietary lipids determine the quality of fodder, which in turn influences the quality of animal products. ω-3 (omega-3) fatty acid is naturally found in plants, particularly fodder crops, and it has a direct impact on the quality of meat (Wright 2013).
- *Minerals and vitamins*: Forage provides vital minerals and vitamins that support animal health, vitality, and reproduction while also enhancing the nutritional value of the finished product. They regulate immune system functions, skin tone, coat quality, body metabolism, and reproduction; they are required in moderation. For this reason, the bio-fortification of fodder crops is a result of the current dynamic movement in scientific study towards micronutrients (Zhu et al. 2009). In addition to being necessary micronutrients, zinc (Zn), iron (Fe), and selenium (Se) also serve as precursors for vital enzyme functions that improve animal digestibility. For instance, a modest amount of zinc and selenium tends to boost the immunity and health of cattle; on the other hand, a large amount might be hazardous, waste resources, and put an extra financial strain on livestock breeders. Livestock absorb vitamins via the ingestion of fodder and feed, which includes both the water-soluble and fat-soluble (A, D, E, and K) vitamins. The body of the calf may retain the fat-soluble material for use at a later time. Moreover, pro-vitamins, or precursors of vitamins, including carotene, are found in plant tissues that are soluble in fat. The quantity of carotene in immature forage leaves is greater than in mature leaves, but exposure to sunshine and high temperatures quickly degrades the carotene (Ren et al. 2020).

- *Plant secondary metabolites*: According to Yang et al. (2018), plant secondary metabolites (SMs) are essential for defense against environmental stressors and pathogenic diseases. The influence of SMs depends on their chemical composition, the amount of food they contain, and how much the animal eats (Acamovic and Brooker 2005). The primary SM compounds present in fodder crops are tannins and phenols; these compounds impact animal productivity and feed digestibility (Hassanpour et al. 2011). Improving the range of crop appropriateness for animal development and production, as well as the nutritional content of their by-products, is thus a crucial feature of fodder crops. By interacting with the amino acids in the protein, the elevated pro-anthocyanidin or CTs lessening the digestibility of the protein. Additionally, CTs induce a total loss of nutritional content by either minimizing or inhibiting absorption in the animal's digestive tract (Yusiati et al. 2018).

7 Forage Breeding

A subfield of agricultural science known as "forage breeding" is devoted to the enhancement and development of plants cultivated particularly for animal feed. Enhancing these plants' resilience, quality, yield, and nutritional content is the aim of fodder breeding in order to promote productive and sustainable livestock production.

Plant breeding is evolution driven by humans (Chand et al. 2022). A population of plants makes up a pasture or hay field. Individual plants of the same fodder species can differ significantly in traits when they are cross-pollinated (such as alfalfa, ryegrass, and switchgrass); this is usually not the case for self-pollinated (lespedeza), vegetatively propagated (bermudagrass), and apomictic (dallisgrass) species. Depending on their environment, breeding facility display distinct phenotypes (Casler and Vogel 2020).

The equation can be used to characterize each phenotype (P), which is the genetic expression of a genotype (G) as influenced by its environment (E):

$$P = G + E + GxE(\text{Interaction Effect})$$

In diverse conditions, such different field sites, genetically identical plants, like those of vegetatively generated cultivars of bermudagrass, might vary in size and other traits.

- *Conventional and Advanced Breeding Approaches*

Crop genetic advancement is fundamentally driven by plant genetic diversity. The amount consumed and the feed's capacity to provide cattle with calories, protein, minerals, and vitamins determine any feed's practical value (Ludemann et al. 2015). Grazing tests, both in vitro and in vivo, are useful tools for choosing better crop species according to their yield and nutritional value. A few key agronomic traits that are linked to distinct gene(s), apomixes, self-incompatibility, low to high levels of polyploidy, high genotypic and phenotypic heterogeneity at the individual level, and other factors make breeding programmes for forage crop improvement difficult and challenging (Rauf et al. 2021).

- *Reduction of ANFs*

  The palatability of the forage crop may be impacted by anti-nutritional factors including plant secondary chemicals (Barry et al. 2001). There is a lot of genetic heterogeneity in ANF fodder crops. To improve feedstuff palatability, breeding variety for high digestible protein and low ANF content is necessary. According to Aouida et al. (2019), there is variability in the amounts of cyanogenic glucoside and cyano-amino acids in cultivars and wild accessions of common vetch.

  ANF is a protease inhibitor found in grass pea that influences protein digestibility and is in charge of monogastric animals' slower development. Grela et al. (2010) found that ODAP and tannin concentration were inversely correlated and that tannin content and blossom colour intensity were positively correlated in 31 European grass pea accessions. White flowers are low in tannin content, while coloured blossoms frequently contain significant levels of tannin. However, the concentration of ODAP in germplasm accessions from the Indian subcontinent was higher (0.70%–2.40%) than in native germplasm from Eastern Asian countries (0.02%–1.20%). The ODAP concentration of grass pea is influenced by both qualitative and quantitative factors (Dixit et al. 2016).

  According to El-Moneim et al. (2000), a number of the grass pea varieties that have been released in India, such as Mahateora (RLS 4595), Prateek (LS 157-14), Ratan (BioL-212), Pusa 24, and LSD1, have excellent yields (1.5 t/ha) and are relatively free of ODAP (0.70%–2.40%) when compared to the native germplasm from Eastern Asian countries (0.02%–1.20%). According to Dixit et al. (2016), the ODAP concentration of grass pea is influenced by both quantitative and qualitative factors. There are several released varieties of grass peas in India with good yields (Dixit et al. 2016).

- *Improving Metabolizable Energy*

  According to Barrett et al. (2015), fodder grasses that produce foliar lipids may increase the energy supply, whereas forage legumes that produce CTs can decrease the rate of protein breakdown. Compared to fodder crops with high DMD and low NDF values, those with low DMD and high NDF resulted to ruminant fullness and decreased metabolizable energy, resulting in animal feeding restriction (Smit et al. 2005). In fodder crops, research by Tas et al. (2006) revealed an inverse association between NDF and leaf blade fructans (also known as water-soluble carbohydrates, or WSC) concentration, but a positive link with CP concentration. In transgenic events compared with FLp 418–20 over the spring and summer, these dietary modifications boosted ME to 1.7 MJ ME/kg of dry matter (kg DM). According to Badenhorst et al. (2018), this was noted in perennial ryegrass.

  However, the genetic variety of pre-existing germplasm for leaf lipid characteristics is low. Forage crops' leaf lipid concentration (7%–8% of DM) must thus be doubled in breeding techniques in order to boost metabolic energy by 10%, which will improve milk fat content (Barrett et al. 2015). In order to reduce photorespiration and maximize energy, a unique protein known as cysteine oleosin (lipid encapsulation protein) was produced by using a synthetic gene (Winichayakul et al. 2022). This protein was constitutively co-expressed with a gene that encodes diacylglycerol acyl-transferase (DGAT) (Chand et al. 2022). This was done using a rice chlorophyll a/b-binding protein and a Rubisco gene-derived green tissue-specific promoter. They observed a 25% increase in biomass in the leaves and roots. According to Bryant et al. (2008), lipid characteristics in forage leaves have the potential to lower animal feed consumption without lowering milk fat content or output, while also increasing farm productivity via effective pasture use. A rise in the concentration of ME in fodder crops, such as perennial ryegrass, was observed to have positive impacts on cattle productivity. The Base 1 M scenario illustrates how productivity per cow would change without impacting the total number of cows if the ME concentration of pasture or forage grass were arbitrarily enhanced by 1 MJ ME/kg DM. Nonetheless, an increase in ME of 0.05 MJ ME/Kg DM of perennial ryegrass can boost farm operation profit by 5% by raising cattle productivity in comparison to the Base 1 M scenario (Ludemann et al. 2015).

Genomics selection (GS) reduces the breeding cycle, maximizes the potential use of low heritable traits, and eventually speeds up the pace of genetic gain in fodder crops by utilising molecular markers to increase selection effectiveness (Kumar et al. 2021). Creating homozygous mutations in polyploids, especially in switchgrass and other forage crops, is a promising biotechnological use of genome editing (Liang et al. 2014).

- *Bioengineering for Metabolite Biosynthetic Pathways*

  Increasing fructan metabolism from sucrose would improve digestibility because fructans are the most major WSC (Chalmers et al. 2005). For example, it has been demonstrated that perennial ryegrass that expresses fructosyl transferase to change fructan biosynthesis can increase ME by up to 1.7 MJ ME/kg DM, decrease NDF concentration in transgenic lines, and improve WSC and in vivo DMD (Badenhorst et al. 2018). Increased digestibility and modifications to the cell wall structure, like adding various lignin monomers and phenylpropanoid derivatives with multiple ester-linked catechols or pyrogallol units, are more strongly correlated, claim Grabber et al. (2019). Additionally, resistant starch benefits human diets by offering more fibre; yet, since it stays undigested in cattle feed for non-ruminants, it is detrimental to animal feed. Therefore, reduced resistant starch will increase the fodder crops' digestibility for animals, including ruminants and non-ruminants (Raigond et al. 2015). As a result, improving digestibility and reducing lignin slightly while sparing biomass and yield potential have been demonstrated by targeting the lignin biosynthesis genes (Halpin 2019). The tall fescue's cloned cDNA sequence, discovered by Chen et al. (2004), encodes a crucial enzyme for lignin biosynthesis—CAD. This transgenic line expresses sense or anti-sense CAD gene constructs, which lower the production of mRNA transcripts and the enzymatic activity of CAD.

  There was 20% and 12% less lignin in the stem and leaves of the T 1 transgenic plant, which had downregulated OMT. This increases the digestibility of the stem by 7% and the leaves by 2%. The cinnamate 4-hydroxylase (C4H) gene was downregulated in lucerne by Reddy et al. (2005) using anti-sense technology. By using the thio-acidolysis method, this resulted in a seven-fold decrease in lignin concentration but no appreciable changes in lignin composition. Interestingly, according to Barros et al. (2019), lucerne is the first genetically modified fodder crop to be commercialized with a lower lignin content and increased digestibility for fodder.

- *Improving Nutrient Contents*
  When the maize SPS gene is overexpressed in lucerne, nodule counts, biomass, and total nitrogenase activity rise, leading to elevated amounts of protein and elemental N (Gebril et al. 2015). More bacterial genes have been cloned and introduced into lucerne, including those for aspartate kinase (AK) and adenylyl sulphate reductase (APR). Aspartate is phosphorylated into asp-phosphate by AK, while sulphur is reduced by APR to form amino acids and other important metabolites. Sulfite is produced by APR from activated sulphate. Tong et al. (2014) reported that these transgenic lines outperformed the wild-type lucerne in cysteine and methionine by 30% and 60%, respectively. According to Hörtensteiner (2009), downregulating the STAY-GREEN (SGR) genes, which encode chloroplast proteins, is thought to be the initial step in destabilising a chloroplast protein complex. Compared to the wild type, lucerne with a downregulated SGR gene (MsSGR) using RNA interference (RNAi) showed increased chlorophyll during senescence, more greenish growth, and improved chlorophyll content (Yang et al. 2018). The nutritional value of fodder crops may therefore be significantly increased by transgenic methods, which incorporate data from food crops.

Current Challenges in Forage Breeding

In comparison to major oilseed or grain crops, pasture crops exhibit significantly lower rates of yield increase through genetic means (Charles Brummer and Casler 2014). This is caused by a number of factors, such as significant genotype-cropping condition and crop utilization interactions, low narrow-sense heritability, longer selection cycles, less opportunities for capitalizing the harvest index, and an outbreeding mating system linked to severe inbreeding depression. Undoubtedly, a significant contributing element is the lack of breeding: in the United States, for instance, several hundred breeders are dedicated to the breeding of maize alone, whereas only a small number of breeders concentrate on forage crops (often including many species). According to psychologists, scarcity tends to alter our thoughts and behaviours, highlighting the necessity of sharpening attention and increasing work productivity (Mullainathan and Shafir 2013).

Species Prioritization

It is astounding how many different species and kinds of legumes and fodder grasses have been produced and used in the EU. Additionally, there is a great deal of variation in the breeding methods of fodder crops. Few forage breeders focus solely on breeding one species; some are now working with over five. Experience has shown that good variants have resulted from both concentrating on a single species and breeding multiple species, which may seem strange at first. A higher species abundance would be expected to compromise the success of breeding efforts for a particular species by implying a smaller scope for such programmes. Experience demonstrates that the breeder's craft partially supersedes scientific theory. The size of the potential seed market for a species' variations is typically connected with breeding efforts; hence, investing in expensive biotechnologies might make sense if one focused on a small number of important crops with a sizable market (Bouton 2007).

In order to create native species cultivars that are more resistant of rainfed farming in dry, hot temperatures than introduced species, diversification of bred species may also become desirable in several US locations (Muir et al. 2005). Although it is debatable whether multi-species mixtures can also demonstrate greater resistance to climatic stresses, their capacity to out-yield pure stands or binary mixtures in a variety of settings may support continued attention to multiple bred species (Van Peer et al. 2004).

It is expected that the shift towards more sustainable crop-livestock systems would lead to a wider production of forage legumes. Replace intensively managed grasses with legume-grass mixtures that receive little to no nitrogen fertilization, or grow these legumes as pure stand, high-protein crops. Breeders' perspective on which species to prioritize will undoubtedly be affected by this. For instance, over the past 10 years, financed programmes and research on crop development at Lodi, Italy, have increasingly focused on legumes (both annual and perennial), as opposed to forage grasses. When compared to other legumes, lucerne (lucerne) can yield more feed protein per unit area (Huyghe 2003).

Matching Selection and Target Environments

The available genetic variation (sA, standard deviation of breeding values), the intensity of selection (i.e., standardized selection differential), the square root of the trait's narrow-sense heritability (h), and the genetic correlation (rg) for the population's (i.e., half-sib family's) response to selection and target environments are the variables that Falconer and Mackay (1996) states affect genetic gain in the target agricultural environments (ΔG for one selection cycle).

$$\Delta G = \mathrm{i\,h\,rg\,sA}$$

Forage breeders find it extremely difficult to get high rg values, particularly for qualities like yield, for two reasons. First, the high costs of experiments resulting from lengthy agricultural cycles and several harvests impede the adoption of selection environments (e.g., swards) that may more closely approximate the target habitats. Second, GE interaction significantly affects the breeding value of forage crop ecotypes and derived material, which provide the essential genetic basis for many breeding programmes, as illustrated by Turesson's (1922) seminal work on cocksfoot. Inadequate breeding progress might result from using a genetic basis that is highly suited to the target surroundings or from selection not matching target conditions (whose detrimental effect increases with distance from unity of rg).

Issues and Plans Concerning Breeding with Diversity

The primary perennial fodder species are susceptible to severe inbreeding depression and exhibit high allogamous behaviour. Red clover, meadow fescue, and ryegrass are examples of diploid species that can be intentionally polyploidized. Unlike most other crops, a large range of potential cultivar types and breeding plants result from these and other factors. Because forage plants are outbreeding and may be found widely in natural or semi-natural grasslands, it is relatively simple for breeders to purposefully introduce foreign germplasm into the breeding population (Sustek-Sánchez et al. 2023).

Making an informed decision can be aided by comprehensive passport data and detailed characterization of genetic resources kept ex situ. Using natural selection can be just as successful in maintaining ecotype populations during in situ maintenance. Many breeding projects have likely neglected the potential value of landraces or other ecotype material for habitats and conditions similar to their evolutionary history as genetic resources. The forages yield responses of landraces of red clover (Boller et al. 2010), white clover (Annicchiarico 1993), lucerne (Annicchiarico 2006), and wild populations of ryegrass (Boller et al. 2009) were equivalent to those of the highest yielding modern kinds. Nonetheless, research that permits an ecological comprehension of the unique adaption pattern of landraces or wild populations is the only way to fully use the diversity of adaptable genes present in these populations (Annicchiarico and Piano 2005). A crucial element in the comprehensive utilization of genetic resources is the appropriate concentration of screening processes on intra-population diversity, which primarily surpasses the between-population variance in terms of forage and seed yield attributes (Julier et al. 2000).

Seed Yielding Ability and Market Exploitation

Boelt and Studer (2010) found that optimizing seed yield potential through efficient utilization is a better strategy for boosting seed output without sacrificing the quality of the feed. However, it seems that markers haven't made a significant impact on the evolution of any forage cultivars thus far. One could wonder why this latency is happening and what can be done about it. The differences between forage crops and the main grains and oilseed crops undoubtedly contribute to the delay in the direct application of markers in breeding programmes. Fixing QTL is harder for cultivars than for inbred lines because most fodder crops are synthetic populations. Until recently, the marker systems used in forage programmes were limited to a small number of markers that were difficult to verify and cost a considerable sum of money when used in a forage breeding programme. Forage yield and other essential variables are genetically complex, therefore in order to discover many significant QTL (generally with relatively minor effects), a thorough genome scan using large marker numbers is typically required. Furthermore, phenotyping the majority of significant features is costly and/or time-consuming and necessitates appropriate GE interaction management. Last but not least, selecting the mapping population is crucial since limited-based material can yield results that are not broadly applicable to the true genetic basis. This is one of the reasons why breeding programmes have not made use of published, effective markers. However, due to shorter linkage disequilibrium, broad-based material makes QTL detection more difficult (Annicchiarico et al. 2015a).

However, the terrain of markers is changing quickly. Single nucleotide polymorphism (SNP)-based high-density marker arrays and sequencing-based marker techniques have been developed and applied to lucerne (Li et al. 2014) and forage grasses (Fè et al. 2014). Numerous SNP markers (approximately 7000–11,000 in several lucerne mapping populations: Annicchiarico et al. 2015b) are now available at a very low cost per data point (Elshire et al. 2011) because of genotyping-by-sequencing technology. Marker assessment will be applied to forage programmes as soon as it is cost-effective. Nevertheless, the majority of those studies were not conducted in a way that would have made it easy to apply the results to the production of synthetic cultivars. Translating QTL data for selection is challenging since mapping populations are typically not descended from or closely related to breeding populations. This raises questions about the QTL-marker phase and the magnitude of the QTL effect in populations with distinct genetic backgrounds. The effective implementation of a significant QTL-based selection project to increase the yield of white clover seeds serves as one example of an exception (Barrett et al. 2009). This finding implies that, provided strong marker–trait correlations are found and a simple marker assay is created, additional direct applications of markers for selection ought to be feasible. Long-term advances of genome sequencing initiatives (like the ongoing lucerne project) and thorough gene annotation should lead to further progress in this approach. As a result, populations would become fixed for these alleles, or at the very least, nearly every member of the synthetic population would display resistance (Brummer and Monteros 2014).

Positive outcomes have also been obtained for lucerne candidate parents' breeding values prediction (Annicchiarico et al. 2015b). This is particularly significant for replacing the laborious stage of parent selection based on half-sib progeny yield responses. Perennial ryegrass is being explored for another strategy that links characteristics to SNP allele frequencies of full-sib populations (Byrne et al. 2013). The third essential precondition is comprehensive phenotyping in order to collect high-quality data on a fairly sized population. If models that cover plant material from diverse genetic pools can be built, it may be desirable to launch a coordinated worldwide study aimed at gathering data across multiple distinct cropping conditions and pooling resources for marker development. Although it's unclear how widely prediction models can be used, their high cost suggests that collaboration across groups should be encouraged. The creation of genomic selection models that are tailored to particular target environments with varying types and amounts of notable stress, or alternatively, appropriate for breeding for broad adaptation to disparate environments is another potential benefit of this perspective, which is currently being investigated on lucerne in the ERA-NET project REFORMA (Annicchiarico et al. 2015a).

Historical Context and Achievements in Forage Breeding

With a long history, forage breeding has produced several noteworthy successes that have benefited agriculture tremendously. Cattle have been fed forages for countless years. Legumes like alfalfa and clover as well as grasses like timothy were among the earliest species to be domesticated (Nelson et al. 2017). Selection was based on outside characteristics such as palatability and yield prior to the application of formal scientific processes. Organized fodder breeding was made possible by the application of Mendelian genetics to plant breeding in the late nineteenth and early twentieth centuries (Charnley 2011). Early breeders became aware of the benefits of hybrid vigour and the potential for selective breeding to raise fodder quality and production (Ijaz et al. 2021).

8 Emerging Technologies in Forage Breeding

The area of fodder breeding is changing as a result of emerging technology, becoming more productive and efficient. With the use of these tools, breeders may create fodder kinds more quickly and precisely with enhanced features. Important new technologies consist of marker-assisted selection and genomic selection. Genomic Selection (GS) selects the finest individuals without requiring lengthy field testing by using genome-wide markers to forecast breeding candidates' performance.

Promotes faster breeding by permitting selection on the basis of genetic potential at an early stage (Sinha et al. 2023). Marker-assisted selection (MAS) uses certain DNA markers associated with desired characteristics to direct the selection procedure (Sun et al. 2024). Enhancing breeding programmes' efficacy by focusing on certain genes linked to qualities, yield, and disease resistance (Hasan et al. 2021). Details about an integrated approach for achieving high quality forage yield are given in (Fig. 1).

High throughput phenotyping (HTP) quickly evaluates plant characteristics in the field by using cutting-edge imaging technology, including drones, satellites, and ground-based sensors. It offers comprehensive information on biomass, growth, stress reactions, and other significant characteristics (Feng et al. 2021). Near-infrared reflectance spectroscopy (NIRS) non-destructively assesses characteristics of fodder quality, including digestibility, fibre content, and protein content. Facilitates quick and precise evaluation of fodder quality, assisting in the choice of better cultivars (Yegon 2023).

Enables targeted gene alteration to add or improve desired features, such enhanced nutritional content or resistance to drought. Offers the chance to create fodder crops with qualities that are difficult to accomplish via traditional breeding (Sami et al. 2021). Involves grafting foreign genes into fodder crops in order to give new features, such as improved growth or insect resistance. Although transgenic technologies are often scrutinized by regulators, they may provide major improvements in fodder breeding (Chand et al. 2022). To find trends and forecast breeding results, it analyses sizable data sets created from genetic, phenotypic, and environmental data.

Improves breeding programme decision-making by offering insights into intricate trait interactions and environmental reactions (Morota et al. 2018). Provides a single platform for managing breeding initiatives by combining a variety of data

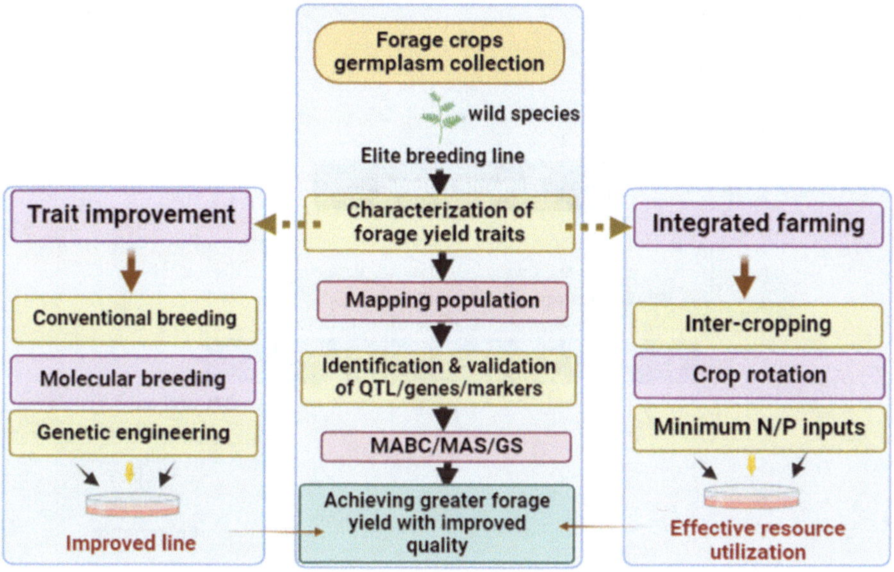

Fig. 1 An integrated approach for achieving high quality forage yield

sources and technologies. Makes data analysis, decision-making, and breeding progress monitoring easier (Gaur et al. 2019).

Emerging Technologies Advancing Forage

- *Flow Cytometry*

For the analysis of huge and complex grass genomes, flow cytometry (FC) has emerged as a revolutionary tool (Doležel et al. 2012). By breaking the genome down into individual chromosomes, it offers opportunity to simplify. Through flow sorting, FC makes it possible to purify many copies of the desired chromosomes. For a karyotype to be successful, each chromosome must be precisely separated from one another (Fig. 2). Given that chromosomes in plant species frequently have similar sizes and proportions of DNA, this is challenging. For example, in meadow fescue and ryegrass, just one of the seven chromosomes can be resolved (Kopecký et al. 2013), while only one of the 21 chromosomes in wheat can be recognized (Vrána et al. 2000). This limitation can be overcome by improving the tools and methods or by separating the target chromosomes from the remaining chromosomes in modified karyotypes (Doležel et al. 2012). The expanded usage of cytogenetic stocks, including deletions, translocations, alien chromosomes, and chromosomal arm additions, has substantially fuelled recent advances in chromosome flow sorting. Using such stocks in barley, wheat, rye, maize, and pea could split crop genomes into separate chromosomes (Suchankova et al. 2006). Forage and turf grasses, however, are limited in their ability to create such cytogenetic stocks due to the flexibility of their genomes, which includes a fairly lax chromosome pairing system and a

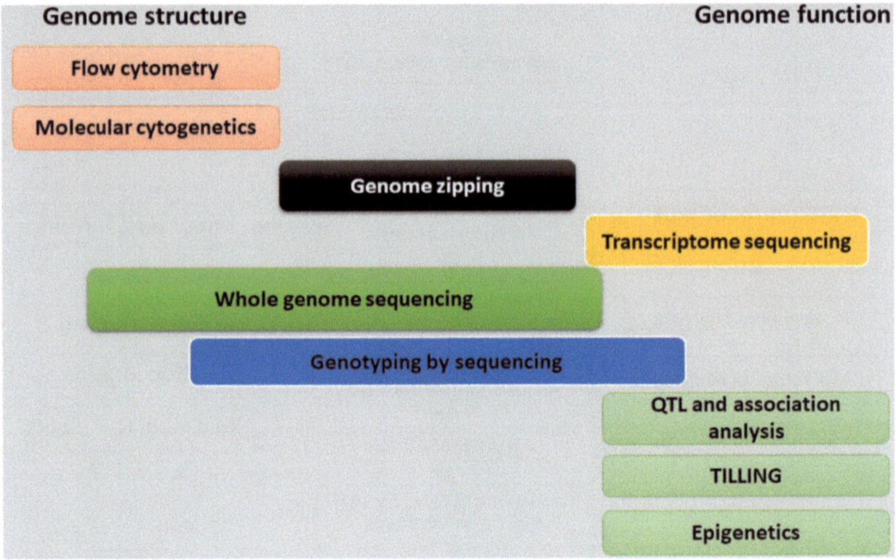

Fig. 2 Emerging technologies advancing forage and turf grass genomics

tolerance for mild aneuploidy. Therefore, using different methods is the only way to separate individual chromosomes from one another in grasses. The most promising involves sorting chromosomes according to the fluorescence signals of different DNA probes and using fluorescent in situ hybridization in suspension (FISHIS) (Giorgi et al. 2013).

GAA microsatellites have already been used with great effectiveness to sort all 14 chromosomes in durum wheat. For all animals with huge and complicated genomes, this represents a breakthrough. Approaches have been identified that can significantly accelerate and streamline physical mapping and genome assembly processes, as emphasized by Doležel et al. (2012). An alternative strategy involves the use of synthetic polyamide probes for chromosome labeling, which bind specifically to target DNA sequences without requiring prior denaturation of the DNA (Gygi et al. 2002). The use of flow-sorted chromosomes has several intriguing uses. In addition to being often utilized in clinical cytogenetics and supporting the physical mapping and sequencing of plant genomes, flow cytogenetics was involved in the early stages of the human genome sequencing project. Sorted chromosomes were a great template for physical mapping with fluorescence in situ hybridization (FISH) (Kubaláková et al. 2003), PCR (Kejnovský et al. 2001), and DNA hybridization (Arumuganathan et al. 1994).

- *Molecular Cytogenetics*

The first steps in determining the chromosomal level genome structure are cytogenetics and karyotyping (Fig. 1). The most recent cytogenetic methods, such as genomic in situ hybridization (GISH) and fluorescence in situ hybridization (FISH), offer great ways to physically map different DNA sequences directly onto chromosomes that have been mounted on a microscope slide. FISH has been used in plants to create physical maps by determining the orientation and order of contigs, estimating the size of contig gaps, and anchoring genetic linkage groups to certain chromosomes (Han et al. 2011). Additionally, it made it possible to track the genesis and development of karyotypes as well as conduct in-depth examinations of chromosomal structure (Mandakova and Lysak 2008). There are very few studies on the use of FISH to identify certain chromosomal sequences in fodder and turf grasses. Cytogenetic markers, such as ribosomal DNAs, are widely used in karyotyping. The differences in the quantity and positions of rDNA loci in plants have been the subject of extensive investigation. Variation in rDNA locus distribution within grass species was found by Książczyk et al. (2010). Harper et al. (2004) estimated parental lineages using rDNA on polyploid fescues. The use of rDNA for such investigations is limited due to the small number of loci and should only function as an indicator or support of earlier obtained results. Tandem organized repeats and anonymous BAC clones were identified on multiple chromosomes in earlier research (Kopecký et al. 2013). This made it possible to thoroughly karyotype ryegrass and fescue species, and it would be useful to use FISHIS to flow sort individual chromosomes in these species. Given its ability to physically map particular DNA sequences, such as cDNAs found by NGS and tandem repeats unique to a species, FISH presents a wealth of possibilities for use in grass genome research. By doing so, these sequences will be arranged in a thorough manner along the chromosomes, making it possible to accurately karyotype prospective breeding lines and cytogenetic stocks (Lewis 2021).

- *Transcriptome Sequencing*

  Even with NGS technology, accessing the DNA sequence of the huge, complex genomes of allogamous turf grass and forage species that are extremely heterozygous is challenging. Transcriptome sequencing, in which only expressed genes are targeted and highly repetitive, non-transcribed genomic areas are eliminated, is an effective method for reducing complexity (Grabherr et al. 2011). Nevertheless, because outbreeding crop species have high degrees of polymorphism, there are many pseudogenes and multigene families, and turf grass and feed species frequently exhibit polyploidy, de novo assembly of short transcriptome reads remains a significant difficulty. Strategies to avoid fragmentation have been developed using de novo assembly methods based on the de Bruijn graph (Ruttink et al. 2013). The transcriptome is also extremely active and constantly evolving. As such, the image of the genes expressed in a specific tissue at a particular developmental stage that can be obtained from mRNA sequencing may be incomplete. Therefore, it could be necessary to merge sequencing libraries from different tissue types that were obtained at different times in order to determine the entire transcriptome. Maher et al. (2009) have shown that transcriptome sequencing is a potent tool for characterizing variations in gene expression, identifying exon/intron boundaries, and researching alternative splicing and post-transcriptional alterations. Entire RNA sequencing (RNA-seq) offers the chance to more precisely characterize several RNA populations, including micro RNAs, transfer RNAs, and ribosomal RNAs, in addition to mRNA transcripts. RNA-seq has shown to be a helpful technique for the high-throughput detection of gene-associated single nucleotide polymorphisms (SNPs) in pasture and turf grasses. According to Ruttink et al. (2013), these SNPs are crucial for genetic research and breeding techniques that use genomics.

Thus far, efforts to measure the transcriptome of turf and fodder grasses depend on estimates of gene content derived from conserved synteny to sequenced Poaceae species. Zhang et al. (2012) estimate that 72%–87% of grass genes are collinear, and the remaining 13%–28% are species-specific genes. All non-collinear genes, however, share traits with pseudogenes (Wicker et al. 2011). An estimated 33,000 genes make up the meadow fescue genome, based on chromosome 4F sequencing work utilizing the GenomeZipper technique (Kopecký et al. 2013). This is akin to the 32,000 genes that were formerly believed to comprise the whole genome of barley (Mayer et al. 2011). But 30,400 genes were estimated to be present in the barley genome based on the most current sequence assembly (Mayer et al. 2012). In contrast, estimates of the number of genes in the wheat A-genome (around 28,000) fall into a similar range (Hernandez et al. 2012).

- *"Zipping" Chromosomes and Whole Forage and Turf Grass Genomes*

The Poaceae genomes' remarkable degree of preserved synteny and the increasing quantity of entire genome sequences that are becoming accessible (International Rice Genome sequencing Project 2005; Mayer et al. 2012) and the development of increasingly extensive transcriptome data for turf grass and fodder (Ruttink et al. 2013) all contributed to the creation of genome zippers. When sequencing Poaceae genomes, a technique known as "genome zipping" is employed to locate syntenic areas and organize syntenic blocks on a genetic linkage map. Also, it addresses species-specific local rearrangements and produces an ordered gene model at the best resolution possible by utilizing the genomic data from sequenced Poaceae genomes. Nevertheless, when different reference genomes are merged side by side, chances present themselves to overcome constraints brought about by species-specific structural rearrangements, gene loss, or translocations. According to Pfeifer et al. (2013), regional synteny breakdown is a major reason why single genome approaches to comparative genomics frequently fail. The orientation and gene order of syntic areas have been successfully determined using these synteny-based gene models for both individual flow-sorted chromosomes (Hernandez et al. 2012) and complete genomes (Mayer et al. 2011). According to Pfeifer et al. (2013), perennial ryegrass showed the example of "genome zipping" in forage and turf grasses, with about 32% of the original gene set anchored in silico in the genome. Combining a finer resolution of the marker scaffold with more comprehensive transcriptome data, as demonstrated by Ruttink et al. (2013), may enable further advances. Since a related percentage of genes can be secure and between 72% and 87% of grass genomes are collinear (Zhang et al. 2012), in silico gene location and order prediction using the GenomeZipper may be able to replace genetic linkage mapping for genes in conserved genome regions. As demonstrated by the perennial ryegrass GenomeZipper (Fig. 1) (Duhoux and Delye 2013) methods that elucidate genome structure can also enhance tools that explain genome function.

Furthermore, it is very challenging to address gene order information from centromeric areas via recombination-based linkage mapping since recombination is inhibited close to genetic centromeres (Huo et al. 2011). The creation of a comprehensive reference genome sequence will also benefit from the GenomeZipper. Gene-rich sections of the genome can be positioned by effectively determining the orientation and order of the genomic sequence in conserved genome regions. Single flow-sorted chromosomal sequencing can further increase this technology's resolution (Kopecký et al. 2013). The "genome zipping" notion will come in handy for many more poorly characterized species of turf grass and fodder in the future. For example, "genome zipping" for Timothy, Orchard grass, or Bluegrass will speed up the process of fine-mapping and map-based cloning of genes or QTL of interest by facilitating the in silico estimation of the genome location of unmapped genes and enabling the production of gene-based markers in specific target regions. According to Pfeifer et al. (2013), the technology used to produce GenomeZippers may be more readily transferred to other species, such fescues, for which a comparatively greater number of DNA sequence resources are available. NGS-based transcriptome sequencing and a marker scaffold connecting assembled genes to syntenic genome blocks will enable the rapid production of high-resolution genome drafts in the future. Recall that "genome zipping" is restricted to conserved genome segments; intriguing behavioural traits exclusive to a species may be found in the estimated 13%–28% of non-syntenic regions. Genotyping turf grass and fodder species via sequencing Genotyping by sequencing (GBS) is an additional rapid and dependable technique for reduced-representation sequencing of multiplexed materials that combines genotyping and the detection of genome-wide molecular markers (Poland and Rife 2012). GBS pools up to 384 individuals on a single sequencing lane by utilizing NGS's capacity to extract enormous amounts of DNA sequencing reads from the ends of genomic restriction fragments (Elshire et al. 2011). Methylation-sensitive restriction enzymes can be used to mostly avoid repetitive genomic areas, which increases SNP calling accuracy and speeds up bioinformatics analysis (Gore et al. 2009). High-density genome-wide marker profiles can be produced with GBS since it is inexpensive per data point and versatile in terms of the percentage of missing values to be sequenced and the required amount of markers to be sequenced (Beissinger et al. 2013).

- *Whole Genome Sequencing*

Establishing whole genome sequences is the final step towards unlocking the genomes of turf grass and fodder (Fig. 1). It is possible to access the entire gene set, their regulatory regions, and the variety in DNA sequence that occurs inside a genome with a high-quality reference genome sequence. Technological advances of great magnitude have expedited the whole genome sequencing of several grass species, thereby enabling the reading of extensive and intricate genomes of turf grass and fodder (Mayer et al. 2012; Zhang et al. 2012). The International Rice Genome Sequencing Project (2005) has made this possible with the conclusion of the rice genome sequencing project. Shotgun sequencing methods still result in severely fragmented assemblies, nevertheless, because grass genomes are large and highly repetitive, often combined with polyploidy. To produce a high-quality genome draft quickly, it may be necessary to combine physical mapping with whole genome shotgun sequencing of individual flow sorted chromosomes, as done, for instance, in wheat. Furthermore, in order to prevent repetitive elements and duplications during de novo assembly, the sequencing method's design must have different insert sizes for mate pair and paired-end libraries. With the potential to greatly improve the assembly process, third-generation sequencing technology allowed for a read length of several (ten) thousand bp. Currently, de novo assembly is being greatly improved by extended sequencing reads from Pacific Biosciences' SMRT technology, but at the expense of a higher sequencing error rate (Nagarajan and Pop 2013). However, errors can be corrected with a correction algorithm and short, high-fidelity sequencing reads, yielding long single-molecule sequencing reads with up to 99.9% base-call accuracy. Completing genome sequences at a faster rate and with greater precision is expected to be possible with such hybrid assembly procedures, which combine various DNA sequencing technologies.

Genomic DNA is broken up into 10-kbp pieces using a method known as "Moleculo Technology," which is then clonally amplified, sheared, barcode-labelled, and sequenced using Illumina technology. Each 10 kb fragment's resulting short sequencing reads are combined independently, allowing for the full reconstruction of every fragment based on short read data. This considerably enables traversing vast repeat areas during the assembly of de novo genome sequencing or it permits the combining of many lengthy reads into a single haplotype. In addition to encouraging the sequencing of more genomes in many species, the combination of better bioinformatics pipelines, more reasonably priced computing power, and higher sequencing throughput at reduced costs will also make it possible to re-sequence complete genomes or genomic subsets of multiple individuals within species or populations (Huang et al. 2013). This makes it possible to identify millions of SNPs and variation patterns across the genome, even with only a 5 to 20× raw data coverage. For example, comparing the gene diversity of cultivars and ecotypes will enable the identification of possible regions chosen during domestication through the use of such high-density genome-wide marker profiles. These "selective sweeps" may be linked to significant biological characteristics that help characterize the genomes of turf grass and fodder. According to Schneider and Dekker (2012) and Eid et al. (2009), with the advancement of third-generation sequencing technology, it will be feasible to sequence a single intact DNA strand effectively and in real time through the use of protein nanopores or polymerase base extension. The amplification-induced bias is eliminated in this way. The ability to sequence the original template strand directly, as opposed to an amplification product, would eventually, considerably enhance single-contig chromosomal assembly and long-range haplotyping, improving the resolution to discover rare DNA mutations.

- *QTL and Genes Controlling Important Traits*

There are an increasing number of publications about QTL analysis, genetic linkage maps, and DNA markers in species of fodder and turf grass. Moreover, systemic trends exist that will advance our understanding of the genome's functioning

(Fig. 1). The integration of marker and QTL data from mapping populations of the same species, or even other species, is made possible by the increasing usage of DNA markers in germplasm (King et al. 2013). Shinozuka et al. (2012) released the first large-scale QTL meta-analysis based on this technique employing an in silico comparison mapping, which proved to be a very promising method for identifying potential cross-species genes. In the future, as more QTL studies become available, QTL meta-analysis will likely become more crucial for integrating and processing QTL data effectively. Long-term QTL research has mostly concentrated on a small number of monogenic and polygenic features, primarily in ryegrasses and fescues. Leading dates and vernalization responses traits related to seed yield (Wang et al. 2011); resistance to abiotic stresses (Bartos et al. 2011); and tolerance to diseases like crown rust, bacterial wilt, dollar spot, or grey leaf (Dracatos et al. 2008) are some of the most well-known examples. But more and more QTL investigations are being conducted for a wide range of variables relevant to morphology, physiology, and tolerance/resistance (Shinozuka et al. 2012). The majority of current efforts to find marker–trait relationships rely on analysis of populations with two parents. Although biparental designs offer a high statistical power for detecting QTLs, the genetic resolution of the QTLs they discover is modest. Furthermore, in a biparental population, the maximum number of distinct alleles that are segregating is four. On the other hand, QTL that include an infinite number of distinct alleles can be found in breeding material using association mapping techniques. According to Nordborg and Weigel (2008), association studies make use of historical recombinations that preserve linkage disequilibrium (LD) between a polymorphic marker and a particular phenotypic trait. The connection is dependent on the degree of LD in the population structure and is statistically difficult to find, despite the very high genetic resolution of these designs (Auzanneau et al. 2007). According to Asp et al. (2011), the early successful reports of association mapping in forage and turf grasses necessitated candidate genes for either biologically or genetically well-defined functions, such as lignin biosynthesis or oligosaccharide metabolism, or for traits like vernalization and flowering time. Yu et al. (2013) reported that complex stress-related features, such as drought tolerance, have been identified by candidate gene-based association mapping.

- *Genome-wide Association Studies*

  Establishing high-resolution marker profiles and applying GWAS will be made possible by recently developed genotyping pipelines, which can be either sequencing- or DNA array-based (Byrne et al. 2013; Altaf et al. 2024). The pipelines will function in tandem with extensive and whole genome SNP resources from forthcoming genome (re)sequencing initiatives. Atwell et al. (2010) claim that GWAS has significantly increased our knowledge of how the genome functions and sheds light on the allelic architecture of complex traits in model plant species like rice and Arabidopsis. These ground-breaking investigations shed light on the optimal design of GWAS as well as the remaining difficulties, such as accurate association detection and understanding of the underlying molecular mechanisms. Gene regulatory networks, functional genome annotation, sequence-based GWAS, and meta-analysis of GWAS are common approaches to address these problems (Nuzhdin et al. 2012). There are a few important factors regarding GWAS in species of turf grass and fodder that require more thought. First, compared to inbreeding species, LD decays much more quickly in allogamous species (Brazauskas et al. 2013; Yamada et al. 2009a, b), necessitating a much greater marker density to capture the LD. Second, there may be additional opportunities and difficulties since the ideal experimental design may differ from the GWAS design in inbreeding species. For instance, precise phenotypic values can be attained by clonally propagating genotypes for comprehensive phenotyping in different conditions. On the other hand, an established plant's genotype cannot be maintained or transmitted through seeds because of heterozygosity and self-incompatibility, and its phenotypic may vary with time. Likewise, populations with allogamous associations might require new statistical models that effectively take into account the structural dependence, which is crucial for the statistical power and resolution of GWAS (Korte et al. 2012). Last but not least, GWAFFs offer the chance to link allele frequencies to characteristics like biomass or seed yield that are best accurately reported at the population level in field swards. This is essential for the effective use of potent genomics-assisted breeding principles, like genomic selection, in the current allogamous forage and turf grass species breeding programmes (Fig. 3).

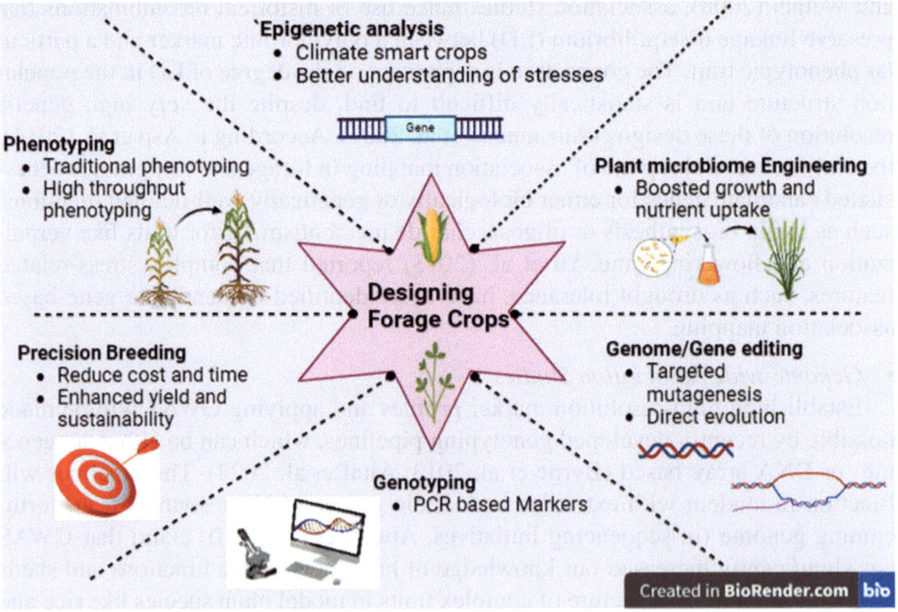

Fig. 3 Cutting-edge novel plant breeding techniques and inventive technology with the ability to solve the food security issues

- *Functional Genomics Resources*

Reverse genetics techniques to find and describe intriguing phenotypes resulting from alteration of a specific gene will surely be stimulated by the availability of whole genome sequences for a growing number of grass species and the capacity to link a genomic region to a particular trait (Jung et al. 2008). It is not easy to create functional genomics resources for turf grasses and fodder, nevertheless. Sequence-specific nucleases can be utilized in genetically modifiable animals to undergo particle bombardment or transformation mediated by Agrobacterium to cause desired gene alterations and double-strand breaks. Voytas (2013) states that three different forms of sequence-specific nucleases have been employed in plants thus far: modified homing endonucleases, or meganucleases; zinc-finger nucleases (ZFNs); and transcription activator-like effector nucleases (TALENs). By precisely aiming at the desired genes, these nucleases allow for site-specific genome editing. Gene silencing by RNA interference (RNAi) and artificial micro RNA (AMRNA) approaches is less successful than nuclease-based approaches, which often cause a partial and unstable reduction in gene expression over generations (Bedell et al. 2012). Other targeted gene inactivation methods, like transfer DNA (Alonso and Ecker 2006) or transposable elements (Krishnan et al. 2009), require hundreds of thousands of individuals to be close to saturation, have insertion biases, or rely on Agrobacterium-mediated transformation for small genomes like Arabidopsis. Assessing a gene's function in a species resistant to persistent genetic transformation can be done effectively via virus-induced gene silencing (Becker and Lange 2010).

Targeting generated local lesions in genomes is another possible strategy (TILLING). The utility of TILLING as a functional genomics methodology has been recognized by model plant species such as Arabidopsis (McCallum et al. 2000) and key crop species such as rice (Till et al. 2007), wheat (Slade et al. 2005), maize (Till et al. 2004), and barley (Mejlhede et al. 2006). It causes chemically induced mutations over the whole genome, with the frequency and density of the caused mutations varying according to the strength of the chemical treatment. It is possible to determine a clear correlation between a mutation and its related phenotype if the mutation occurs in a critical region of a certain gene. TILLING is a non-transgenic method that can target any gene and is suitable for fodder and turf grass species, requiring relatively modest population densities (Wang et al. 2012). Furthermore, a particularly effective technique for finding mutations is low-cost high-throughput sequencing of multidimensional DNA pools followed by SNP analysis (Tsai et al. 2013). One significant limitation, though, is that the induced mutation in the phenotypic cannot be expressed without self-pollination, therefore only self-compatible plant material can be employed.

- *Epigenetics*

Even while GWAS has been successful in identifying highly heritable qualities (Huang et al. 2013), the genetic variations that have been found have very minor effects on more complicated behaviours (Ingvarsson and Street 2011). In many instances, their combined influence only contributes to a minor portion of the noted phenotypic variance, prompting enquiries into the heredity that remains unaccounted for (Vineis and Pearce 2010). Future GWAS may need to focus on identifying associations whose effects are environment-dependent because epigenetics may play a significant role in the heritability that cannot be explained (Brachi et al. 2011). Since some of the most significant features, such vernalization, are partially epigenetically controlled, the study of epigenetics in forage and turf grass species has hardly been touched and is probably going to gain more traction (Pennisi 2013). Mapping the DNA methylome will be accelerated even further by sequencing-based technologies to examine whole genome histone modifications (ChiP-Seq), which will replace traditional microarray-based techniques (ChIP-chip). It is anticipated that this field of increasing attention will yield new findings and a better comprehension of the fundamental ideas of genomic imprinting (Fig. 3).

9 Initiatives for Open and Collaborative Breeding

Engages farmers and other interested parties in the breeding process to guarantee that novel fodder cultivars satisfy regional requirements and preferences. Promotes the uptake and effectiveness of enhanced fodder crops (Mekonnen et al. 2022). Encourages global researchers and breeders to share genetic resources, information, and instruments. Promotes cooperation between stakeholders to quicken innovation and the creation of new forage kinds (Ngunga et al. 2021).

10 Considerations in Breeding for Resilience and Adaptation

Resilience is the capacity of a system to tolerate perturbation and rebuild itself in the face of change while essentially preserving its identity, function, structure, and feedback loops. Resilience is frequently attributed to diversity, especially in the agricultural sector. An established ecological theory states that more species contributing to a given ecosystem function stabilizes the function. Diversity in and of itself may not increase resilience, but the range of reactions that exist within a single function may be very important when dealing with variation and change. Such response diversity is indicative of the potential to adapt to different contexts and perform effectively under a variety of situations. It's important to understand the intricate relationships that exist between plants and their surroundings. Breeding programmes need to take into account the soil types, local climate, and management techniques (Tracy et al. 2018). Complex breeding techniques and cutting-edge statistical methods are needed to balance the selection for different characteristics, such as yield, quality, and stress tolerance (Behera et al. 2024).

Including farmers in the breeding process rises the likelihood that new varieties will be accepted and suit real-world demands. A useful resource for understanding regional circumstances and preferences is participatory breeding (Lammerts van Bueren et al. 2018). In addition to taking soil health, carbon sequestration, and biodiversity into account, breeding for resilience should also take these larger ecosystem services into account (Rao et al. 2015). It is critical to create cultivars that can adjust to potential future climatic conditions. The qualities essential for long-term resilience may be identified with the use of climate modelling and predictive breeding (Snowdon et al. 2021).

11 Socio-economic and Environmental Impacts of Forages

Forages are vital to rural lives, sustainable agriculture, and the health of ecosystems (Rao et al. 2015). By creating jobs in planting, harvesting, and processing, forage farming promotes rural employment. Farmers may make money from forage crops, particularly in mixed agricultural systems where cattle play a significant role (Dhyani et al. 2007). Better forages promote the health and productivity of animals, which enhances the production of meat, milk, and wool. Growing forages on-site lowers production costs and boosts farm profitability by reducing dependency on bought feed (Chand et al. 2022). By boosting the sustainability and productivity of livestock systems, which provide necessary food items like milk and meat, forages contribute to food security. Including forages in diversified agricultural systems makes them more resistant to climatic and economic shocks, which helps to maintain stable food supply (Muir et al. 2017).

By including forages in crop rotations, farms may increase their resistance to weather and market variations while also diversifying their revenue sources and farm productivity. This reduces risk. When primary crops are damaged by unfavourable weather, forages may step in as a dependable supply of food (Dumont et al. 2022). The addition of organic matter to the soil via forages improves its water-holding ability, fertility, and structure. By fixing atmospheric nitrogen, legume-rich forages improve soil fertility and lessen the demand for artificial fertilizers (Silveira and Kohmann 2020).

The wide root systems of forage crops contribute to soil stabilization, which lessens soil erosion caused by water and wind. By continuously covering the soil, forages shield it from erosion and nutrient runoff (Horrocks et al. 2019). By enhancing soil infiltration and water retention, forages lessen surface runoff and boost groundwater recharge.

During dry spells, production may be sustained by deep-rooted forage species because they can get water from deeper soil layers (Palacios et al. 2008). By providing a range of species with habitats, fodder crops help to promote biodiversity on farms. Blooming forages aid in crop pollination and pest management by providing a home for pollinators and beneficial insects (Crotty et al. 2015). By storing carbon in their biomass and roots, forages help to slow down global warming by storing carbon in the soil. Forages contribute to a decrease in greenhouse gas emissions from agricultural systems by improving soil health and lowering the requirement for synthetic fertilizers (Sarkar et al. 2020). Various forage systems may disrupt the cycles of pests and diseases, eliminating the need for chemical pesticides. By offering homes to agricultural pests' natural predators, forages improve biological control (Murdoch et al. 2003). In areas with a shortage of arable land, the production of food crops must be balanced with the allotment of land for fodder crops. Sustainable grazing methods are crucial since excessive grazing may cause land degradation (Karlsson and Röös 2019). To maintain economic viability, farmers must have access to markets for livestock and forage products. To sustain forage-based systems, sufficient funding must be allocated to research, extension services, and infrastructure (Kumar and Faruqui 2009). Water-efficient species and methods are crucial since fodder crops may demand substantial amounts of water in some places. The necessity for optimum management techniques is highlighted by the fact that poorly managed forage systems may lead to nutrient runoff and water contamination (Duan et al. 2023).

12 Future Directions and Perspectives

With a number of new trends and initiatives positioned to strengthen the role of forages in sustainable agriculture, the future of forage research and development is bright. Making better and more accurate use of whole-genome information to select for complex characteristics (Simeão Resende et al. 2014). Using gene editing techniques to precisely introduce certain features, such enhanced nutritional content or disease resistance (Haq et al. 2022). The use of robots, sensors, and drones enables the rapid and precise evaluation of plant characteristics under field conditions. Furthermore, the integration of phenotypic and genomic data enhances the accuracy and efficiency of breeding and selection processes (Gebremedhin et al. 2019). Using big data analytics and digital technologies to make well-informed choices on breeding, usage, and forage management (Bestelmeyer et al. 2024). Using mobile technology, farmers may obtain real-time data and management guidance (Bestelmeyer et al. 2024). Using IoT devices and smart sensors to monitor and maximize plant health, soil moisture, and nutrient levels in forage production. To increase productivity and save labour expenses, automate processes like fertilization, harvesting, and watering (Boursianis et al. 2022).

13 Conclusion

The future of forage breeding is vital for advancing sustainable agriculture, particularly in addressing challenges like climate change, environmental degradation, and the growing global demand for food. Emerging technologies such as genomic selection, CRISPR gene editing, and high-throughput phenotyping are transforming the development of forage varieties. These innovations not only enhance the nutritional quality of forages but also improve their resilience to diseases, pests, and extreme weather conditions similar to heat and drought. Integrating forages into agroecological practices—such as cover cropping, crop rotation, and agroforestry—contributes to greater ecosystem resilience, increased biodiversity, and improved soil health. Precision agriculture and digital tools further optimize forage management and production. To ensure extensive adoption of these innovations, strong socioeconomic and policy support, including farmer engagement, capacity building, and market development, is essential. International collaboration in research and knowledge transfer is also vital to ensure that the benefits of forage breeding are shared globally. By combining technological advancements with sustainable practices and socioeconomic strategies, the future of forage breeding clutches the potential to significantly enhance food security, rural livelihoods, and environmental health.

Author Contributions S.Z., J.I., M.A.A., R.R., and B.A.A., designed and conceived the chapter idea. S.Z., J.I., M.A.A., R.I., wrote the original draft. S.K., S.I., Z.U., G.M., R.I., H.S., Z.U., and A.I., reviewed and edited the manuscript and provided valuable Insights. All authors have made valuable revisions and edited the manuscript and approved the final version.

References

Aasim M, Şahin Demirbağ N, Khawar KM, Kendir H, Özcan S (2011) Direct axillary shoot regeneration from the mature seed explant of the hairy vetch (Vicia villosa Roth). Arch Biol Sci 63(3):757–762

Aboagye IA, Oba M, Castillo AR, Koenig KM, Iwaasa AD, Beauchemin KA (2018) Effects of hydrolyzable tannin with or without condensed tannin on methane emissions, nitrogen use, and performance of beef cattle fed a high-forage diet. J Anim Sci 96(12):5276–5286

Acamovic T, Brooker JD (2005) Biochemistry of plant secondary metabolites and their effects in animals. Proceedings of the Nutrition Society 64(3):403–412

Alonso JM, Ecker JR (2006) Moving forward in reverse: genetic technologies to enable genome-wide phenomic screens in Arabidopsis. Nat Rev Genet 7(7):524–536

Altaf MT, Tatar M, Ali A, Liaqat W, Mortazvi P, Kayihan C et al (2024) Advancements in QTL mapping and GWAS application in plant improvement. Turk J Bot 48(7):376–426

Annicchiarico P (1993) Variation for dry matter yield, seed yield and other agronomic traits in Ladino white clover landraces and natural populations. Euphytica 71(1):131–141

Annicchiarico P (2006) Diversity, genetic structure, distinctness and agronomic value of Italian lucerne (Medicago sativa L.) landraces. Euphytica 148:269–282

Annicchiarico P, Piano E (2005) Use of artificial environments to reproduce and exploit genotype× location interaction for lucerne in northern Italy. Theor Appl Genet 110(2):219–227

Annicchiarico P, Barrett B, Brummer EC, Julier B, Marshall AH (2015a) Achievements and challenges in improving temperate perennial forage legumes. Crit Rev Plant Sci 34(1–3):327–380

Annicchiarico P, Nazzicari N, Li X, Wei Y, Pecetti L, Brummer EC (2015b) Accuracy of genomic selection for alfalfa biomass yield in different reference populations. BMC Genomics 16:1–13

Aouida M, Rook F, Bianca Maimann A, Sánchez-Pérez R, Abid G, Fauconnier ML, Jebara M (2019) Polymorphisms in cyanogenic glucoside and cyanoamino acid content in natural accessions of common vetch (Vicia sativa L.) and selection for improved agronomic performance. Plant Breeding 138(3):348–359

Arumuganathan K, Martin GB, Telenius H, Tanksley SD, Earle ED (1994) Chromosome 2-specific DNA clones from flow-sorted chromosomes of tomato. Mol Gen Genet MGG 242:551–558

Asp T, Byrne S, Gundlach H, Bruggmann R, Mayer KF, Andersen JR et al (2011) Comparative sequence analysis of VRN1 alleles of Lolium perenne with the co-linear regions in barley, wheat, and rice. Mol Gen Genomics 286:433–447

Atwell S, Huang YS, Vilhjálmsson BJ, Willems G, Horton M, Li Y et al (2010) Genome-wide association study of 107 phenotypes in Arabidopsis thaliana inbred lines. Nature 465(7298):627–631

Auzanneau J, Huyghe C, Julier B, Barre P (2007) Linkage disequilibrium in synthetic varieties of perennial ryegrass. Theor Appl Genet 115(6):837–847

Ayyadurai P, Priya RS, Jegathjothi N, Gokila J (2013) A review on optimizing forage quality through management. Int J Agric Sci Resour 3:173–180

Badenhorst PE, Panter S, Palanisamy R, Georges S, Smith KF, Mouradov A, Mason J, Spangenberg GC (2018) Molecular breeding of transgenic perennial ryegrass (*Lolium perenne* L.) with altered fructan biosynthesis through the expression of fructosyltransferases. Mol Breed 38(2):21

Barrett BA, Faville MJ, Nichols SN, Simpson WR, Bryan GT, Conner AJ (2009) Breaking through the feed barrier: Options for improving forage genetics. Crop Science 49(1):1–12

Bartoš M, Janeček Š, Klimešová J (2011) Effect of mowing and fertilization on biomass and carbohydrate reserves of Molinia caerulea at two organizational levels. Acta Oecologica 37(4):299–306

Barrett BA, Faville MJ, Nichols SN, Simpson WR, Bryan GT, Conner AJ (2015) Breaking through the feed barrier: options for improving forage genetics. Anim Prod Sci 55(7):883–892

Barros J, Temple S, Dixon RA (2019) Development and commercialization of reduced lignin alfalfa. Curr Opin Biotechnol 56:48–54

Barry TN, McNeill DM, McNabb WC (2001) Plant secondary compounds; their impact on forage nutritive value and upon animal production. In: 19th International Grassland Congress, Sao Pedro, Brazil. Fundacao Estudos Agrarios Luiz Queiroz (FEALQ), Piracicaba

Becker A, Lange M (2010) VIGS—genomics goes functional. Trends Plant Sci 15:1–4

Bedell VM, Wang Y, Campbell JM, Poshusta TL, Starker CG, Krug RG II et al (2012) In vivo genome editing using a high-efficiency TALEN system. Nature 491(7422):114–118

Behera PP, Singode A, Bhat BV, Ronda V, Borah N, Verma H, Sarma RN (2024) Genetic gains in forage sorghum for adaptive traits for non-conventional area through multi-trait-based stability selection methods. Front Plant Sci 15:1248663

Beissinger TM, Hirsch CN, Sekhon RS, Foerster JM, Johnson JM, Muttoni G et al (2013) Marker density and read depth for genotyping populations using genotyping-by-sequencing. Genetics 193(4):1073–1081

Bestelmeyer BT, McCord SE, Browning DM, Burkett LM, Elias E, Estell RE, Williamson J (2024) Fulfilling the promise of digital tools to build rangeland resilience. Front Ecol Environ 22(5):27–36

Bittner S (2006) When quinones meet amino acids: chemical, physical and biological consequences. Amino Acids 30(3):205–224. https://doi.org/10.1007/s00726-005-0298-2

Bochalya RS, Gupta AK, Mehta S, Kumar D, Menia M, Bhagat S et al (2021) Conservation agriculture: an option to mitigate the adverse effect of climate change: a review. Pharma Innov J 10(7):51–65

Boelt B, Studer B (2010) Breeding for grass seed yield. In: Fodder crops and amenity grasses, pp 161–174. https://doi.org/10.1007/978-1-4419-0760-8_7

Boller B, Peter-Schmid MK, Tresch E, Tanner P, Schubiger FX (2009) Ecotypes of Italian ryegrass from Swiss permanent grassland outperform current recommended cultivars. Euphytica 170:53–65

Boller B, Posselt UK, Veronesi F (2010) Fodder crops and amenity grasses, vol 5. Springer, New York

Boursianis AD, Papadopoulou MS, Diamantoulakis P, Liopa-Tsakalidi A, Barouchas P, Salahas G et al (2022) Internet of things (IoT) and agricultural unmanned aerial vehicles (UAVs) in smart farming: a comprehensive review. Internet Things 18:100187

Bouton J (2007) The economic benefits of forage improvement in the United States. Euphytica 154:263–270

Brachi B, Morris GP, Borevitz JO (2011) Genome-wide association studies in plants: the missing heritability is in the field. Genome Biol 12:1–8

Brazauskas G, Xing Y, Studer B, Schejbel B, Frei U, Berg PR, Lübberstedt T (2013) Identification of genomic loci associated with crown rust resistance in perennial ryegrass (Lolium perenne L.) divergently selected populations. Plant Sci 208:34–41

Brummer EC, Monteros MJ (2014) Marker-assisted selection. Legume Perspectives 4:17–18

Bryant J, Lopez-Villalobos N, Holmes C, Pryce J, Rossi J, Macdonald K (2008) Development and evaluation of a pastoral simulation model that predicts dairy cattle performance based on animal genotype and environmental sensitivity information. Agric Syst 97(1–2):13–25. https://doi.org/10.1016/j.agsy.2007.10.007

Byrne S, Czaban A, Studer B, Panitz F, Bendixen C, Asp T (2013) Genome wide allele frequency fingerprints (GWAFFs) of populations via genotyping by sequencing. PLoS One 8(3):e57438

Capstaff NM, Miller AJ (2018) Improving the yield and nutritional quality of forage crops. Front Plant Sci 9:535

Casler MD, Vogel KP (2020) Forage breeding. In: Forages: the science of grassland agriculture, vol 2. John Wiley & Sons, Ltd., pp 553–565. https://doi.org/10.1002/9781119436669.ch30

Chalmers J, Lidgett A, Cummings N, Cao Y, Forster J, Spangenberg G (2005) Molecular genetics of fructan metabolism in perennial ryegrass. Plant Biotechnol J 3(5):459–474. https://doi.org/10.1111/j.1467-7652.2005.00148.x

Chand S, Indu, Singhal RK, Govindasamy P (2022) Agronomical and breeding approaches to improve the nutritional status of forage crops for better livestock productivity. Grass Forage Sci 77(1):11–32

Charles Brummer E, Casler MD (2014) Cool-season forages. In: Yield gains in major US field crops, vol 33. John Wiley & Sons, Ltd., Oxford, pp 33–51

Charnley B (2011) Agricultural science, plant breeding and the emergence of a mendelian system in Britain, 1880–1930. University of Leeds

Chen L, Auh CK, Dowling P, Bell J, Lehmann D, Wang ZY (2004) Transgenic down-regulation of caffeic acid O-methyltransferase (COMT) led to improved digestibility in tall fescue (Festuca arundinacea). Funct Plant Biol 31(3):235–245. https://doi.org/10.1071/FP03254

Christensen RG, Eun JS, Yang SY, Min BR, MacAdam JW (2017) In vitro effects of birdsfoot trefoil (Lotus corniculatus L.) pasture on ruminal fermentation, microbial population, and methane production. Prof Anim Sci 33(4):451–460

Crotty FV, Fychan R, Scullion J, Sanderson R, Marley CL (2015) Assessing the impact of agricultural forage crops on soil biodiversity and abundance. Soil Biol Biochem 91:119–126

Deveci M, Aşçı ÖÖ, Demirkol G, Karakuş YS (2016) The importance of forage crops for sustainable agriculture. Int J Ecosyst Ecol Sci 6(3):341–346

Dhyani SK, Samra JS, Handa AK (2007) Forestry to support increased agricultural production: focus on employment generation and rural development. Agric Econ Res Rev 20(2):179–202

Dickson J (2022) Forage quality, animal performance and behaviour of bred beef heifers grazing stockpiled perennial and annual forages in the late fall/early winter in Manitoba. Winnipeg, MB, Canada: University of Manitoba

Dixit GP, Parihar AK, Bohra A, Singh NP (2016) Achievements and prospects of grass pea (Lathyrus sativus L.) improvement for sustainable food production. Crop J 4(5):407–416

Doležel J, Vrána J, Šafář J, Bartoš J, Kubaláková M, Šimková H (2012) Chromosomes in the flow to simplify genome analysis. Funct Integr Genomics 12:397–416

Dracatos PM, Cogan NOI, Dobrowolski MP, Sawbridge TI, Spangenberg GC, Smith KF, Forster JW (2008) Discovery and genetic mapping of single nucleotide polymorphisms in candidate genes for pathogen defence response in perennial ryegrass (Lolium perenne L.). Theor Appl Genet 117(2):203–219

Duan C, Yu C, Shi P, Huangqing D, Zhang X, Dai E (2023) Assessing trade-offs among productive, economic, and environmental indicators of forage systems in southern Tibetan crop-livestock integration. Sci Total Environ 876:162641

Duhoux A, Delye C (2013) Reference genes to study herbicide stress response in Lolium sp.: up-regulation of P450 genes in plants resistant to acetolactate-synthase inhibitors. PLoS One 8(5):e63576

Dumont B, Franca A, Lópezi GF, Mosnier C, Pauler CM (2022) Diversification increases the resilience of European grassland-based systems but is not a one-size-fits-all strategy. Grass Forage Sci 77(4):247–256

Eid J, Fehr A, Gray J, Luong K, Lyle J, Otto G et al (2009) Real-time DNA sequencing from single polymerase molecules. Science 323(5910):133–138

El-Moneim AMA, Van Dorrestein B, Baum M, & Mulugeta W (2000) Improving the nutritional quality and yield potential of grasspea (Lathyrus sativus L.). Food and Nutrition Bulletin 21(4):493–496

Elshire RJ, Glaubitz JC, Sun Q, Poland JA, Kawamoto K, Buckler ES, Mitchell SE (2011) A robust, simple genotyping-by-sequencing (GBS) approach for high diversity species. PLoS One 6(5):e19379

Entz MH, Bullied WJ, Katepa-Mupondwa F (1995) Rotational benefits of forage crops in Canadian prairie cropping systems. J Prod Agric 8(4):521–529

Entz MH, Baron VS, Carr PM, Meyer DW, Smith SR Jr, McCaughey WP (2002) Potential of forages to diversify cropping systems in the Northern Great Plains. Agron J 94(2):240–250

Falconer DS, Mackay TFC (1996) Introduction to quantitative genetics (4th ed.). Harlow, England: Longman (Pearson Education Ltd.)

FAOSTAT F (2020) Statistics, food and agriculture organization of the United Nations. Food Agric Organ. United Nations

Fè D, Ashraf B, Byrne S, Czaban A, Roulund N, Lenk I, Asp T (2014) Prospects for introducing genomic selection into forage grass breeding. In: The future of European grasslands. Aberystwyth University, Aberystwyth, p 830

Feng L, Chen S, Zhang C, Zhang Y, He Y (2021) A comprehensive review on recent applications of unmanned aerial vehicle remote sensing with various sensors for high-throughput plant phenotyping. Comput Electron Agric 182:106033

Gaur PM, Samineni S, Thudi M, Tripathi S, Sajja SB, Jayalakshmi V, Dixit GP (2019) Integrated breeding approaches for improving drought and heat adaptation in chickpea (*Cicer arietinum* L.). Plant Breed 138(4):389–400

Gebremedhin A, Badenhorst PE, Wang J, Spangenberg GC, Smith KF (2019) Prospects for measurement of dry matter yield in forage breeding programs using sensor technologies. Agronomy 9(2):65

Gebril S, Seger M, Villanueva FM, Ortega JL, Bagga S, Sengupta-Gopalan C (2015) Transgenic alfalfa (Medicago sativa) with increased sucrose phosphate synthase activity shows enhanced growth when grown under N2-fixing conditions. Planta 242(4):1009–1024. https://doi.org/10.1007/s00425-015-2342-0

Giorgi D, Farina A, Grosso V, Gennaro A, Ceoloni C, Lucretti S (2013) FISHIS: fluorescence in situ hybridization in suspension and chromosome flow sorting made easy. PLoS One 8(2):e57994

Grabber JH, Davidson C, Tobimatsu Y, Kim H, Lu F, Zhu Y, Ralph J (2019) Structural features of alternative lignin monomers associated with improved digestibility of artificially lignified maize cell walls. Plant Sci 287:110070

Grabherr MG, Haas BJ, Yassour M, Levin JZ, Thompson DA, Amit I et al (2011) Full-length transcriptome assembly from RNA-Seq data without a reference genome. Nat Biotechnol 29(7):644–652

Grela ER, Rybiński W, Klebaniuk R, & Matras J (2010) Morphological characteristics of some accessions of grass pea (Lathyrus sativus L.) grown in Europe and nutritional traits of their seeds. Genetic Resources and Crop Evolution 57(5):693–701

Gore MA, Chia JM, Elshire RJ, Sun Q, Ersoz ES, Hurwitz BL et al (2009) A first-generation haplotype map of maize. Science 326(5956):1115–1117

Gygi MP, Ferguson MD, Mefford HC, Lund KP, O'Day C, Zhou P et al (2002) Use of fluorescent sequence-specific polyamides to discriminate human chromosomes by microscopy and flow cytometry. Nucleic Acids Res 30(13):2790–2799

Habermann E, Dias de Oliveira EA, Contin DR, Delvecchio G, Viciedo DO, de Moraes MA, de Mello Prado R, de Pinho Costa KA, Braga MR, Martinez CA (2019) Warming and water deficit impact leaf photosynthesis and decrease forage quality and digestibility of a C4 tropical grass. Physiol Plant 165(2):383–402. https://doi.org/10.1111/ppl.12891

Hassanpour S, MaheriSis N, & Eshratkhah B (2011) Plants and secondary metabolites (Tannins): A Review

Halpin C (2019) Lignin engineering to improve saccharification and digestibility in grasses. Curr Opin Biotechnol 56:223–229

Han Y, Zheng D, Vimolmangkang S, Khan MA, Beever JE, Korban SS (2011) Integration of physical and genetic maps in apple confirms whole-genome and segmental duplications in the apple genome. J Exp Bot 62(14):5117–5130
Haq SIU, Zheng D, Feng N, Jiang X, Qiao F, He JS, Qiu QS (2022) Progresses of CRISPR/Cas9 genome editing in forage crops. J Plant Physiol 279:153–860
Harper JA, Thomas ID, Lovatt JA, Thomas HM (2004) Physical mapping of rDNA sites in possible diploid progenitors of polyploid Festuca species. Plant Syst Evol 245:163–168
Hasan N, Choudhary S, Naaz N, Sharma N, Laskar RA (2021) Recent advancements in molecular marker-assisted selection and applications in plant breeding programmes. J Genet Eng Biotechnol 19(1):128
Hatfield RD, Kalscheur KF (2020) Carbohydrate and protein nutritional chemistry of forages. In RF Barnes, CJ. Nelson, KJ Moore, & M. Collins (Eds.), Forages: The science of grassland agriculture (Vol. 2, pp. 595–607). Hoboken, NJ, USA: Wiley (John Wiley & Sons)
Hernandez P, Martis M, Dorado G, Pfeifer M, Gálvez S, Schaaf S et al (2012) Next-generation sequencing and syntenic integration of flow-sorted arms of wheat chromosome 4A exposes the chromosome structure and gene content. Plant J 69(3):377–386
Horrocks CA, Arango J, Arévalo A, Nuñez J, Cardoso JA, Dungait JAJ (2019) Smart forage selection could significantly improve soil health in the tropics. Sci Total Environ 688:609–621
Hörtensteiner S (2009) Stay-green regulates chlorophyll and chlorophyll-binding protein degradation during senescence. Trends Plant Sci 14(3):155–162. https://doi.org/10.1016/j.tplants.2009.01.002
Howard A (2020) The soil and health. A Distant Mirror, La Vergne
Huang X, Lu T, Han B (2013) Resequencing rice genomes: an emerging new era of rice genomics. Trends Genet 29(4):225–232
Huo N, Garvin DF, You FM, McMahon S, Luo MC, Gu YQ et al (2011) Comparison of a high-density genetic linkage map to genome features in the model grass Brachypodium distachyon. Theor Appl Genet 123:455–464
Huyghe C (2003) Les fourrages et la production de protéines. Fourrages 174:145–162
Ijaz S, Khan AH, Sarwar MK (2021) Conventional breeding strategies and programs for winter fodders. In: Sustainable winter fodder, pp 3–13. https://doi.org/10.1201/9781003055365-2
Ingvarsson PK, Street NR (2011) Association genetics of complex traits in plants. New Phytol 189(4):909–922
Jonker A, Yu P (2017) The occurrence, biosynthesis, and molecular structure of proanthocyanidins and their effects on legume forage protein precipitation, digestion and absorption in the ruminant digestive tract. Int J Mol Sci 18(5):1105
Julier B, Huyghe C, Ecalle C (2000) Within-and among-cultivar genetic variation in alfalfa: forage quality, morphology, and yield. Crop Sci 40(2):365–369
Jung KH, An G, Ronald PC (2008) Towards a better bowl of rice: assigning function to tens of thousands of rice genes. Nat Rev Genet 9(2):91–101
Kaithwas M, Singh S, Prusty S, Mondal G, Kundu SS (2020) Evaluation of legume and cereal fodders for carbohydrate and protein fractions, nutrient digestibility, energy and forage quality. Range Manag Agrofor 41(1):126–132
Karlsson JO, Röös E (2019) Resource-efficient use of land and animals—environmental impacts of food systems based on organic cropping and avoided food-feed competition. Land Use Policy 85:63–72
Katoch R (2022) Nutritional quality management of forages in the Himalayan region, pp 99–149. https://doi.org/10.1007/978-981-16-5437-4
Katoch R (2023) Techniques in forage quality analysis. Springer, Singapore
Kejnovský E, Vrána J, Matsunaga S, Souček P, Široký J, Doležel J, Vyskot B (2001) Localization of male-specifically expressed MROS genes of Silene latifolia by PCR on flow-sorted sex chromosomes and autosomes. Genetics 158(3):1269–1277
King J, Thomas A, James C, King I, Armstead I (2013) A DArT marker genetic map of perennial ryegrass (Lolium perenne L.) integrated with detailed comparative mapping information;

comparison with existing DArT marker genetic maps of Lolium perenne, L. multiflorum and Festuca pratensis. BMC Genomics 14:1–7

Kopecký D, Martis M, Číhalíková J, Hřibová E, Vrána J, Bartoš J et al (2013) Flow sorting and sequencing meadow fescue chromosome 4F. Plant Physiol 163(3):1323–1337

Korte A, Vilhjálmsson BJ, Segura V, Platt A, Long Q, Nordborg M (2012) A mixed-model approach for genome-wide association studies of correlated traits in structured populations. Nat Genet 44(9):1066–1071

Krishnan A, Guiderdoni E, An G, Hsing YIC, Han CD, Lee MC et al (2009) Mutant resources in rice for functional genomics of the grasses. Plant Physiol 149(1):165–170

Książczyk T, Taciak M, Zwierzykowski Z (2010) Variability of ribosomal DNA sites in Festuca pratensis, Lolium perenne, and their intergeneric hybrids, revealed by FISH and GISH. J Appl Genet 51:449–460

Kubaláková M, Valárik M, Barto J, Vrána J, Cíhalíková J, Molnár-Láng M, Dolezel J (2003) Analysis and sorting of rye (Secale cereale L.) chromosomes using flow cytometry. Genome 46(5):893–905

Kumar S, Faruqui SA (2009) Production potential and economic viability of food-forage based cropping system under irrigated conditions. Indian J Agron 54(1):36–41

Kumar N, Rana M, Kumar B, Chand S, Shiv A, Wani SH, Kumar S (2021) Genomic selection for wheat improvement. In: Wani SH, Mohan A, Singh GP (eds) Physiological, molecular, and genetic perspectives of wheat improvement. Springer, Singapore, pp 175–207

Lammerts van Bueren ET, Struik PC, van Eekeren N, Nuijten E (2018) Towards resilience through systems-based plant breeding. A review. Agron Sustain Dev 38:1–21

Lee MR (2014) Forage polyphenol oxidase and ruminant livestock nutrition. Front Plant Sci 5:694. https://doi.org/10.3389/fpls.2014.00694

Lee MA (2018) A global comparison of the nutritive values of forage plants grown in contrasting environments. J Plant Res 131(4):641–654. https://doi.org/10.1007/s10265-018-1024-y

Lee MRF, Theobald VJ, Tweed JKS, Winters AL, Scollan ND (2009) Effect of feeding fresh or conditioned red clover on milk fatty acids and nitrogen utilization in lactating dairy cows. J Dairy Sci 92(3):1136–1147. https://doi.org/10.3168/jds.2008-1692

Lewis NM (2021) The economic consequences of reciprocal translocations (RTs) in global pig breeding and the benefits of using a novel "multiprobe" approach as a standard screening methodology. Master's thesis, University of Kent, United Kingdom

Li X, Han Y, Wei Y, Acharya A, Farmer AD, Ho J et al (2014) Development of an alfalfa SNP array and its use to evaluate patterns of population structure and linkage disequilibrium. PLoS One 9(1):e84329

Liang C, Wang Y, Zhu Y, Tang J, Hu B, Liu L, Ou S, Wu H, Sun X, Chu J, Chu C (2014) OsNAP connects abscisic acid and leaf senescence by fine-tuning abscisic acid biosynthesis and directly targeting senescence-associated genes in rice. Proc Natl Acad Sci USA 111(27):10013–10018. https://doi.org/10.1073/pnas.1321568111

Ludemann CI, Eckard RJ, Cullen BR, Jacobs JL, Malcolm B, Smith KF (2015) Higher energy concentration traits in perennial ryegrass (Lolium perenne L.) may increase profitability and improve energy conversion on dairy farms. Agric Syst 137:89–100

Maher CA, Kumar-Sinha C, Cao X, Kalyana-Sundaram S, Han B, Jing X et al (2009) Transcriptome sequencing to detect gene fusions in cancer. Nature 458(7234):97–101

Maheswari VU, Reddy PS, Kumar CV, Eswari KB (2024) Character association studies for yield attributing traits across different seasons in pearl millet [Pennisetum glaucum (L.) R. Br]. International Journal of Plant & Soil Science 36(9):368–377

Mandakova T, Lysak MA (2008) Chromosomal phylogeny and karyotype evolution in x= 7 crucifer species (Brassicaceae). Plant Cell 20(10):2559–2570

Mayer KF, Martis M, Hedley PE, Šimková H, Liu H, Morris JA et al (2011) Unlocking the barley genome by chromosomal and comparative genomics. Plant Cell 23(4):1249–1263

Mayer KF, Waugh R, Langridge P, Close TJ, Wise RP, Graner A et al (2012) A physical, genetic and functional sequence assembly of the barley genome. Nature 491:711–716

McCallum CM, Comai L, Greene EA, Henikoff S (2000) Targeting Induced Local Lesions IN Genomes (TILLING) for plant functional genomics. Plant Physiol 123(2):439–442

Mejlhede N, Kyjovska Z, Backes G, Burhenne K, Rasmussen SK, Jahoor A (2006) EcoTILLING for the identification of allelic variation in the powdery mildew resistance genes mlo and Mla of barley. Plant Breed 125(5):461–467
Mekonnen K, Bezabih M, Thorne P, Gebreyes MG, Hammond J, Adie A (2022) Feed and forage development in mixed crop–livestock systems of the Ethiopian highlands: Africa RISING project research experience. Agron J 114(1):46–62
Melo CD, Maduro Dias CS, Wallon S, Borba AE, Madruga J, Borges PA et al (2022) Influence of climate variability and soil fertility on the forage quality and productivity in Azorean pastures. Agriculture 12(3):358
Moore VM, Schlautman B, Fei SZ, Roberts LM, Wolfe M, Ryan MR, Lorenz AJ (2022) Plant breeding for intercropping in temperate field crop systems: a review. Front Plant Sci 13:843065
Morota G, Ventura RV, Silva FF, Koyama M, Fernando SC (2018) Big data analytics and precision animal agriculture symposium: machine learning and data mining advance predictive big data analysis in precision animal agriculture. J Anim Sci 96(4):1540–1550
Muir JP, Taylor J, Interrante SM (2005) Herbage and seed from Texan native perennial herbaceous legumes. Rangel Ecol Manag 58(6):643–651
Muir JP, Tedeschi LO, Dubeux JC, Peters M, Burkart S (2017) Enhancing food security in Latin America with forage legumes. Arch Latinoam Prod Anim 25(3–4):113–131
Mukhtiar A, Pickova J, Taufiq A, Muhammad L, Abid F, Muhammad J (2016) Oxidation of lipids in foods. Sarhad J Agric 32(3):230–238. https://doi.org/10.17582/journal.sja/2016.32.3.230.238
Mullainathan S, Shafir E (2013) Scarcity: Why having too little means so much. New York, NY, USA: Henry Holt and Company
Murdoch AJ, Njuguna JM, Lukuyu B, Musembi F, Mwangi DM, Maina JM et al (2003) Integrated pest management options to improve maize forage yield and quality for small-scale dairy farmers in central Kenya. Asp Appl Biol 70:71–78
Nagarajan N, Pop M (2013) Sequence assembly demystified. Nat Rev Genet 14(3):157–167
Nelson CJ, Moore KJ, Collins M (2017) Forages and grasslands in a changing world. Forages 1:3–17
Newman Y, Lambert B, Muir J (2024) Defining forage quality (SCS-2006–09). Gainesville, FL, USA: University of Florida IFAS Extension
Ngunga D, Nzogela B, Maleko D, Mwendia SW, Mwilawa A, Omore AO, Notenbaert AMO (2021) Forage seed stakeholder workshop (12 May 2021): Addressing obstacles for a thriving forage seed sector in Tanzania. Nairobi, Kenya: International Livestock Research Institute (ILRI)
Niyigena V, Ashworth AJ, Nieman C, Acharya M, Coffey KP, Philipp D et al (2021) Factors affecting sugar accumulation and fluxes in warm-and cool-season forages grown in a silvopastoral system. Agronomy 11(2):354
Nordborg M, Weigel D (2008) Next-generation genetics in plants. Nature 456(7223):720–723
Nuzhdin SV, Friesen ML, McIntyre LM (2012) Genotype–phenotype mapping in a post-GWAS world. Trends Genet 28(9):421–426
Palacios MP, Mendoza-Grimón V, Fernández F, Fernandez-Vera JR, Hernandez-Moreno JM (2008) Sustainable reclaimed water management by subsurface drip irrigation system: a study case for forage production. Water Pract Technol 3(2):2008049
Parveen I, Threadgill MD, Moorby JM, Winters A (2010) Oxidative phenols in forage crops containing polyphenol oxidase enzymes. J Agric Food Chem 58(3):1371–1382. https://doi.org/10.1021/jf9024294
Pennisi E (2013) Plant biology. Evolution heresy? Epigenetics underlies heritable plant traits. Science 341(6150):1055
Pfeifer M, Martis M, Asp T, Mayer KF, Lübberstedt T, Byrne S et al (2013) The perennial ryegrass GenomeZipper: targeted use of genome resources for comparative grass genomics. Plant Physiol 161(2):571–582
Poland JA, Rife TW (2012) Genotyping-by-sequencing for plant breeding and genetics. The Plant Genome 5(3):92–102
Project, I. R. G. S (2005) The map-based sequence of the rice genome. Nature 436(7052):793–800

Raigond P, Ezekiel R, Raigond B (2015) Resistant starch in food: a review. J Sci Food Agric 95(10):1968–1978

Rao IM, Peters M, Castro A, Schultze-Kraft R, White D, Fisher M, Rudel TK (2015) LivestockPlus: the sustainable intensification of forage-based agricultural systems to improve livelihoods and ecosystem services in the tropics. Trop Grassl Forrajes Trop 3(2):59

Rauf S, Ortiz R, Malinowski DP, Clarindo WR, Kainat W, Shehzad M et al (2021) Induced polyploidy: a tool for forage species improvement. Agriculture 11(3):210

Reddy MS, Chen F, Shadle G, Jackson L, Aljoe H, Dixon RA (2005) Targeted down-regulation of cytochrome P450 enzymes for forage quality improvement in alfalfa (Medicago sativa L.). Proc Natl Acad Sci USA 102(46):16573–16578

Ren H, Feng Y, Liu T, Li J, Wang Z, Fu S et al (2020) Effects of different simulated seasonal temperatures on the fermentation characteristics and microbial community diversities of the maize straw and cabbage waste co-ensiling system. Sci Total Environ 708:135113

Ruckle ME, Meier MA, Frey L, Eicke S, Kölliker R, Zeeman SC, Studer B (2017) Diurnal leaf starch content: an orphan trait in forage legumes. Agronomy 7(1):16

Ruttink T, Sterck L, Rohde A, Bendixen C, Rouzé P, Asp T et al (2013) Orthology guided assembly in highly heterozygous crops: creating a reference transcriptome to uncover genetic diversity in Lolium perenne. Plant Biotechnol J 11(5):605–617

Sami A, Xue Z, Tazein S, Arshad A, He Zhu Z, Ping Chen Y, Jin Zhou K (2021) CRISPR–Cas9-based genetic engineering for crop improvement under drought stress. Bioengineered 12(1):5814–5829

Sanz-Sáez Á, Erice G, Aguirreolea J, Muñoz F, Sánchez-Díaz M, Irigoyen JJ (2012) Alfalfa forage digestibility, quality and yield under future climate change scenarios vary with Sinorhizobium meliloti strain. J Plant Physiol 169(8):782–788. https://doi.org/10.1016/j.jplph.2012.01.010

Sarkar R, Corriher-Olson V, Long C, Somenahally A (2020) Challenges and potentials for soil organic carbon sequestration in forage and grazing systems. Rangel Ecol Manag 73(6):786–795

Schneider GF, Dekker C (2012) DNA sequencing with nanopores. Nat Biotechnol 30(4):326–328

Shinozuka H, Cogan NO, Spangenberg GC, Forster JW (2012) Quantitative Trait Locus (QTL) meta-analysis and comparative genomics for candidate gene prediction in perennial ryegrass (Lolium perenne L.). BMC Genet 13:1–12

Šidlauskaitė G, Kemešytė V, Toleikienė M, Kadžiulienė Ž (2022) Plant diversity, functional group composition and legumes effects versus fertilisation on the yield and forage quality. Sustainability 14(3):1182

Silveira ML, Kohmann MM (2020) Maintaining soil fertility and health for sustainable pastures. In: Management strategies for sustainable cattle production in southern pastures. Academic Press, London, pp 35–58

Simeão Resende RM, Casler MD, de Resende MDV (2014) Genomic selection in forage breeding: accuracy and methods. Crop Sci 54(1):143–156

Sinha D, Maurya AK, Abdi G, Majeed M, Agarwal R, Mukherjee R, Chen JT (2023) Integrated genomic selection for accelerating breeding programs of climate-smart cereals. Genes 14(7):1484

Slade AJ, Fuerstenberg SI, Loeffler D, Steine MN, Facciotti D (2005) A reverse genetic, nontransgenic approach to wheat crop improvement by TILLING. Nat Biotechnol 23(1):75–81

Smit HJ, Tas BM, Taweel HZ, Tamminga S, Elgersma A (2005) Effects of perennial ryegrass (Lolium perenne L.) cultivars on herbage production, nutritional quality and herbage intake of grazing dairy cows. Grass and Forage Science, 60(3):297–309

Snowdon RJ, Wittkop B, Chen TW, Stahl A (2021) Crop adaptation to climate change as a consequence of long-term breeding. Theor Appl Genet 134(6):1613–1623

Stokes SR, Prostko EP (2024) Understanding forage quality analysis (L-5198). College Station, TX, USA: Texas A&M AgriLife Extension Service

Suchankova P, Kubalakova M, Kovarova P, Bartos J, Cihalikova J, Molnar-Lang M et al (2006) Dissection of the nuclear genome of barley by chromosome flow sorting. Theor Appl Genet 113:651–659

Sun L, Lai M, Ghouri F, Nawaz MA, Ali F, Baloch FS et al (2024) Modern plant breeding techniques in crop improvement and genetic diversity: from molecular markers and gene editing to artificial intelligence—a critical review. Plants 13(19):2676

Suresh Kumar S (2023) Influence of potassium fertilizer application timing on cotton production as related to soil potassium on US Coastal Plain soils. [Likely thesis] Auburn, AL, USA: Auburn University

Sustek-Sánchez F, Rognli OA, Rostoks N, Sõmera M, Jaškūnė K, Kovi MR et al (2023) Improving abiotic stress tolerance of forage grasses–prospects of using genome editing. Front Plant Sci 14:1127532

Tas BM, Taweel HZ, Smit HJ, Elgersma A, Dijkstra J, Tamminga S (2006) Effects of perennial ryegrass cultivars on milk yield and nitrogen utilization in grazing dairy cows. Journal of dairy science 89(9):3494–3500

Till BJ, Reynolds SH, Weil C, Springer N, Burtner C, Young K et al (2004) Discovery of induced point mutations in maize genes by TILLING. BMC Plant Biol 4:1–8

Till BJ, Cooper J, Tai TH, Colowit P, Greene EA, Henikoff S, Comai L (2007) Discovery of chemically induced mutations in rice by TILLING. BMC Plant Biol 7:1–12

Tong Z, Xie C, Ma L, Liu L, Jin Y, Dong J, Wang T (2014) Coexpression of bacterial aspartate kinase and adenylylsulfate reductase genes substantially increases sulfur amino acid levels in transgenic alfalfa (Medicago sativa L.). PLoS One 9(2):88–310

Tracy BF, Foster JL, Butler TJ, Islam MA, Toledo D, Vendramini JMB (2018) Resilience in forage and grazinglands. Crop Sci 58(1):31–42

Tsai H, Missirian V, Ngo KJ, Tran RK, Chan SR, Sundaresan V, Comai L (2013) Production of a high-efficiency TILLING population through polyploidization. Plant Physiol 161(4):1604–1614

Turesson G (1922) The genotypical response of the plant species to the habitat. Hereditas 3(3):211–350

Van Peer L, Nijs I, Reheul D, De Cauwer B (2004) Species richness and susceptibility to heat and drought extremes in synthesized grassland ecosystems: compositional vs physiological effects. Funct Ecol 18:769–778

Vineis P, Pearce N (2010) Missing heritability in genome-wide association study research. Nat Rev Genet 11(8):589–589

Voytas DF (2013) Plant genome engineering with sequence-specific nucleases. Annu Rev Plant Biol 64(1):327–350

Vrána J, Kubaláková M, Simková H, Číhalíkovái J, Lysák MA, Dolezel J (2000) Flow sorting of mitotic chromosomes in common wheat (Triticum aestivum L.). Genetics 156(4):2033–2041

Wang J, Baillie RC, Cogan NOI, McFarlane NM, Dupal MP, Smith KF, Forster JW (2011) Molecular genetic marker-based analysis of species-differentiated phenotypic characters in an interspecific ryegrass mapping population. Crop Pasture Sci 62(10):892–902

Wang TL, Uauy C, Robson F, Till B (2012) TILLING in extremis. Plant Biotechnol J 10(7):761–772

Watts KA (2009) Carbohydrates in forage: what is a safe grass. In: Pagan JD (ed) Advances in equine nutrition, vol IV. Nottingham University Press, Nottingham, pp 29–42

Wicker T, Mayer KF, Gundlach H, Martis M, Steuernagel B, Scholz U et al (2011) Frequent gene movement and pseudogene evolution is common to the large and complex genomes of wheat, barley, and their relatives. Plant Cell 23(5):1706–1718

Winichayakul S, Macknight R, Le Lievre L, Beechey-Gradwell Z, Lee R, Cooney L, ... Roberts N (2022) Insight into the regulatory networks underlying the high lipid perennial ryegrass growth under different irradiances. PloS one 17(10):e0275503

Wright AM (2013) Effect of forage type and corn supplementation on animal performance and meat quality. Clemson University

Yamada T, Spangenberg G, Barrett B, Baird I, Woodfield D (2009a) White clover seed yield: a case study in marker-assisted selection. In: Molecular breeding of forage and turf. Springer, New York, pp 241–250

Yamada T, Spangenberg G, Smith KF, Dobrowolski MP, Cogan NO, Spangenberg GC, Forster JW (2009b) Utilizing linkage disequilibrium and association mapping to implement candidate gene based markers in perennial ryegrass breeding. In: Molecular breeding of forage and turf. Springer, New York, pp 259–274

Yang L, Wen KS, Ruan X, Zhao YX, Wei F, Wang Q (2018) Response of plant secondary metabolites to environmental factors. Molecules 23(4):762

Yegon DK (2023) Evaluation of portable near-infrared spectrometer for rapid and non-destructive determination of quality and authenticity of baobab (Adansonia Digitata l.) fruit pulp in Kenya. Doctoral dissertation, JKUAT-CoANRE

Yu X, Bai G, Liu S, Luo N, Wang Y, Richmond DS et al (2013) Association of candidate genes with drought tolerance traits in diverse perennial ryegrass accessions. J Exp Bot 64(6):1537–1551

Yusiati LM, Kurniawati A, Hanim C, & Anas DM (2018, February) Protein binding capacity of different forages tannin. In IOP Conference Series: Earth and Environmental Science (Vol. 119, No. 1, p. 012007). IOP Publishing

Zhang G, Liu X, Quan Z, Cheng S, Xu X, Pan S et al (2012) Genome sequence of foxtail millet (Setaria italica) provides insights into grass evolution and biofuel potential. Nat Biotechnol 30(6):549–554

Zhang Y, Ran Y, Nagy I, Lenk I, Qiu JL, Asp T, Jensen CS, Gao C (2020) Targeted mutagenesis in ryegrass (Lolium spp.) using the CRISPR/Cas9 system. Plant Biotechnol J 18(9):1854

Zhu YG, Pilon-Smits EA, Zhao FJ, Williams PN, Meharg AA (2009) Selenium in higher plants: understanding mechanisms for biofortification and phytoremediation. Trends in plant science 14(8):436–442

MIX
Papier aus verantwortungsvollen Quellen
Paper from responsible sources
FSC® C105338

If you have any concerns about our products,
you can contact us on
ProductSafety@springernature.com

In case Publisher is established outside the EU,
the EU authorized representative is:
Springer Nature Customer Service Center GmbH
Europaplatz 3, 69115 Heidelberg, Germany

Printed by Libri Plureos GmbH
in Hamburg, Germany